PROPERTY OF  W9-BQS-784

T-REX CORPORATION

~~1850 CENTENNIAL~~

~~PARK DRIVE~~

~~RESTON, VA 20191~~

11495 Sunset Hills Rd, Suite 220

Reston, VA 20190

Successful Proposal Strategies for Small Businesses

Using Knowledge Management to Win Government, Private-Sector, and International Contracts

Fourth Edition

For a complete listing of the *Artech House Technology Management and Professional Development Library,* turn to the back of this book.

Successful Proposal Strategies for Small Businesses

Using Knowledge Management to Win Government, Private-Sector, and International Contracts

Fourth Edition

Robert S. Frey

ARTECH
HOUSE

BOSTON | LONDON
artechhouse.com

Library of Congress Cataloging-in-Publication Data
Frey, Robert S.
  Successful proposal strategies for small businesses: using knowledge
  management to win government, private-sector, and international
  contracts/ Robert S. Frey—4th ed. (Artech House technology
  management and professional development library)
  Includes bibliographical references and index.

  ISBN 1-58053-957-2

  1. Proposal writing for grants—United States.   2. Small
  business—United States—Finance.
  I. Title. II. Series.

HG177.5.U6F74 2005
658.8'04—dc22   2004062876

British Library Cataloguing in Publication Data
Frey, Robert S.
  Successful proposal strategies for small businesses: using knowledge
  management to win government, private-sector, and international
  contracts.—4th ed. (Artech House technology management and
  professional development library)
  1. Proposal writing for grants—United States   2. Small
    business—United States—Finance
  I. Title
  658.8'04

  ISBN 1-58053-957-2

Cover design by Igor Valdman

© **2005 ARTECH HOUSE, INC.**
685 Canton Street
Norwood, MA 02062

International Standard Book Number: 1-58053-957-2

10 9 8 7 6 5 4 3 2 1

For

my friend and fellow teacher, Dr. Terry C. Tarbell,

an outstanding business developer who sees the effort through from target

identification and qualification, capture management and proposal development,

and oral presentation to contract award.

And in memory of

Jonathan L. Friedman

my colleague and friend who died on Saturday, April 24, 2004.

With love and respect for the dignity that he brought to life.

Contents

Chapter 11

Pursuing international business and structuring international proposals   255

Chapter 12

Proposal production and publication   291

Acknowledgments

THE FOURTH EDITION OF THIS BOOK HAS BEEN DEVELOPED, expanded, and refined during the past 15 years. I would like to express my appreciation once again to Mr. Bruce Elbert, series editor of the Artech House Technology Management and Professional Development Library. He must also be recognized and thanked for relentlessly yet good-naturedly stretching the book into new areas. This work has been enhanced greatly through his efforts, and I am genuinely appreciative.

The graphic-arts talent of Ms. Lisa Richard must also be recognized. Lisa generated the graphics for all four editions of this book. She can be contacted at LOTSLR@aol.com. Ms. Sarah A. Fowlie developed the companion CD-ROM for the second, third, and fourth editions of this volume. Sarah is a freelance computer trainer in desktop-publishing applications, graphics applications, databases, and Web development. She can be contacted at safowlie@yahoo.com.

Finally, the business and editorial acumen of Mrs. Terry Raezer Frey contributed significantly to the value, consistency, and readability of this work. And for her love and unconditional day-to-day support, I am both a better author and better person.

Introduction

FUNDAMENTALLY, *Successful Proposal Strategies for Small Businesses: Using Knowledge Management to Win Government, Private-Sector, and International Contracts, Fourth Edition,* and its companion CD-ROM are highly accessible, self-contained desktop references developed to be informative, practical, and easy to use. They help small and mid-sized businesses, as well as nonprofit organizations and public-sector agencies, achieve effective, efficient, and disciplined business development, proposal development, and knowledge management (KM) processes. These, in turn, contribute to increased contract and grant awards and enhanced levels of revenue. Using this book, any small company or organization with a viable product or service can learn how to gain and keep a client's attention, even when working with only a few employees. Entrepreneurs can use the book to assist in establishing best-of-breed business development, proposal development, knowledge management, and publications infrastructures and processes within their organizations. In many ways, a small company's future performance in the marketplace will be a direct result of how effectively it chooses to implement disciplined business development, proposal development, and KM processes and methodologies, as well as the modes of thinking presented in this work.

Reviews of previous editions of this book have been published in such prestigious forums as *Business Week* (New York); *Minorities and Women in Business* (Washington, D.C.); *Turning Point* magazine (Los Angeles); *Canada One Magazine* (on the Web); *E-merging Business* magazine (Pacific Palisades, California); *Small Business Advisor* (Los Angeles); and *Women's Business of South Florida* (Hollywood, Florida). In addition, Amazon.com lists exemplary reviews of the book's various editions.

Successful Proposal Strategies for Small Businesses, Fourth Edition, provides effective, field-tested guidance for small businesses (SB), disadvantaged business enterprises (DBEs), minority business enterprises (MBEs), women-owned business enterprises (WBEs), veteran-owned firms, and other organizations to plan, organize, manage, and develop effective, highly competitive responses to federal, state, and local government requests for proposals (RFPs) or requests for solution (RFSs), private-sector solicitations, and international tenders. Entrepreneurs, business development staff, capture managers, proposal managers, proposal writers, proposal specialists, and coordinators will benefit from applying the structured processes illustrated in this work.

Specific focus is on small and midsized business enterprises and exploring the important human and organizational dynamics related to the proposal life cycle that contribute directly to winning new contracts. Step-by-step, *Successful Proposal Strategies for Small Businesses, Fourth Edition,* clearly maps and details every stage of the contractor proposal response life cycle. This work shows how to maximize small business strengths and leverage knowledge and intellectual capital in order to conduct client- centered marketing and produce benefits-focused, requirements-driven proposals and oral presentations that respond fully to client success criteria and critical issues. This new edition is also valuable for educators in preparing grant proposals and in teaching proposal development courses in business curricula in colleges, universities, and distance-learning programs. Importantly, a full section is devoted to successful grant proposal management. In addition, and very importantly, as the U.S. federal government evolves and sells its services to other branches of government, government staff can also leverage the proven and agile marketing, knowledge management, proposal development, and communications strategies presented in this edition.

Salient among the new edition's contributions to this field is its focus on the proposal as a sales document and on demonstrating how structured and repeatable KM processes, approaches, and automated tools directly benefit companies' proposal and oral presentation development efforts.

The book places the proposal response process within the larger context of small companies' overall strategic and mission planning, as well as business development and corporate communication and image management activities. An extremely comprehensive and expanded listing of small business Web-based resources, as well as business and proposal-related acronyms, is also provided both in the book and on the accompanying CD-ROM. The CD-ROM also includes fully updated, useful, and timesaving proposal- and marketing-related templates, along with planning and review tools.

Among the highly beneficial aspects of this book's fourth edition are significant additions and expanded treatment of topics that include (1) next generation governmentwide acquisition contract (GWAC) vehicles, (2) performance-based acquisition (PBA), (3) Mentor-Protégé programs, (4) President's Management Agenda (PMA) and e-government initiatives, (5) the increasing importance of oral presentations in federal procurements, (6) exit strategies from the Small Business Administration (SBA) 8(a) program, (7) benefits of KM to proposal development, (8) leading-edge developments in federal civilian and defense electronic acquisition (including all of the latest major federal e-business and e-commerce Web sites), (9) growing importance of OMB Circular A-76 Studies for increased efficiency and lower costs, (10) the pivotal role of the capture manager or campaign manager in the proposal process, and (11) fact-based storytelling as a powerful framework for conveying proposal solutions.

Successful Proposal Strategies for Small Businesses, Fourth Edition, gives both the big picture and the down-in-the-trenches perspective about marketing and proposal development, management, production, and infrastructure support in a rapidly evolving global economy. The book discusses how marketing and proposal life cycles can and should mesh with operational, management, and infrastructure support activities within a small company and shows how human and organizational dynamics drive successful marketing and proposal processes.

Unlike most books, cassettes, CDs, videotapes, and training seminars on developing proposals, *Successful Proposal Strategies for Small Businesses, Fourth Edition,* focuses on the special constraints and strengths of small businesses as they relate to the proposal process. Many of the best-known proposal seminars, for example, are designed for large businesses competing on massive defense and aerospace hardware and systems procurements. Marketing and proposal development in a small business environment—particularly in the support services arena—presents special challenges in terms of support infrastructure, staffing levels, depth of expertise, bid and proposal resources, and business culture. Meeting these distinctive challenges is the purpose of this new edition.

The late Vince Lombardi, legendary coach of the Green Bay Packers, is reputed to have said that he longed to "lie exhausted in victory." That is, to expend the very best effort, to harness the talent and spirit within, and to channel that immense power toward a very specific goal. In Lombardi's thoughts, that goal was victory in the early Super Bowl competitions of the National Football League. In my own thoughts, that goal is to bring all the knowledge, experience, initiative, and positive emotion—the passion—I can into producing a winning proposal.

Unlike many other professions, proposal preparation in the contractor arena for federal, state, local, private-sector, and international opportunities occurs in very discrete and often overlapping bundles of intense activity. There is a clear beginning, middle, and end to the preparation process. Often in a mere span of 5 to 45 days and nights, a host of technical and programmatic information, cost strategies, and marketing intelligence must be condensed, distilled, and fitted together into a set of polished documents. Considering the length of time required to bring journals and books to press, it is astounding that such a choreographed process of information retrieval and management, assembly, and packaging must unfold in the space of only a week or several weeks! And yet for those small and large businesses that compete in the contracting marketplace, it is a matter of survival.

Successful proposal preparation is built largely upon a winning attitude, commitment, attention to detail, teamwork at all levels, communication, emotional and physical endurance, and adequate and well-timed allocation of company human and material resources. To be sure, success also depends upon marketing intelligence about the customer and your competition, informed and timely bid–no bid decisions, planning, scheduling, and superior information management. But my experience has suggested that what makes the difference once a company decides to respond to an RFP or SF330 synopsis lies in the area of human and organizational dynamics rather than in technical and strategic excellence alone. Can a diverse group of technical, management, and support people work together effectively for protracted periods of time—including nights, weekends, and holidays—to produce a winning document? Will company management commit the best technical talent, lease or acquire adequate computer or publishing equipment, make dedicated work space available for the proposal team, or allocate bonus monies to reward the above-and-beyond efforts of particular people?

To lie exhausted in victory. Plans and milestone schedules, bullet drafts and storyboards, writing and editorial guidelines, action item lists, internal review cycles, and document configuration management schemas all come down to one thing—getting a winning proposal assembled, out the door, and delivered before the established due date. While I was coordinating a $100 million Air Force proposal for a Virginia-based contractor, the entire marketing and proposal life cycle came down to one overcast Saturday in December, not long before the holidays. Thoughts were not on marketing target identification, intelligence gathering, teaming arrangements, RFP analysis, outline development, program pricing, or Red Team review comments. Rather, there were 150 copies of various volumes that had to be photoreproduced and put into three-ring notebooks, with multiple foldout pages inserted in each one, and an overnight carrier office nearby that was scheduled to close promptly at 5 P.M. Just the night before, several members of the proposal team had worked into the early morning hours. People were exhausted from several weeks of grueling schedules, missed meals, and no recreation, taping boxes shut at breakneck speed, loading them into several cars, and making multiple trips to the shipping office. When that effort was over, I, along with several members of my staff, felt too tired to move. And yet, there was a palpable feeling of accomplishment, a feeling of victory.

For those full-time professionals in the proposal development business, proposals must become a way of life if we are to survive and grow in our careers. Alternative strategies for time management, stress management, family life, and personal pursuits must be developed and nurtured. In ways analogous to military service, the proposal professional must adjust quickly despite tiredness, personal and family concerns, time of day or night, and level of pressure. But the possibility of personal satisfaction from performing proposal work well can be second to none.

Chapter 1

Competitive proposals and small business

Successful Proposal Strategies for Small Businesses: Using Knowledge Management to Win Government, Private-Sector, and International Contracts, Fourth Edition, is designed to provide entrepreneurs, as well as beginner and experienced proposal managers, capture managers, proposal writers, proposal specialists and coordinators, and business development staff with a useful resource for planning, organizing, managing, and preparing effective responses to U.S. federal government requests for proposals (RFPs), requests for solutions (RFSs), and architect-engineer (A-E) standard form (SF) 330s. (Architectural and engineering firms submit SF330s routinely to establish their credentials with client organizations.) There is also significant attention devoted to responding to U.S. private-sector solicitations and international tenders.

This book illustrates the close relationship between the federal acquisition process and the response life cycle that unfolds within the contractor community. The specialized statutory and regulatory structure that currently governs and dominates the federal acquisition process and the contractor proposal process is summarized. Important and exciting new directions in federal electronic commerce (EC) following the issuance of George W. Bush's President's Management Agenda (PMA) and the passage of the Federal Acquisition Streamlining Act (FASA) and the Federal Acquisition Reform Act (FARA) are highlighted. Ethical business acquisition practices are emphasized, and effective long-term marketing and customer-relationship building approaches are presented.

Small businesses are confronted with distinctive opportunities and constraints in the federal marketplace. *Successful Proposal Strategies for Small Businesses* focuses specifically on small business enterprises, exploring the important human and organizational dynamics related to the proposal life cycle that can facilitate success in acquiring new business. Thinking to win is a crucial aspect in the world of federal, private-sector, and international procurement.

Salient points in the contractor proposal response life cycle are discussed in detail, as are the major components of the proposal documents and the client's RFPs. The role of a small company's proposal manager is explored at length, and valuable knowledge management (KM) activities in support of the proposal process are described. Effective proposal-writing techniques are provided along with successful proposal publication and production scenarios. Proposal and marketing cost-tracking, control, and recovery strategies are reviewed; and select client and competitor information and intelligence sources for the U.S. government, U.S. private-sector, and international opportunities are enumerated (Appendix C). Guidance for planning and producing compliant and responsive SF330s is presented. And structuring proposals for international and U.S. private-sector clients is discussed as well. Finally, to support the users of *Successful Proposal Strategies for Small Businesses*, a lengthy and expanded listing of proposal, business, and acquisition-related acronyms is provided as are definitions of select terminology (Appendix D).

No one person or methodology can offer absolutely definitive step-by-step instructions to win federal, private-sector, or international proposals. There are no shortcuts to building and growing an entire business development infrastructure to market clients, develop long-term professional relationships, and win new business. In recognition of the hard work, right thinking, informed decisions, careful planning, and exacting execution of proper proposal techniques, this book is offered as a starting point in proposal literacy. We hope that it serves as a users' manual, consulted frequently for suggestions and guidance throughout the proposal planning and response process. Best wishes for successful proposals in your company's future!

1.1 Overview

Winning. The federal competitive procurement process [1] is absolutely binary—contractors either win or lose with their proposals. With the exception of multiple-award situations, there are no rewards for coming in second. To allocate your company's bid and proposal (B&P), marketing, and internal research and development (IR&D) funds to pursue procurements for which there is only a marginal probability of winning is, at best, questionable business planning. Federal agencies often have a variety of domestic, as well as overseas,[1] contractor or vendor firms from which to select a specific supplier of goods or services. At a minimum, you have to know your potential client and his or her requirements, as well as hopes, fears, and biases; and, in turn, your client must be made aware of your company's particular technical capabilities, relevant contractual experience, managerial experience, available human talent, and financial stability in the context of an ongoing marketing relationship. One or two briefings from your company to top-level government agency administrators will most likely be insufficient to secure new business in the competitive federal marketplace. This applies to the state, municipality, and U.S. private-sector marketplaces as well. Organizations, in general, procure goods and services from companies that they have come to know and trust and that have demonstrated an ongoing interest in an organization's technical, operational, programmatic, and profitability issues. Increasingly, client organizations expect your company to share both technological and cost risks for a given program.

Many small contracting firms that provide goods and services to the federal government are primarily or even solely dependent upon federal contracts for their survival and growth. Consequently, proposal development, management, design, and preparation are the most important business activities that your company performs. Proposal development and writing are more than just full-time jobs. It can be a 12- to 16-hour-a-day, 6- or 7-day-a-week effort just to keep from falling hopelessly behind [2]. Proper, intelligent planning and preparation will certainly make proposal development more manageable. Your company should not start developing a proposal unless it intends to win. An exception to this guideline is if

---

1   Competition is growing from Japanese, Taiwanese, Canadian, Western European, and emerging Eastern European nations for U.S. government contracts.

your company wants to submit a proposal on a particular procurement in order to gain experience in assembling proposals or to gain recognition from the government as a potential supplier [3]. The American Graduate University suggests that as many as three-quarters of the proposals received by government procuring agencies are deemed to be nonresponsive or inadequate [4]. If your company competes heavily in the federal marketplace, then proposals are your most important product.

It does not matter how large your company is. For example, let us assume that yours is a company with $12 million posted in revenue during the last fiscal year. To simply maintain revenues at that level during the next fiscal year, you will burn $1 million each month in contract backlog, as shown in Figure 1.1. That means that you must win $1 million each month in new or recompete business just to keep the revenue pipeline full. Yet winning $1 million per month in new or recompete business will not allow your company to grow revenuewise at all! To put that $1 million of business per month in appropriate context—your company would have to bid $3 million per month in proposals and have a win ratio of 33% to bring in that level of revenue. And $3 million worth of proposals translates into identifying two to three times that amount in potential marketing opportunities that then have to be qualified and pursued. Many times, release schedules for procurement opportunities slip, or funding is withheld, or the specific requirements get rolled into a larger procurement. As a result, what appears to be a solid lead in January has evaporated by June. See Figure 1.2 for an illustration of this pipeline process. Note that business development has *bookings* goals; operating groups have *revenue* goals. The same applies for a company with $1.2 billion of posted revenue.

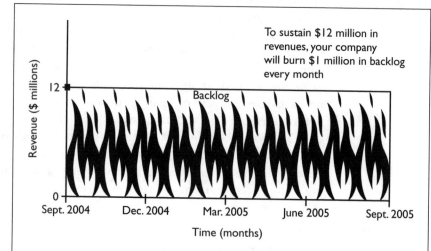

Figure 1.1
Contract backlog burn rate.

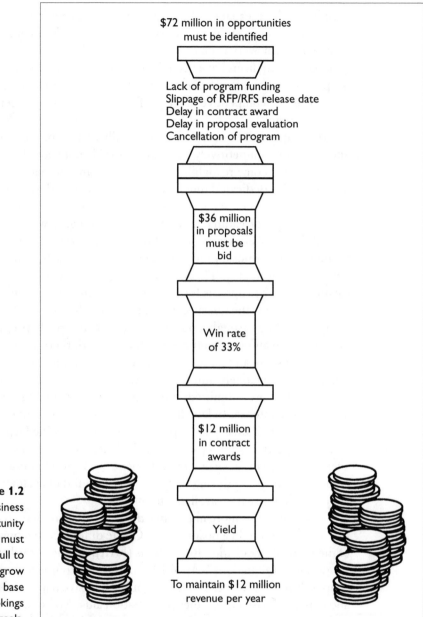

Figure 1.2
Your business
opportunity
pipeline must
remain full to
maintain and grow
your revenue base
and meet bookings
goals.

"Without a plan, the proposal process will be chaotic and the product, at best, will be inferior" [5]. Gone is the time of last-minute, haphazard proposal preparation by a few individuals working in isolation from senior, in-house review, such resources as a proposal knowledge base, and other corporate or divisional guidance. Your company simply cannot compete effectively with the many U.S.-based and overseas contracting firms if every proposal you submit is not your finest effort. Your company will, of course, not win every procurement—25% to 40% is a reasonable win ratio, although certain firms have been documented to win 60% or more of their proposals consistently—but you must strive to have each and every proposal be in the competitive range [6] from a technical, management, and cost standpoint.

It is important to note that a technically sound, programmatically effective, and competitively priced proposal is not enough. With content and cost must come readability, appearance, and format. And these elements require dedicated time to accomplish. Cover design, page formatting, editing, generating graphics, word processing and desktop publishing, proofreading, and photoreproducing, as well as collating, assembling, and electronic uploading to client Web sites are all vital steps in the overall proposal preparation cycle. Put yourself in the role of a government evaluator. That person, along with his or her colleagues, has to look at many proposals for each procurement. Why should I do business with you? is the question they are asking themselves. Would you enjoy struggling through a poorly written, amateurishly prepared document in the evening or on the weekend? Indeed, there are increasing numbers of small and large businesses chasing fewer and fewer federal dollars. Even relatively minor procurements are resulting in 50 or more proposals that are submitted. For major procurements, the competition is even more intense. In late 2003, the federal government received 430 proposals from industry in response to one high-profile Department of Commerce services bid opportunity. Debriefings across a wide variety of agencies suggest that evaluators are spending 15 to 30 minutes on each company's proposal during the preliminary round of evaluation. There simply is no more time available to them.

As a result, it is more imperative than ever that your company's proposal stand out in a positive way. Create difference! Section 52.215-7 of the Federal Acquisition Regulations (FAR), "Unnecessarily Elaborate Proposals or Quotations" (April 1984), cautions contractors not to submit proposals that contain elaborate artwork, expensive paper and bindings, and expensive visual and other presentation aids. To be sure, certain federal agencies, as well as state and local organizations that follow the FAR, will look unfavorably on any proposal documents that go beyond basic typewriter-level presentation values. Yet your competitors are spending tens of thousands of dollars both in-house and through professional proposal consulting firms to prepare full-color, graphics-intensive, high-impact proposal documents and multimedia oral presentations. The challenge is to know your client well enough to sense what level of proposal media and presentation style they will respond to favorably. Some clients stipulate that proposals should demonstrate the quality of deliverable document that your company is capable of producing. You will have to balance the perception of cost consciousness in your proposal documents and electronic submittals with the genuine need to make your volumes establish the standard of presentation excellence among those submitted for a particular procurement.

1.2 From set-asides to full-and-open competition

There are no "bluebirds" in full-and-open competition. The gravy train—to the extent that it still persists in the business world—is over! A reorientation and fundamental transformation of your company's collective thinking and attitude will be required to begin the challenging shift from 8(a) [7] set-aside procurement to competitive federal business acquisition. The term *8(a)* refers to the U.S. Small Business Administration's (SBA; http://www.sba.gov) program to assist qualified small and woman- and minority-owned businesses during their early years of operation. Program participation is divided into two stages: developmental and transitional. The developmental stage lasts 4 years and the transitional stage spans 5 years. Established in July 1953 under the Eisenhower administration, the SBA provides financial, technical, and programmatic assistance to help Americans start, operate, and grow their businesses. The 8(a) program, named for Section 8(a) of the Small Business Act, is the most well-known element of the Minority Enterprise Development (MED) program. Since 1953, the SBA has provided assistance to more than 20 million small business owners. The SBA provides financial assistance through its participating lenders in the form of loan guarantees, not direct loans. In 2000, for example, the SBA backed more than $12.3 billion in loans to small businesses. The SBA's FY2004 budget request stands at $797.9 million.

Currently, Hector V. Barreto, Jr., a Hispanic-American with a long history of small business support, administers the SBA. His predecessor, Aida Alvarez, suggested that diversity, technology, and globalization were critical to small business success in the twenty-first century.

Today, the SBA has a leadership role in implementing President Bush's Small Business Agenda to the benefit of all 23 million small businesses nationwide. Among the core strategic programmatic goals for the SBA are "championing small business interests" and "empowering entrepreneurs," as presented in the FY2004 *SBA Budget Request & Performance Plan* [8]. According to the SBA Performance Scorecard, fully 23% of federal prime contracts will go to small businesses in FY2004 [9].

Importantly, 40% of all subcontracting procurement dollars are targeted for small business as well in FY2004 [10]. The Performance Scorecard is based on the goal hierarchy established by the Government Performance Results Act (GPRA). In accordance with GPRA, SBA identifies its overarching strategic goals, the outcome goals associated with each general goal, and the specific performance goals and performance indicators that are used to measure how programs contribute toward outcome goals [11].

A new offering from SBA is the online 8(a) Procurement Academy (http://www.sba.gov/gcbd/accessing_contracts/), which offers CD and Web-based training to all 8(a) firms. The course has five modules: (1) Getting Started, (2) Identify Your Customer, (3) Develop Your Market Strategy, (4) Compete for Contract, and (5) Perform for Success.

The SBA's national Small Business Week honors the contributions of the nation's small business owners. Winners are selected on their record of stability, growth in employment and sales, sound financial reports, innovation, and the company's response to adversity, as well as community service.

Changing attitudes can be a difficult and lengthy process. The process of change must begin and be fostered on an ongoing basis by senior management, through such forums as marketing meetings, acquisition team meetings, proposal kickoff meetings, your company's internal newsletter or communications vehicle (intranet-based or hard copy), and project management mentoring programs. The thinking and processes that proved so successful and comfortable during the early developmental 8(a) years of your company's history are often precisely the very things that thwart your company's potential for growth in the competitive arena. Many entrepreneurial companies are characterized by informal business organizations and cultures that are functional and effective for small companies only.

However, if companies are successful and grow in terms of revenue, equity, human resources, and client and technical knowledge, they will reach a point at which an informal culture and organization are thoroughly inadequate. This is particularly apparent in the areas of planning, management structure and processes, internal and external communications, and support infrastructure. Successful, growing companies should reorganize, bring in new senior operations and business development executives as appropriate, and develop and implement a strategic planning process [12]. Management responsibilities should be delegated downward and outward so that a small company's organizational structure is not so "sharply hierarchical" [13]. Empowerment is critical to the lowest possible level within your organization. Senior management should actively encourage all staff to identify and understand problems and then propose positive, team-oriented solutions. When implemented, this approach helps to leverage everyone's talents and domain-specific expertise and contributes to the collective knowledge base of the company.

Dedicated effort in accordance with a well-defined, yet flexible, plan, broad-based and in-depth knowledge of your clients and competitors, superlative performance on past and present projects, and a formalized company organization and communication network all contribute to successful proposals. In an extremely important quantitative study conducted by Price Waterhouse from 1990 to 1992, it was determined that companies that exhibit superior performance as measured by competitive contract awards managed "their business acquisition as a formal, disciplined process. These companies view business acquisition as a structured set of interrelated activities to win contract awards. The superior performers continuously improve the methods they use to pursue opportunities" [14].

In an effort to help ensure the long-term success of participants in the 8(a) program, the SBA has established goals for the percentage of 8(a) and non-8(a) contracts that a company should pursue, with the number of non-8(a) contracts increasing each year. A company is put on a schedule during its final 5 transition years in the 8(a) program, with non-8(a) business targeted to be 55% to 75% of company revenues by the final year prior to graduation into the arena of full and open competition [15]. Your company will need to develop an exit strategy from the 8(a) program if you are to survive and prosper. That strategy might include developing a balanced albeit aggressive mix of government and commercial contracts, establishing long-term business partnerships with major prime contractors, and mergers and acquisitions. Small businesses that have been successful in leveraging partnerships with major prime contractors into increased revenue point to their flexible pricing strategies, solid past performance record, and sufficient financial backing [16]. The important point is to begin a structured, ongoing planning process to envision business life beyond the 8(a) program several years prior to graduation. One extremely valuable approach is to conduct semiannual focused planning sessions, during which specific actions required for successful 8(a) exit are identified, documented, communicated, and acted upon. You will want to assess the following parameters:

- Business development structure and processes;
- Accounting system and accounting practices;
- Project management approaches and processes;
- Procurement processes;
- Training requirements;

- Delegation of authority inside your company;
- Internal lines of communication;
- Recruitment capacity and processes;
- Profitability goals;
- Core competencies;
- Required external certifications and accreditations (e.g., SEI CMMI®, ISO, IEEE).

It is important to point out cogent examples of companies that have successfully migrated out of the 8(a) program and flourished in the open competitive arena. Native-American-owned, Virginia-based CEXEC. graduated from the 8(a) program in 1993. Today, CEXEC employs 250 people and staffs projects in 18 states. Headed by President and CEO Douglas C. Rhoades, CEXEC received the Department of the Interior's Minority Enterprise Contractor of the Year Award in 1997, 4 years after graduation. User Technology Associates Group (UTA), led by Yong K. Kim, graduated successfully in 1997 and proceeded to attain revenues of $80 million in 2000 with a professional staff of 900. UTA was the number one fastest growing company in the Washington, D.C., area with 13,500% growth. UTA was acquired by DigitalNet Holdings, Inc., of Herndon, Virginia.

Aegir Systems, which provides engineering and technical services, was founded in 1981 by Ella D. Williams. It now employs more than 75 professionals with $4.5 million in sales. This Oxnard, California–based firm graduated from the 8(a) program in 1995. Ms. Williams is quoted as saying that without the SBA 8(a) program, there is "no way I would be in business today." Hispanic-American William Soza, president and founder of SOZA & Company, participated in the 8(a) program from 1989 to 1997. This Virginia-based company provides information technology, management consulting, and financial services. Today, SOZA & Company is a wholly owned subsidiary of Perot Systems Government Services, Inc.

And, finally, Sytel in Bethesda, Maryland, a woman-owned firm that provides e-business (eB) services, networking, and information technology support, graduated successfully from the 8(a) program in June 1998. The company currently employs 300 professionals. Sytel is one of only 53 companies worldwide to be inducted into the *INC.* 500 Hall of Fame. Jeannette Lee White, Sytel's president and CEO, was named by *Success* magazine as one of the top entrepreneurs in the United States.

1.3 Small business constraints

In terms directly relevant to proposal design, development, and preparation, many small businesses must navigate effectively amidst a narrow opportunity pipeline, very limited B&P funds, a lack of depth in human resources, a small business base, a contract backlog deficit, a low level of contractual experience, a lack of name recognition in the marketplace, and line of credit challenges. A small business base, for example, can lead to higher indirect costs, which in turn can place a company at a competitive disadvantage during procurement efforts. An insufficient staff can translate to few or no people dedicated to the tasks of advanced and strategic planning, ongoing marketing and relationship building with particular client organizations, proposal operations, proposal reviews, proposal editing and proofreading, proposal publication, oral-presentation preparation and delivery, and postproposal marketing. Staffing challenges emerge quickly as full-time project managers work 40 billable hours each week for their client and then additional time to serve as proposal managers, proposal reviewers, or presentation team members. (In predominately service-oriented contracting firms, the company's overall profitability is affected by the degree to which its personnel are fully billable. Transfer ratios, that is, billable time versus total time worked, must remain very high.) And thin contractual experience can lead to low or marginal scores received for "Past Performance" or "Relevant Experience" sections of the RFP, RFS, and the SF330.

1.4 Maximizing small business strengths

America's 23 million small businesses employ more than 50% of the private workforce. According to the SBA, a record 898,000 new small businesses recorded initial employment during 1998. And women are creating new businesses and new jobs at double the national rate, and own nearly 40% of all firms in the United States. In 1994, small businesses were awarded 28% of the $160 billion worth of available federal government contract awards [17].

Small businesses (corporations that employ 500 people or less) constitute 99.7% of U.S. employers. "These companies have certain competitive advantages: They are lean in terms of administration, they can position themselves in a market niche that large corporations cannot fill, and they can offer superior service to customers" [18]. On June 24, 1998, the White House announced new federal rules to give minority firms an edge when bidding for federal government contracts while respecting a 1995 U.S.

Supreme Court ruling that limited affirmative action programs. Under these new rules, small firms that are certified to be disadvantaged by the U.S. SBA will receive a price break of 10% in calculating the lowest bidder for government contracts [19].

It is the policy of the United States, as stated in the Small Business Act, that all small businesses have the maximum practicable opportunity to participate in providing goods and services to the government. To ensure that small businesses get their fair share, the SBA negotiates annual procurement preference goals with each federal agency and reviews each agency's results. The SBA is responsible for ensuring that the statutory government-wide goals are met in the aggregate. The current statutory goals are as follows:

- 23% of prime contracts for small businesses;
- 7% of prime and subcontracts for small disadvantaged businesses;
- 5% of prime and subcontracts for women-owned small businesses;
- 3% of prime contracts for Historically Underutilized Business Zone (HUBZone; see Appendix D) small businesses;
- 3% of prime and subcontracts for service-disabled veteran-owned small businesses.

Small businesses have the potential to respond rapidly to emerging business opportunities because they have fewer layers of management approval in the decision-making chain. Company policies can be modified quickly to meet client requests and requirements [20]. Small businesses can carefully control their growth in terms of acquiring technical talent (subject matter experts) and penetrating new market sectors. The opportunity for excellent in-house communications throughout the network of authority exists with small businesses. Small companies are ideally positioned to develop, right from the start, open-architecture internal automated KM systems for maintaining and searching staff résumés, project summaries, proposal modules, lessons learned, success stories, and marketing opportunities. And because of the staffing deficit, people tend to become cross-trained and proficient in a wide variety of tasks. More people are given the opportunity to understand the big picture of the proposal life cycle and of specific business targets. Conversely, in large, multidivision corporations, very few staff fully understand the multidimensional complexities of massive procurement targets. On large procurements, for example, some major firms devote one or more staff exclusively to handling subcontractor résumés. In a small firm, that level of work breakdown is simply not possible.

1.5 SBIR and STTR programs

Important mechanisms for generating revenue in the small business community for those firms with strong scientific or engineering capabilities include the three-phase Small Business Innovation Research (SBIR) program (see http://www.sba.gov/sbir/indexsbir-sttr.html) and the Small Business Technology Transfer (STTR) pilot program (see http://www.sba.gov/sbir/indexsbir-sttr.html). Enacted on July 22, 1982, as part of the Small Business Innovation Development Act (P.L. 97-219), the SBIR program encourages small businesses with fewer than 500 employees to explore their technological potential and provides the incentive to profit from its commercialization. This legislation requires federal agencies to set aside special funding for relevant small business research and development (R&D). Worldwide commercial rights to any patents normally will go to the small company. Public Law 106-55 reauthorized the SBIR program through 2008. The SBIR program is highly competitive and merit based—it is in no way an assistance program for small businesses. It is, however, open only to small American-owned and independently operated businesses and is not intended for nonprofit organizations.

The risks associated with conducting significant R&D efforts are often beyond the economic and resource capabilities of small businesses. SBIR, in effect, protects small businesses and enables them to compete on the same level as larger businesses. In FY2001, the SBIR program produced 3,215 Phase I awards and 1,533 Phase II awards for approximately $1.5 billion. Since its beginning, the SBIR program has awarded more than $4 billion to various companies, thus allowing them to test high-risk theories and develop innovative technologies. Of note is that companies with 10 or fewer employees have won more than one-third of all SBIR awards to date, and the SBIR program has consistently awarded more than 10% of its funds to minority-owned firms. Five of the 10 SBIR federal agencies required to participate in the SBIR program based on their R&D budgets make more than 90% of the awards annually—Department of Defense (DoD), Department of Health and Human Services (DHHS), Department of Energy (DOE), NASA, and the National Science Foundation (NSF). Other agencies include the Department of Agriculture (USDA), Department of Education, Department of Transportation (DOT), EPA, and the Department of Commerce.

The Small Business Administration's Office of Technology has the responsibility for coordinating the SBIR and STTR programs. Upon request, SBA will mail you a quarterly Presolicitation Announcement (PSA) that lists the agencies that will make SBIR offerings in the next fiscal quarter, along with the release, closing, and award announcement dates. The quarterly PSA also provides a one-line statement for each SBIR topic. However, this PSA is not a substitute for the agency SBIR solicitations themselves. DoD will automatically send its solicitations to all companies on the SBA PSA mailing list, but no other agency does. You will have to request SBIR solicitations from each agency in which your company is interested each year. Some agencies, such as DoD and DHHS, issue more than one pamphlet or bulletin each year in which SBIR solicitations are published. Each participating federal agency publishes an extremely helpful volume of abstracts that summarizes the SBIR proposals that it funds each year. To obtain a copy, you can contact the SBIR program manager in each agency in which your company has interest.

The SBIR program has three phases: Phase I, Feasibility Study; Phase II, Full-scale Research; and Phase III, Commercialization. More than 15,000 small businesses have received at least one Phase I SBIR award. Funding at Phase I extends up to $100,000, and finances up to 6 months of research. Phase II funding can reach $750,000, and research can span 2 years. Only Phase I winners are considered for Phase II. With the DOE, for example, success ratios have been about 12% in Phase I and 45% in Phase II. Some states, such as Kentucky through its SBIR Bridge Grant program established in 1988, as well as Alaska and Delaware, assist small firms to continue product development research projects begun under federal Phase I SBIR awards.

As a result of the Small Business Innovation Research Program Reauthorization Act of 2000, the SBA's Office of Technology established the *Federal and State Technology Partnership* program, or FAST. FAST was designed to strengthen the technological competitiveness of small businesses in all 50 states as well as the District of Columbia and U.S. territories. FAST is a competitive grants program that allows each state to receive funding that can be used to provide an array of services in support of the SBIR program.

In most cases, Phases I and II provide full allowable costs and a negotiated fee or profit for the small company. Phase III involves no SBIR funding, although the SBA has developed the Commercialization Matching System to help SBIR awardees locate funding sources for finalizing their innovations. The small business might procure funding for Phase III of the SBIR process from commercial banks (e.g., the Small Business Investment Corporation, the equity investment arm of the Bank of America Community Development Banking Group); venture capitalists; private-sector nonprofit small business lending groups such as the Development Credit Fund, Inc. (DCF) of Baltimore, Maryland (http://www.developmentcredit.com), or the Maryland-based Council for Economic and Business Opportunity (CEBO); large companies; or other non-SBIR federal agency funding. Other help is available for companies entering the commercialization phase of the SBIR program. For example,

Dawnbreaker (http://www.dawnbreaker.com), a professional services firm located in Rochester, New York, has assisted more than 1,500 DOE SBIR awardees since 1989. Dawnbreaker staff work with DOE SBIR Phase II grantees to develop a strategic business plan and to prepare the firm for commercialization. Dawnbreaker then hosts two Commercialization Opportunity Forums in Washington, D.C., each year to showcase firms that are ready to enter the third SBIR phase [21].

More than 39% of Phase II SBIR projects will result in a commercialized product. Successfully commercialized SBIR projects include DR-LINK, a natural language retrieval system based on linguistic technology that processes ongoing news streams or information from databases. Patent examiners at the U.S. Patent and Trademark Office now use the DR-LINK system daily to search for prior art in more than 4,000 published databases. This SBIR effort was funded through DARPA, the Defense Advanced Research Projects Agency. SBIR support from the National Science Foundation, DOE, and DoD allowed AstroPower of Newark, Delaware, to develop superior thin-layer silicon- and gallium-arsenide technology with optical and speed advantages in photovoltaic devices. The NSF and National Institutes of Health (NIH) have made major SBIR contributions to research conducted by Martek Corporation in Columbia, Maryland. This biotechnology firm discovered how to make a critical ingredient needed by infants that is normally found in mothers' milk. Medical and health organizations in Europe and the World Health Organization (WHO) have recommended that this critical ingredient be added to infant formulas to offset nutrient deficiencies in babies that are not breast-fed. And Scientific Computing Components, in New Haven, Connecticut, has leveraged SBIR funds to produce a number of breakthroughs in commercial software related to high-performance computing.

Although the format for SBIR Phase I proposals varies across the participating federal agencies, a typical format is as follows:

- Cover sheet;
- Project abstract, anticipated benefits, and key words;
- Identification of the problem or opportunity;
- Background information;
- Phase I technical objectives;
- Phase I work plan;
- Related work;
- Relationship to future research and development;
- Potential postapplications;

- Key personnel;
- Facilities and equipment;
- Consultants;
- Current and pending support;
- Budget;
- Previous SBIR awards.

The STTR pilot program's role is to foster the innovation necessary to meet America's scientific and technical challenges in the twenty-first century. Central to the program is the expansion of the public- and private-sector partnership to include joint venture opportunities for small businesses. Each year because their R&D budgets exceed $1 billion, DoD, DOE, DHHS, NASA, and NSF are required under the STTR to reserve a portion of their R&D funds for award to small businesses and nonprofit research institution (e.g., colleges and universities) partnerships. At least 40% of the work must be performed by the small business. The nonprofit research institution must be based in the United States, and can be a nonprofit college or university, domestic nonprofit research organization, or Federally Funded R&D Center (FFRDC). Like SBIR, STTR is a competitive, three-phase program coordinated by the SBA and announced through the PSA process. Award thresholds are $100,000 for Phase I and $500,000 for Phase II. In FY2001, the STTR program distributed $78 million across 224 Phase I awards and 113 Phase II awards.

1.6 Organizing your company to acquire new business

To support your company's efforts to acquire, that is, win new and follow-on business, consider forming, and then actively supporting, a business development or advanced planning group.[2] In many smaller firms, marketing and proposal efforts are handled exclusively through each division or line organization. One division may or may not be aware of duplication of marketing efforts with other divisions, related contractual experience performed by another division, or human talent and subject-

---

2   The name of these business and planning groups varies from company to company. Some are called business development groups (BDGs), advanced planning groups, strategic planning and business development groups (SP&BDGs), special programs groups, and marketing departments, for example.

matter expertise in another division, for example. The formation of a centralized corporate BDG should not preclude a given division's involvement in its own business planning and proposal development. Rather, the BDG can serve to focus, channel, and support divisional business-related activities in accordance with your company's formalized mission statement [22] and strategic plan. Because of its corporate vantage, the BDG can help identify and make available the appropriate human talent, material resources, and information resident throughout your entire company in order to pursue a business opportunity. The functional charter of the BDG can also extend to include the following closely related activities:

- Strategic, business, and marketing planning;
- Business opportunity/Federal Business Opportunities (FBO) and other opportunity tracking and reporting;
- Intelligence gathering: marketing support and client contacts;
- Formalizing the process of establishing business objectives, gathering data, analyzing and synthesizing data, prioritizing, and action planning;
- Acquisition team formation and guidance;
- Bid–no bid decision-making coordination;
- Teaming agreement coordination;
- Cost strategizing for proposals;
- Company information management, distribution, and archiving (including proposal data center, or automated knowledge base);
- Proposal and documentation standards development and dissemination;
- Proposal coordination and production;
- Oral presentation development and coaching;
- Company image development and public relations;
- Corporate communications (such as newsletters, trade shows, and advertising);
- Marketing and proposal management training.

To ensure adequate connection with and visibility from senior management, a full-scale BDG should be under the leadership of a vice president (VP) for business development. The most appropriate candidate for this pivotal position is an individual with an advanced and relevant technical degree coupled with at least 5 years of clearly demonstrated competitive business development experience and success with "closing the deal" in the federal government arena. Ensure that the candidates you consider can bring a successful track record of identifying, qualifying, and winning major-impact programs. Contacts alone are a necessary but insufficient gauge of a business developer's successful performance. This individual must also possess or develop a solid knowledge of your company's internal operations and capabilities, as well as the buying habits and specific sell points of your clients.

Even under the constraints faced by very small companies (less than $5 million in annual revenues), this VP functions most effectively when he or she is not obligated to be an on-the-road company marketeer as well as an internal business-development planner, organizer, and administrator. (In addition, an administrative assistant seems absolutely essential to enhance the functionality of the VP position.) Under this VP's guidance would be two primary functional groups: external and internal sales support (see Figure 1.3). The external sales support element might include a full-time corporate marketeer(s) as well as a key division manager(s). External sales efforts would involve direct client contact and interaction at multiple technical and administrative levels. It would also extend to relationship building with potential major teaming partners and specialty subcontractors appropriate to your company's lines of business (LOBs) or product lines.

The internal sales support element might be subdivided further into proposal development and production, KM, and corporate communications, as depicted in Figure 1.4. Proposal development and production should logically include publication of the proposal documents. Proposal design and development become extremely challenging if they do not include oversight of the priorities and resources of the publications group. The function and focus of both the external, as well as internal, sales support elements are to

Figure 1.3
A cooperative sales approach ensures effective information exchange.

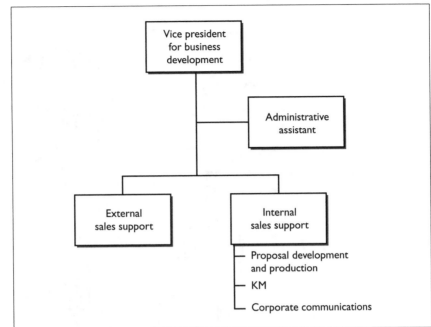

Figure 1.4
Suggested BDG
organization.

project a professional, client-oriented corporate image. Understanding your client's business, technical requirements, hopes, fears, and biases and demonstrating that understanding in every proposal you prepare is absolutely critical to your success in the federal marketplace in 2005 and beyond (see Figure 1.5). Superior proposals should complement the public relations and advertising activities of your firm.

Effective interaction and cooperative information exchange and knowledge transfer between the external and internal sales support staff are absolutely critical to proposal success. Neither group succeeds on its own. Marketing intelligence learned from client and competitor interaction must be infused into the proposal during the important bid–no bid decision-making process, prekickoff meeting planning sessions, the kickoff meeting itself, the formal internal review steps, and postproposal activities, such as preparation for oral presentations and Final Proposal Revision (FPR). Clear, formalized, well-supported internal sales support processes contribute to winning new business. Too many times, companies view the identification of marketing opportunities as paramount to their success. As important as these leads can be, unless there is a well-defined, structured set of processes and methodologies internal to your company that harness the necessary human talent, information technology resources, dedicated space, and B&P monies, the leads do not result in contract awards. They result in also-rans (proposals submitted that do not win).

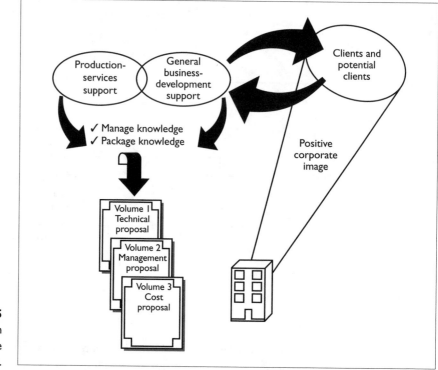

Figure 1.5
Projecting an
appropriate
corporate image.

It has been this author's observation that small companies tend to undergo oscillations, and even convulsions, between centralized business development and control, and decentralized divisional business activities. Instead of using the BDG as a vital corporate support structure, some companies prefer to dissolve or de-scope the BDG, despite the 1- to 2-year extraordinary level of effort generally required to establish the BDG in the first place. Full-time corporate business development staff can be significantly more effective than employees assigned to BDG support on a part-time basis only. A certain level of centralized oversight and control is highly beneficial and cost-effective. Continual restructuring and refocusing of marketing activities result in the diversion of valuable time and energy into counterproductive internal organizational challenges. Building an effective and responsive business development infrastructure takes time—on the order of 2 or 3 years. Senior management must be attentive to, yet patient with, this dynamic process.

Figure 1.6 illustrates a proven growth paradigm for small businesses that contract with the U.S. federal government. Several key areas associated with thriving within the federal market space require additional exploration, including business culture, management, planning, marketing, and financial/infrastructure.

- Diversify customer base–early!
- Win every GWAC, ID/IQ, and Schedule possible!
- Must have an aggressive customer satisfaction program.
- Stretch your entire organization— pursue and bid at least twice as many opportunities as you can comfortably support.
- Attract and retain really good people—don't get stuck on the wrong side of the 80/20 rule!

Figure 1.6
Proven growth paradigm.

- *Business culture.* Your small company's business culture must be simultaneously employee-centric and customer-focused. Leveraging a corporate value system that builds on trust within the executive leaders, your organization should be driven to performance excellence and the advancement of your customer's mission.

- *Management.* Accountability and decisiveness are two critical ingredients for successful management of your organization. There must be a system for total accountability. Along with accountability must come authority and genuine "skin in the game." Key people, who should be seasoned performers, must feel emotionally and financially engaged with the business outcomes of your organization. Decisiveness includes the willingness and capacity to take measured risks.

- *Planning.* Every organization functions best with a strategic plan (see Section 1.7) that derives from a corporate vision and mission statement. It is vital to review actual progress against goals, objectives, strategies, and action plans on a regular basis. Goals generally have 3- to 5-year time horizons, whereas objectives have 1-year time frames. All goals and objectives should be meaningful, measurable, and achievable.

- *Marketing.* The keys in the marketing area are staff and focus. You must stay focused on your company's core competencies when pursuing bid opportunities. Avoid the "shotgun approach" that leads to dilution of B&P funds and human resources. In addition, ensure that there is a business development leader as well as a proposal development leader. The proposal development leader should have a strong command of proposal processes as well as knowledge management practices.

- *Financial/infrastructure.* Your organization must develop excellent banking relationships. Invest in robust accounting systems because timely and accurate financial status reporting is essential. Integrate systems, including a human resources information system (HRIS), recruitment, accounting, and procurement. Pursue external quality assurance certification and assessment (e.g., ISO 9001:2000, SEI CMMI®). Consider hiring a chief administrative officer to integrate and manage the infrastructure people and processes.

1.7 Effective strategic and mission planning

Strategic and mission planning are crucial considerations to ensure a planned pattern of growth for your company and to focus your bid–no bid decision making. According to Thompson and Strickland, a strategic plan is "a comprehensive statement about the organization's mission and future direction, near-term and long-term performance targets, and how management intends to produce the desired results and fulfill the mission, given the organization's overall situation" [23]. Strategic planning works best (that is, is more widely accepted and more easily implemented) when it is interactive to a certain extent, not merely formulated by a separate group who creates the plan with everyone else then told to follow it. The starting point for the strategic planning process is understanding the fundamental goal of your company. No matter how small your company may be, this goal should be identified, understood, developed, communicated, and revisited. Important questions to keep in mind when formulating a strategic plan for your company are listed next [24]:

- Where is your company now in terms of market sector or LOB (that is, defining your business)?
- What is your company's mission in the marketplace?
- What is your company's distinctive (core) competence?
- Who are your company's principal clients (i.e., market segmentation)?
- What are your principal services, products, or knowledge lines?
- What is your company's brand?
- Where does your company want to be businesswise in 5 years? Develop a vision of the future state of your business.

- What are your company's internal strengths and weaknesses (for example, business, technical, knowledge base, internal business processes, public relations, and fiscal capability)?
- What external business or economic opportunities, threats, or challenges (e.g., the entry of new competitors into the marketplace, adverse government policies, adverse demographic changes, and global security issues) face your company now [25]? These are often referred to as SWOTs, for strengths, weaknesses, opportunities, and threats.
- Who are your primary competitors? Understanding how your competition thinks and acts is critical to your success (competitor analysis) [26].
- How are your company's knowledge assets being monitored, developed, and applied?
- How do you identify the domains of knowledge that matter to the competitive strategy of your company? In effect, how do you find the seeds of innovation and sustained competitive advantage in a dynamic global marketplace?

Strategic planning is not a static process; rather, it should be a dynamic, living activity. Business conditions and priorities change over time and technologies refresh constantly, and your strategic plan should reflect this evolution. Provide the necessary resources and steering mechanisms to implement your strategic plan, and revisit and revise the plan regularly. The mission statement is a close adjunct of the strategic plan.

Formulating, articulating, and implementing a meaningful strategic plan is critical for obtaining lines of credit through banking institutions and demonstrating to government auditing agencies that planned business growth over a period of time justifies an expansion in business base and therefore a long-term reduction in company overhead costs. Strategic planning also serves as the overarching guidance for your company's marketing activities, which is the focus of our next chapter. In addition, strategic planning provides formalized, informed guidance for making bid–no bid decisions on specific proposal opportunities. If responding to an RFP and winning that piece of business does not mesh with your company's principal lines of business and mission in the marketplace, then submitting a proposal is not appropriate. Just because your company has the technical capability to perform most of the scope of work does not mean that you should invest the time, energy, and resources into preparing a proposal.

1.8 Converting knowledge into proposal success

The question becomes not whether, but how quickly companies will move to embrace the knowledge model of the corporation.
—Richard C. Huseman and Jon P. Goodman "Realm of the Red Queen," The Corporate Knowledge Center @ EC², p. 15 [27]

A recent Baldridge Foundation survey of 700 *Fortune* 500 chief executive officers (CEOs) determined that knowledge management came in second as the top challenge facing companies—right behind globalization. Dr. Karl E. Sveiby—former executive chairman and co-owner of Ekonomi+Teknik Förlag, one of Scandinavia's biggest publishing companies in the trade press and business press sector, and now an international consultant on knowledge management—defines the term *knowledge management* as the art of creating value by leveraging an organization's intangible assets, or intellectual capital [28]. "Intellectual capital is the most important source for sustainable competitive advantages in companies," suggests Professor Johan Roos of the Imagination Lab Foundation in Lausanne, Switzerland, and formerly of the International Institute for Management Development (IMD) in Switzerland [29]. "The knowledge of any organization is found in its people, substantiated in its business processes, products, customer interactions, and lastly its information systems" [30]. In *Working Knowledge: How Organizations Manage What They Know* (Harvard Business School Press, 1997), Tom Davenport and Laurence Prusak emphasize that "[k]nowledge is as much an act or process as an artifact or thing" [31]. Knowledge is an asset that flows [32].

At the most fundamental level—that of a mathematical equation—KM is a dynamic combination of structured processes and automated tools multiplied by executive-level leadership and vision—and leveraged exponentially by passionate people and communities of practice (CoPs) (see Figure 1.7):

$$KM = [(\text{Processes} + \text{Tools}) \times \text{Vision}]^{\text{People}}$$

For practical purposes, KM is a holistic cluster of sustainable, proactive, conscious, and comprehensive organizational and business activities that encompasses enterprisewide processes, techniques, and professional practices and interactions. These dynamic activities are directed toward enhanced decision support throughout the entire organization, informed business actions, and rapid prototyping of technical solutions, as well as rapid prototyping of proposal documents, improved client support, and

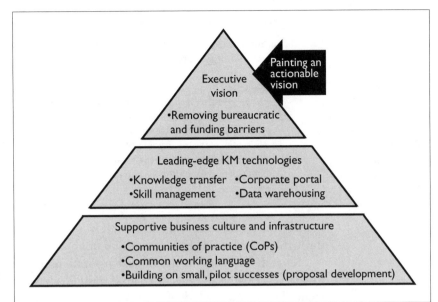

Figure 1.7
Developing a
sustainable KM
solution.

product innovations. Specific process elements in KM span (1) creating, (2) identifying, (3) collecting, (4) indexing, (5) codifying, (6) organizing, (7) evaluating, (8) making visible and actionable, (9) accessing, (10) leveraging, and (11) optimizing institutional knowledge for conservation, distribution, and reuse. When actively supported and valued by forward-looking senior management, a culture of mutual trust, the passion of people at all levels, and technology-based user-driven solutions and tools, KM encourages your organization's people to create, share, and benefit from their own, as well as your stakeholders' and clients', knowledge. KM also encapsulates individual and project team [e.g., communities of practice (CoPs)] knowledge into transferable corporate assets that offer potential market value and provide valuable decision support. Data must be converted into information, which is transformed into knowledge, which, in turn, must grow into authentic understanding and action. Practical benefits of KM are presented in Figure 1.8.

Implementing business processes that actively promote knowledge valuation, development, transfer, management, and congealment will help American companies maximize both intellectual and structural capital. Traditional command-and-control management functions must now give way to mapping and acquiring intellectual capital, communicating a clear vision for the firm, developing and implementing a business lexicon with shared meanings and nuances to foster generalized understanding of strategic goals, and facilitating the rapid assessment and multidirectional flow of knowledge throughout the organization. Caveats and pronouncements

Enterprisewide, ongoing application of intellectual capital to achieve organizational mission and goals

Benefits
•Improves decision support through just-in-time intelligence
•Enhances mission performance
•Facilitates informed business actions
•Fosters rapid prototyping of proposals and technical solutions
•Encourages improved client support
•Supports risk-mitigation initiatives

Figure 1.8
KM provides the ability to anticipate and adapt to change.

issued from senior management will not be dispersed and internalized throughout the knowledge organization or knowledge enterprise.

Since 1991, major international corporations and organizations have proactively established executive-level positions in knowledge management. These firms include Arthur Andersen, Booz Allen Hamilton, Canadian Imperial Bank of Commerce (CIBC), Coopers & Lybrand, Dow Chemical, and IBM Corporation, as well as Monsanto, Skandia AFS, Texas Instruments, U S West, and Xerox [33]. Other companies and organizations at the vanguard of recognizing and acting upon the reality, and the value, of intellectual capital and intangible assets (e.g., corporate-brain power, organizational knowledge, client relationships, innovation ability, and employee morale) include General Electric, Hughes Space & Communications, PricewaterhouseCoopers, and Altria Group, Inc., which have launched considerable efforts to understand and enhance intellectual capital management.

In addition, Buckman Laboratories International, a U.S.-based biotechnology firm, has created a Knowledge Sharing Department. Buckman Labs provides incentives in the form of financial rewards and management positions to those employees who contribute to its knowledge-sharing culture. Chevron has engineered a best practice database to capture and make available the company's collective experience with drilling conditions and innovative solutions to technical problems on site. And Sweden's Celemì company published the world's first audit of intangible assets in its 1995 annual report. Celemì's Intangible Assets Monitor focuses on its customers (image-enhancing customers, from whom testimonials are valuable; brand names; trademarks); their internal organization (patents, computer systems, management infrastructure); and their staff (competencies, flexibility). Skandia published the first-ever annual report supplement on intellectual capital. Pfizer of Switzerland has created competence models for recruiting executives that include knowledge building and sharing as important criteria. WM-data of Sweden, a fast-growing information technology (IT) company links nonfinancial indicators to strategy, and considers financial ratios of little use for management. And Sweden's leading telecommunications company, Telia Sonera, has published an annual Statement of Human Resources for each year of the past decade. This statement includes a profit-and-loss account that visualizes human resources costs and a balance sheet that shows investments in human resources [34].

"The business problem that knowledge management is designed to solve is that knowledge acquired through experience doesn't get reused because it isn't shared in a formal way.... But the ultimate goal isn't creating a departmental island of success recycling. It's giving the organization the capacity to be more effective every passing day with the gathering of institutional memory the way human beings have the capacity to become more effective and mature every day with the accumulation of thoughts and memories" [35]. A representative knowledge maturity model is presented in Figure 1.9. From direct experience in a small information technology business, we know this to be the case. KM processes such as a grassroots birds-of-a-feather program and corporate-level centers for excellence built around core competencies, have been implemented not merely to create internal hubs of institutional contract-related experience and technical know-how, but rather to facilitate the cost-effective, time-efficient rapid prototyping of proposals and presentations related to business development with a minimum of full-time staff. Until very recently, most companies and organizations—large and small—knew how to produce goods or services only. Now they must learn how to apply the right knowledge within appropriate infrastructures, processes, technologies, and geographies. The U.S. federal government's Federal Acquisition Regulations (FAR) still employ only the terms supplies and services (FAR Part 8) in reference to contracting. There is no mention of knowledge as a value-added product or service. On the other hand, International Accounting Standard (IAS) 38 specifically recognizes intangible assets, such as brand names, patents, rights, customer lists, supplier relationships, designs, and computer software.

During a speech before the American Accounting Association's annual meeting held in Dallas, Texas, in August 1997, Michael Sutton—chief accountant of the U.S. Securities and Exchange Commission (SEC)—noted that "historically, accounting has been strongly influenced by the reporting needs of a manufacturing-based economy" [36]. In April 1996, the SEC had convened a symposium on intangible assets in Washington, D.C., during which invited participants from prestigious business, academic, and government organizations discussed issues related to the measurement of intangible assets by preparers of financial reports, concerns about disclosures related to intangible assets, and the experience of U.S.

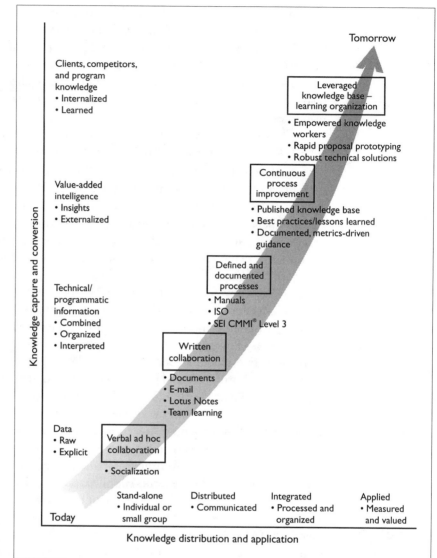

Figure 1.9
Knowledge
maturity model.

and overseas trendsetters with regard to the accounting and disclosure of intangible assets. The SEC, however, has not yet provided any guidelines or issued any directives vis-à-vis intangible assets for direct application in American corporations. Baruch Lev, professor of accounting and finance at New York University's Stern School of Business, concludes that "[n]early 40% of the market valuation of the average company is missing from its balance sheet" [37]. Clearly, your company's success in this new millennium will depend increasingly upon the quality of your people in terms of the level of knowledge, information, and know-how at their fingertips as well as your management's ability to systematically foster highly adaptive and openly collaborative business processes that leverage this collective knowledge and enable intelligent decision making (see Figure 1.10).

KM has become a significant enterprisewide initiative within a growing number of civilian and defense agencies in the U.S. federal government. The General Services Administration (GSA) appointed the federal government's first chief knowledge officer (CKO), Dr. Shereen Remez, in mid-1999. Major KM initiatives are also under way within the Department of the Navy (DoN), U.S. Army and Army Corps of Engineers (USACE), the Federal Aviation Administration (FAA), Federal Highway Administration (FHWA), General Accounting Office (GAO), Department of State (DOS), and many others. According to a recent study published by the Input research firm, the "U.S. Government spending on Knowledge Management (KM) products and services will increase at a compound annual growth rate of 9 percent, from $820 million in GFY 2003 to nearly 1.3 billion by GFY 2008." Many agencies are expected to make significant

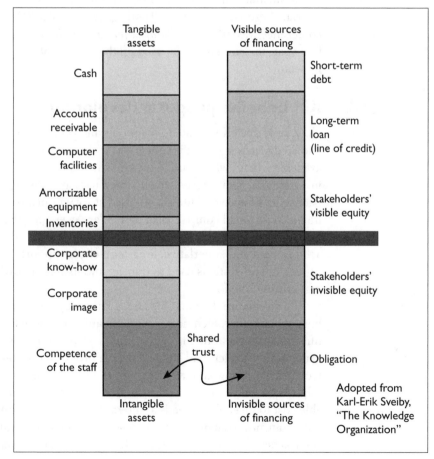

Figure 1.10
Balance sheet of the knowledge organization.

investments in KM in the next 5 years, including the Departments of Homeland Security, State, and Justice. The Federal Chief Information Officers Council (the CIO Council) established the Knowledge Management Working Group as an interagency body to bring the benefits of the government's intellectual assets to all federal organizations, customers, and partners (see http://www.km.gov). Comprised of the CIOs of 29 federal agencies and departments and representatives from the Office of Management and Budget (OMB), the CIO Council works to improve agency practices related to the modernization, sharing, and performance of federal government information assets and resources. The CIO Council, whose charter was codified formally through the E-Government Act of 2002, also is responsible for implementing elements of the Government Paperwork Elimination Act (GPEA), Government Performance and Results Act (GPRA), Federal Information Security Management Act of 2002 (FISMA), and the Information Technology Management Reform Act of 1996 (Clinger-Cohen Act). Importantly, the federal government sponsored the first major 4-day KM e-government (e-Gov) conference in Alexandria, Virginia, in April 2000 and again in April 2001. The second conference was titled *The Catalyst for Electronic Government*. In 2004, its fifth year, the conference focused on the rapid evolution of KM in the public sector.

1.8.1 KM benefits proposal development

Now, how does KM relate to proposals and proposal management? Just as knowledge management is concerned with which knowledge should be available where, when, and in which form within an organization, proposal management—at its core—is concerned with assembling, synthesizing, and packaging knowledge into a fact-based story within a very limited time frame. With small companies in particular, leveraging the collective intellect and institutional knowledge is integral to business development achievement and superlative, long-term client support. Improving the processes and systems used to transfer and manage knowledge is an ongoing, critical function.

Proposals are, first and foremost, knowledge products that include a host of marketing, technical, programmatic, institutional, pricing, and certification information. Through a choreographed process of knowledge generation, transfer, and congealment, a proposal is designed to sell both technical and programmatic capabilities of a company to accomplish all required activities (or provide specified products) on time and at a reasonable cost to meet client needs and requirements (see Figure 1.11). In addition, your proposal must often convey more intangible values and benefits such as peace of mind, problem-solving acumen, fiscal stability, risk

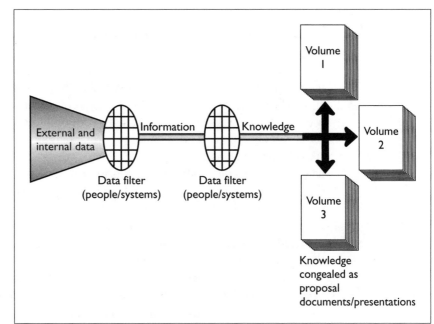

Figure 1.11
The knowledge
transfer process.

mitigation, and so forth. Consider, for example, the Pontiac automobile
television advertisements of several years ago. General Motors was not selling steel, leather, composite materials, and plastics—in effect, the vehicle
per se. Rather, the company was selling excitement (e.g., GM's slogan was
"We build excitement—Pontiac"). The same goes for books. When you
purchase a book, you are not buying wood pulp and soy inks. You are buying insight, a romantic escape, or guidance.

Your company's proposal document(s) is scored, literally, by the client's evaluators against formalized, specific standards in a process analogous to pattern recognition. A proposal is the tangible result of knowledge-
building processes, supported in turn by hard and directed work and
buoyed by a positive collective attitude within your company's proposal
team of knowledge workers, as world-renowned management consultant
Peter Drucker in 1998 termed employees in the new American economy.

In actuality, résumés and project summaries reflect in one form a significant percentage of intellectual capital of an organization, small or large.
Relevant knowledge resident in the employees of the company as gained
through education, training, and professional experience coupled with initiative, dedication, and innovation is highlighted in each résumé. Similarly,
effective project summaries reflect the organizational knowledge of the firm
from technical, managerial, and geographic perspectives.

Crucial to proposal success is the transfer of relevant marketing information, lessons learned from current and past contracts, and technically
and programmatically relevant best practices to appropriate members of

the proposal team. And from there, that information must be interpreted, assigned value, and translated into knowledge that will ultimately be congealed into the actual proposal documents—into narrative, theme statements, graphics, captions, compliance matrices, presentation slides, tabs, and so forth. If marketing and other data and information do not find their way into the knowledge product called a proposal, then a company's proposal will not receive maximal point score. Therefore, the marketing data are valueless. In effect, if intellectual capital measured in client, project, competitor, and political knowledge is not transferred to appropriate staff (knowledge workers) and converted efficiently into hardcopy or electronic proposal documents via established, articulated business processes, your company's proposal win ratios in terms of dollars (x dollars of contracts awarded for every $100 of contract value pursued) and number (x number of proposals won) will be suboptimal.

In your own company, evaluate critical decision points in your business processes. The bid–no bid or downselect decision point in the marketing or proposal life cycle is one salient example within the private sector. Decision quality improves—as measured by leveraged B&P dollars, increased proposal win rates, heightened staff morale, and enhanced revenue stream—when the right knowledge is available and applied when and where it is required. That right knowledge may take the form of customer understanding, knowledge of incumbent staff compensation packages, competitor intelligence, relevant political insight, innovative technical solution, current resource allocation, and so forth. See Table 1.1 for a list of the types of knowledge assets that your organization should identify, assess, archive, and disseminate internally—in effect, value.

With information and knowledge becoming the currency of American business in the twenty-first century, the organizations that can share this new currency ecologically in an effort to arrive at effective and appropriate solutions for both their external and internal clients in the shortest possible time and with the least amount of resources expended will be those firms that occupy positions, or *optima*, above the knowledge landscape (see Figure 1.12). From these vantage points, companies will be less likely to be displaced by new technologies, reconfigured marketing paradigms, social and demographic shifts, fluctuations in consumer confidence, global tensions, and political and legislative climate changes.

1.8.2 Internal and external clients: looking at clients in a whole new way

Let us reconsider the notion of clients. Within a traditional company, clients are perceived as the buyers or procurers of products or services being offered. In a knowledge-oriented company, clients include both external

Company Assets
Company core competencies (technical discussion)
Company bonding, registrations, and certifications
Licenses
Maps (e.g., office locations, project locations, number of projects by state or region)
Customer base
Patents
Trademarks
Company organizational charts
Company best practices
Industry certifications and accreditations [e.g., Software Engineering Institute (SEI) Capability Maturity Model Integration® (CMMI®), ISO 9001:2000, and IEEE]
Facility and personnel security clearances
Industry and business community awards, commendations, and recognitions (e.g., SBA Entrepreneur of the Year, Ernst & Young)
Customer endorsements and commendations
Small/minority business involvement; HUBZone involvement
Corporate-sponsored community service and outreach activities
Company videos
Facilities diagrams
Best management practices (BMPs)
Technical and business articles and books authored by company staff
Interview guides
Vendor manuals
Contractual Resources
Project or product descriptions or summaries
Project performance metrics
Award-fee statistics per project/overarching trends
Project success stories
Customer survey results

Table 1.1
Cross Section of Company Knowledge and Information Assets

Human Resources

Résumés

Personnel diversity data (minorities, women, veterans, handicapped individuals)

Incumbent capture rates

Employee turnover rates

Employee recruitment plan and metrics

Employee retention metrics

Staff by project, office, technical discipline, and degree

Business Development and Proposal Development Assets

Strategic plans and benchmarking data

Market segmentation analyses

Competitor information

B&P spending patterns

Winning proposals

Multimedia presentations to clients

Teaming agreements

Teaming-agreement statements of work (SOWs)

Proposal templates (e.g., executive summaries, cover letters)

Freedom of Information Act (FOIA) documents

Proposal lessons learned (from client debriefings)

Sales volumes by client sector, line of business (LOB), and geographic area

Tactical business plans

Corporate Infrastructure and Process Assets

Documentation capabilities

Business processes (e.g., procurement systems and mechanisms, financial systems, invoicing procedures, cost accounting, scheduling, automated tracking tools)

Project-Specific Assets

Customer evaluation reports (i.e., PPIRS, ACASS, CPARS, PPAIS, and PPIMS ratings)

Award-fee determinations

Table 1.1
(Continued)

Contract deliverables
Service-level agreements (SLAs)
Earned value analysis (EVA) results by project
Work breakdown structures (WBSs)
Project management plans
Computer-Related Assets
Software development capabilities and practices
Software source code for proprietary applications
Nonoperational hardware and software test-bed environments
Computer equipment and resources
Financial Assets
Budget forecasts
Capital expenses
Annual revenues
Relationship Assets
Cooperative agreements with colleges and universities, particularly HBCUs
Strategic business alliances (with such organizations as Microsoft Solutions, Lotus, Dell, and Oracle)
Public Relations Assets
Ads placed in professional and trade journals
Trade-show modules (text and graphics)
External news releases
Engineering/Manufacturing Assets
Engineering white papers (position papers)
Manufacturing defect-rate data
Engineering drawings
Engineering change proposals (ECPs)

Table 1.1
(Continued)

Test procedures
Requirements data
Policies and Plans
Make-or-buy policy
Management plans
Health and safety plans
Technology transfer plans
Professional compensation plans
Design control plans
Software quality assurance (SQA) plans
Phase in–phase out and transition plans
Subcontractor management plans
Configuration management (CM) plans

Table 1.1
(Continued)

Figure 1.12
Strive for optima
above the
knowledge
landscape.

elements, such as buyers, key subcontractors and vendors, regulatory agencies, and government institutions, as well as internal clients. The latter might include professional, project management, and support staff; operational units; sales and marketing elements; financial and accounting resources; senior management; and so forth. The currency of knowledge must flow in meaningful, multidirectional pathways among all of these clients. Each client is critical to the sustainability and performance (i.e., positive energetics) of the company, and each requires, and in turn provides, vital information and knowledge to the other much as enzymes are shared within a living cell to perform a host of vital maintenance, replicative, and growth functions.

Endnotes

1. See Federal Acquisition Regulation (FAR) Part 6.

2. "Do-It-Yourself Proposal Plan," TRW Space and Defense Proposal Operations, Rev. 3, February 1989, p. iii.

3. Stewart, Rodney D., and Ann L. Stewart, *Proposal Preparation*, New York: John Wiley & Sons, 1984.

4. *Proposal Preparation Manual*, Vol. II, Covina, CA: Procurement Associates, Inc., 1989, pp. 1–2.

5. "Do-It-Yourself Proposal Plan," TRW Space and Defense Proposal Operations, Rev. 3, February 1989, p. ii.

6. See FAR 15.609. Competitive range consists of those proposals that have a reasonable chance of being selected for contract award. Win percentages can be presented in two very distinct ways: in terms of numbers of proposals won versus total number submitted and also in terms of dollars awarded versus total potential dollars for all procurements on which your company proposed. For example, if your company has won 35% of its proposals in terms of number and only 12% in terms of dollars, that means that you are winning proposals of low dollar value.

7. 8(a) refers to Section 8(a) of the Small Business Act [15 U.S.C. 637(a)], which established a program that authorizes the SBA to enter into all types of contracts with other agencies and let subcontracts for performing those contracts to firms eligible for program participation based upon the criteria established in 13 CFR 124.101-113. The SBA's subcontractors are referred to as "8(a) contractors." (See FAR Subpart 19.8.) Eligibility in the program focuses on those groups that

have historically been denied access to capital, educational resources, and markets.

In 1998, the SBA listed 6,100 8(a) firms throughout the United States. Of those firms, 45.7% were owned by African-Americans, 22.9% were owned by Spanish-speaking people, and 20.7% were owned by Asians. The remaining companies were owned by Puerto Ricans, Native Americans, and Aleuts. Slightly more than 20% of the 8(a) companies are currently based in the Washington, D.C., area. The SBA allocated $5.2 billion in 1995 in set-aside contracts for minority-owned businesses. [See Munro, Neil, "Clinton Set-Aside Plan Becomes Election-Year Pawn," *Washington Technology*, April 11, 1996, pp. 5, 93; and Munro, Neil, "8(a) Program Survives Republican Attack," *Washington Technology's 8(a) and Small Business Report*, September 12, 1996, p. S-6.]

8. *SBA Budget Request & Performance Plan*, FY2004 Congressional Submission, Fourth Printing, February 10, 2003, p. 1.

9. *SBA Budget Request & Performance Plan*, FY2004 Congressional Submission, Fourth Printing, February 10, 2003, p. 38.

10. *SBA Budget Request & Performance Plan*, FY2004 Congressional Submission, Fourth Printing, February 10, 2003, p. 39.

11. *SBA Budget Request & Performance Plan*, FY2004 Congressional Submission, Fourth Printing, February 10, 2003, p. 2.

12. Cranston, H. S., and Eric G. Flamholtz, "The Problems of Success," *Management Decision*, Vol. 26, September 1988, p. 17.

13. Hartman, Curtis, and Steven Pearlstein, "The Joy of Working," *INC.*, November 1987, p. 62.

14. O'Guin, Michael, "Competitive Intelligence and Superior Business Performance: A Strategic Benchmarking Study," *Competitive Intelligence Review*, Vol. 5, 1994, p. 8.

15. Murray, Bill, "Looking Beyond Graduation," *Washington Technology's 8(a) and Small Business Report*, March 7, 1996, p. S-4.

16. Anderson, Tania, "8(a) Companies Learn the Secret of Partnering," *Washington Technology*, October 10, 1996, p. 22.

17. Kelleher, Kevin, "Feds, State Go On Line with Contracts," *San Francisco Business Times*, Vol. 9, April 1995, p. 3.

18. Fox, Harold W., "Strategic Superiorities of Small Size," *Advanced Management Journal*, Vol. 51, 1986, p. 14.

19. "Disadvantaged Firms to Get Edge in Bids for Federal Contracts," *Baltimore Sun*, June 25, 1998, p. 2A.

20. Hartman, Curtis, and Steven Pearlstein, "The Joy of Working," *INC.*, November 1987, p. 67.

21. Frank, C., "Uncle Sam Helps Small Business Get High Tech to Market," *Missouri TechNet News*, November 4, 1997; http://www.umr.edu/~tscsbdc/sbirsumm.html.

22. A mission statement is a concise, written expression of a company's long-term business, technical, and programmatic goals. It is a narrative outline of "who we are, what we do, and where we are headed" (Thompson, Arthur A., and A. J. Strickland, *Strategy Formulation and Implementation: Tasks of the General Manager*, 4th ed., Homewood, IL and Boston, MA: BPI/IRWIN, 1989, p. 23.). Once formulated, this statement should be communicated publicly to employees, clients, and vendors, among others. One effective mechanism is to add a "pull quote" from your company's mission statement to your marketing brochures, annual reports, company letterhead, or newsletter. Some companies have poster-sized copies of their mission statement displayed prominently in their facilities to clearly convey their identity.

23. Thompson, Arthur S., and A. J. Strickland, *Strategy Formulation and Implementation: Tasks of the General Manager*, 4th ed., Homewood, IL: Irwin, 1989, p. 19.

24. Certain elements in this listing were adapted and modified from Piper, Thomas S., "A Corporate Strategic Plan for General Sciences Corporation," Spring 1989, p. 1 (unpublished).

25. For additional business challenges, as well as opportunities, strengths, and weaknesses, see Thompson, Arthur S., and A. J. Strickland, *Strategy Formulation and Implementation: Tasks of the General Manager*, 4th ed., Homewood, IL: Irwin, 1989, pp. 109–111.

26. Benchmarking is a rigorous process for linking competitive analysis to your company's strategy development. Benchmarking is a method that measures the performance of your best-in-class competitors relative to your industry's key success factors. It also is a mechanism for identifying,

learning, and adapting the business processes by which the best-in-class achieve those performance levels to your company.

"Benchmarking results should provide insight into how well we are doing and specific process changes that will improve our operations." See Leibfried, Kate H. J., and Joe Oebbecke, "Benchmarking: Gaining a New Perspective on How You Are Doing," *Enterprise Integration Services Supplement*, October 1994, pp. 8–9.

27. Huseman, Richard C., and Jon P. Goodman, "Realm of the Red Queen: The Impact of Change on Corporate Structure, Corporate Education, and the Emergence of Knowledge Organizations," The Corporate Knowledge Center @ EC^2, Annenberg Center for Communication, University of Southern California, 1998, p. 15.

28. Angus, Jeff, Jeetu Patel, and Jennifer Harty, "Knowledge Management: Great Concept . . . But What Is It?" *InformationWeek*, No. 673, March 16, 1998.

29. "Community KM Glossary," http://knowledgecreators.com/km/kes/glossary.htm and Karl E. Sveiby, "What Is Knowledge Management?" p. 2; http://www.sveiby.com/articles/KnowledgeManagement.html, April 2001.

The term *intellectual capital* is attributed to economist John Kenneth Galbraith (1969). Intellectual capital includes individual talents and knowledge as well as documented knowledge that spans reports, research papers, patents, software source code, books, articles, and manuscripts. See also Touraj Nasseri, "Knowledge Leverage: The Ultimate Advantage," *Kognos: The E-Journal of Knowledge Issues,* Summer 1996.

30. Roos, Johan, "Intellectual Performance: Exploring an Intellectual Capital System in Small Companies," October 30, 1996.

31. Hoyt, Brad, "What Is KM?" *Knowledge Management News,* 1998; http://www.kmnews.com/Editorial/whatmk.htm.

32. Davenport, Tom, and Larry Prusak, "Know What You Know," http://www.brint.com/km/davenport/cio/know.htm, 2000.

Tom Davenport is a professor and director of the Information Management Program at the University of Texas at Austin. Larry Prusak is managing principal of the IBM Consulting Group in Boston.

33. "Knowledge Management: Collaborating on the Competitive Edge," *CIO White Paper Library,* 2000; http://www.cio.com/sponsors/0600_km.

34. Grundstein, Michel, "Companies & Executives in Knowledge Management," http://www.brint.com/km/cko.htm.

35. Information in this paragraph was drawn from Karl E. Svieby, "What Is Knowledge Management?" pp. 5–6; http://www.sveiby.com/ articles/KnowledgeManagement.html, April 2001.

   See also Karl E. Sveiby, "The Intangible Assets Monitor," http://www.sveiby.com/articles/IntangAss/CompanyMonitor.html, December 20, 1997. Celemì employs metrics such as efficiency (sales per customer, staff turnover), stability (repeat orders, growth in sales per administrative staff), and growth/ renewal (average years of professional competence). Celemì is headquartered in Malmö, Sweden, and is dedicated to creating processes that help companies leverage the power of learning.

36. Sutton, Michael H., "Dangerous Ideas: A Sequel," p. 3; http://www. corpforms99.com/83.html. (Remarks delivered during the American Accounting Association 1997 Annual Meeting in Dallas, Texas, on August 18, 1997.)

37. http://www.martech.co.nz/kbl.html#intro.

   Professor Lev notes, "How ironic that accounting is the last vestige of those who believe that things are assets and that ideas are expendable." See Baruch Lev, "Accounting Needs New Standards for Capitalizing Intangibles," *ASAP: Forbes Supplement on the Information Age,* April 7, 1997; http://www.forbes.com/asap/97/ 0407/034.htm.

Chapter 2

Strategic partnering and subcontracting opportunities

RALPH PETERSON, president and CEO of CH2M Hill Companies, a 7,000-person, 50-year-old firm that provides professional services worldwide in water, environmental, transportation, and industrial facilities, has remarked that "[p]artnerships and alliances are being forged that just a short time ago could not have been imagined" [1]. Strategic alliances can contribute synergy that results in sustained competitive advantage in a rapidly changing global business environment. The stakes are high and the opportunities are real. Many major corporations now maintain their own internal procurement and small business centers or agencies. For example, the Bethesda, Maryland–based space and aeronautics megacorporation, Lockheed Martin, staffs five internal procurement agencies nationwide. "Twenty-two to 28 percent of the company's contracting dollars go to small, disadvantaged and women-owned businesses." And computer manufacturer Gateway, based in South Dakota, awarded several million dollars worth of contracts to more than 1,000 small businesses in 1997 [2].

Hamilton Standard, a division of United Technologies in Connecticut; The National Renewable Energy Laboratory (NREL) in Golden, Colorado; and the Jet Propulsion Laboratory (JPL) in Pasadena, California, are all winners of the Dwight D. Eisenhower Award for Excellence from the U.S. Small Business Administration. The Eisenhower Award is a national program for large prime contractors in the manufacturing, service, R&D, and construction arenas that have excelled in their use of small businesses as subcontractors and suppliers. Divisions of Lockheed Martin in Orlando, Florida, and Cherry Hill, New Jersey, were among the winners of the Eisenhower Award in the manufacturing and service categories [3].

2.1 Subcontracting opportunities and pathways to success

Evidence of new directions in subcontracting is seen in the public/private partnership among the SBA, General Motors, Ford Motor Company, and Chrysler that is designed to improve opportunities for minority-owned small businesses in a major industry. This landmark memorandum of understanding (MOU) marked the first time that a private-sector industrial group of this magnitude and importance has engaged in partnership with the SBA. The MOU represents a sharp increase in the contract dollars flowing to small disadvantaged businesses (SDBs). Subcontracting awards of $3 billion over 3 years are anticipated [4].

In general, proactive small business entrepreneurs are significantly leveraging their limited marketing, financial, and technical resources as well as augmenting their revenue stream by attracting and forming strategic partnerships, as subcontractors and protégés, with large prime contracting companies to pursue new business opportunities and expand work with existing clients. Many procurements let by the U.S. federal government, as well as by the public sector at the state and local levels, stipulate small, small disadvantaged, minority-owned, or women-owned business target thresholds for the resulting contracts. For example, a procurement related to the Everglades by the state of Florida required that a minimum of 16% of the contract revenue be allocated to minority-/women-owned business enterprises (M/WBE).

The opportunities for small business success have never been better! As of June 2004, the federal government's subcontracting goals were as follows:

- 23% for small businesses;
- 7% for SDBs;
- 5% for women-owned businesses;
- 3% for HUBZone small businesses;
- 3% for service-disabled veteran-owned businesses.

These guidelines are of direct benefit to small businesses of all types because they require large federal contractors to proactively team with small businesses to pursue U.S. government contracts.

2.2 Critical success factors

Critical to a small firm's subcontracting success are the following elements: (1) superlative contractual performance on existing or past projects (being the best-in-class); (2) flexible, audited, and approved pricing strategies and structures; (3) the allocation of top-flight, dedicated professional staff to proposal teams with the prime contractor; (4) fiscal strength as measured by such parameters as operating profit, net income, positive cash flow, operating history, and future business assumptions; (5) well-defined and articulated core competencies and product lines; and (6) fair, equitable, and ethical teaming agreements that are always honored. Strategic alliances or partnerships must be managed as enduring business relationships—with mutually compatible objectives, shared risks and resources, real-time knowledge transfer, and mutual trust.

Understand that the prime contractor will have to sell your small company's human talent and knowledge, contractual experience, and fiscal solvency in their proposal to a given public- or private-sector client. Make that task extremely easy for them by having a customer-focused, fact-based story to tell, and make that information available to them in a timely manner during the proposal response life cycle. Far too many small businesses submit poorly written, incomplete materials to the prime contractor for integration into the prime's proposal. This inappropriate practice adds to the prime contractor's B&P costs, and detracts from your working relationship with that prime in the future. You may need to contract freelance or temporary staff to prepare required proposal modules and other materials in an electronic form compatible with the prime contractor's publishing systems. Above all, adhere to the proposal schedule as developed by the prime contractor. These aspects of working with prime contractors are reviewed in Chapter 7.

Whenever possible, small businesses should strive to forge and negotiate exclusive teaming agreements with prime contractors which are active and successful within your company's major LOBs. Cultivate successful experience within the prime's and your target client's technical areas of interest. Understand how to conduct contractual business with the prime and the ultimate client. Demonstrate your company's ability to staff the project with stellar professional staff (through incumbent capture, for example) who are committed to the life of the project and not merely senior staff whose résumés help to win the job, but who leave the project after a minimal amount of time. Build camaraderie in the relationship, and establish a mutually beneficial strategic partnership for the long term.

2.3 Specific strategies for achieving subcontracts

The following strategies can help you achieve subcontracts with prime contractors:

- *Register your small business with the new CCR/PRO-Net portal (http://www.ccr.gov) right away.* As of January 1, 2004, the U.S. Small Business Administration's PRO-*Net* database had been combined with the Department of Defense's (DoD) Central Contractor Registration (CCR) database. *The result?* One Web portal for vendor registration and for searching for small business sources—and an integrated, more efficient mechanism that small businesses can use to market their products and services to the federal government. All of the search options and company information that existed in PRO-*Net* are now found at the CCR's "Dynamic Small Business Search" site. The new CCR/PRO-*Net* portal is part of the Integrated Acquisition Environment (IAE), one of the important e-government initiatives under the President's Management Agenda (PMA) (http://www.whitehouse.gov/omb/budget/fy2002/mgmt.pdf). The CCR is the primary vendor database for the U.S. federal government. This system collects, validates, stores, and disseminates data in support of agency acquisition missions. Both current and potential federal government vendors are required to register with CCR in order to be awarded contracts by the U.S. government. The CCR Web portal is also valuable for searching for small business sources. PRO-*Net* and the Procurement Automated Source System (PASS) database are now defunct.

  Prime contractors use SUB-*Net* (http://web.sba.gov/subnet) to post subcontracting opportunities. These may or may not be reserved for small business, and they may include either solicitations or other notices—for example, notices of sources sought for teaming partners and subcontractors on future contracts. Small businesses can review this Web site to identify opportunities in their areas of expertise. While the Web site is designed primarily as a place for large businesses to post solicitations and notices, it is also used for the same purpose by federal agencies, state and local governments, nonprofit organizations, colleges and universities, and even foreign governments. This Web site has shifted the traditional marketing strategy from the shotgun approach to one that is more focused and sophisticated. Instead of marketing blindly to many prime contractors, with no certainty that any given company has a need for their product or service, small businesses can use their limited resources to identify concrete, tangible opportunities and then bid on them.

- You can search for subcontracting opportunities on SUB-*Net* in several ways. First, if you are looking for opportunities in a particular industry, you can search for all Requests for Proposals and Notices of Sources Sought by the North American Industry Classification System (NAICS) code for that industry. From the main menu, simply click on "Search for Solicitation," then click on "Search the Database" and then enter the six-digit NAICS code for your industry. (Add a zero to the end if the NAICS general industry code contains only five digits.) This is the fastest way for you to see if there is anything on the bulletin board that interests you. If you would rather not worry about NAICS codes, you can search by the generic description or solicitation number, or if you want to look at every opportunity on the bulletin board regardless of industry, you can click on "View all solicitations." This will display everything on the bulletin board for which the bid date has not yet expired. (All notices on this page are sorted in order of bid closing date, and they drop off automatically on the date and hour at which they expire.)

    When you find a solicitation or notice that appeals to you, simply print a hardcopy (for your records) and contact the company that posted the notice for a copy of the complete bid package. Everything posted on SUB-*Net* will have the name, telephone number, fax number, and e-mail address of a person to contact for questions or additional information.

- *Small Business Subcontracting Opportunities Directory.* The SBA maintains a Subcontracting Opportunities Directory (http://www.sba.gov/GC/indexcontacts-sbsd.html). SBA obtains the names and addresses for this directory from subcontracting plans that are submitted to the government when a large business receives a federal contract for more than $500,000 (over $1 million in construction). SBA's Commercial Market Representatives (CMRs) counsel small businesses on how to market their products and services to the prime contractors in this directory. To find your nearest CMR, go to http://www.sba.gov/GC/contacts.html and click on the fourth menu selection.

- In addition, if appropriate, list your firm with the Worldwide Minority Business Network (http://www.mbnet.com/ mbnet_info.htm).

- *Participate in private-sector supplier diversity programs.* Major corporations and organizations ranging from General Dynamics, AT&T, American Express, Marriott, Apple Computer, NBC, and Owens-Illinois to ConocoPhillips, Computer Sciences Corporation, Dial, Continental Airlines, Kodak, Charter One Bank, and the Indiana University at Bloomington maintain and staff robust supplier diversity programs. These programs actively seek, certify, qualify, develop, and use M/WBEs. For example, AT&T now does business with more than 2,100 minority and women-owned businesses (and, during the past 36 years, has provided more than $16 billion to these types of businesses). In 2003, *Hispanic Magazine* selected AT&T as having one of the top 25 minority vendor programs in the country. Also in 2003, Marriott spent more than $150 million with M/WBEs.

- *Register and prequalify your small business with the SADBUS, or Small and Disadvantaged Business Utilization Specialist, in specific federal agencies.* At the state level, the analogous office may be called the "Office of Supplier Diversity and Outreach." And many large corporations such as Swift & Company in Greeley, Colorado, now have small business centers, staffed with Small Business Liaison Officers (SBLOs). Every federal contracting agency and prime contractor doing substantial business with the federal government is obliged to designate an SBLO or SADBUS within the agency or firm. This person is the first point of contact for your small business wishing to make your products and services known to the contracting entity.

- *Create, maintain, and register an easily accessible Web site for your small company.* Prime contractors often perform Internet searches to locate potential teaming partners. Registration refers to listing your site with various Internet-based search engines, such as 123 Submit, Submit It, Add Me, Submit Away, and Submit All.

- *Interact with Small Business Development Centers (SBDCs) in your state (http://www.sba.gov/sbdc/).* The 63 SBDCs with 1,100 service locations nationwide are educational and research resources for small businesses. The Association of Small Business Development Centers (ASBDC) links each SBDC program into a national network. In addition, some states maintain Regional Minority Supplier Development Councils and Regional Minority Purchasing Councils.

- *As fiscally appropriate for your small business, employ reputable consulting firms to open the door to strategic partnership discussions.* Reputable consulting firms include INPUT (http://www.input.com) and Federal Sources, Inc. (FSI) (http://www.fedsources.com), both located in Virginia.

- *Register your company in the Diversity Information Resources (DIR) online M/WBE sourcing portal.* This portal includes more than 9,000 certified M/WBEs. In addition to access to the M/WBE portal, the supplier registration link allows your corporation to establish a supplier registration link from your corporate Web site. (http://www.diversityinforesources.com/)

- *Attend and participate in appropriate conferences and trade shows for your industry.* Small businesses are strongly encouraged to attend and participate in relevant federal government conferences. For example, NASA's Annual Mentor-Protégé Program Conference is the space agency's formal technical and business development program for SDBs and WOBs. Convened most recently under the leadership of Mr. Lamont Hames, NASA's Mentor-Protégé Program manager and chief of staff within the Office of Small and Disadvantaged Business Utilization (OSDBU), small businesses were provided the opportunity to meet senior executive-level NASA associate administrators and managers as well as major prime contractors.

  At the 9th Annual NASA Mentor-Protégé Program Conference held for 2 days in December 2004 in Washington, D.C., more than 200 NASA protégés and other attendees learned about the fundamentals of the NASA Mentor-Protégé Program, principles of effective teaming agreements, and proven approaches for winning new business within the Mentor-Protégé relationship. As a presenter at this particular NASA conference, I know firsthand how much knowledge is shared and learned in this particular forum.

Specifics addressed at the NASA Conference spanned these areas:

- Mentor-Protégé Program policy [located within the NASA Federal Acquisition Regulations (NFAR) at Section 1819.72];
- NASA's Mentor-Protégé (MP) management information systems (MP/MIS) tool;
- Procurement vehicles—GSA Schedule, indefinite delivery/indefinite quality [ID/IQ], 8(a), and so forth;
- MP and the RFP.

In addition, there was an in-depth presentation on the business development life cycle (opportunity identification, qualification, and strategy development), capture management (win strategy, framework, approach, past performance, and cost), and proposal development (roles and accountability, teaming, solution development, and performance-based contracting). Information on important NASA-specific Web sites were provided, including http://prod.nais.nasa.gov/cgi-bin/nens/index.cgi and http://prod.nais.nasa.gov/cgi-bin/nais/forecast.cgi.

- *Secure your company's membership and participate in trade and business associations.* This includes interaction with such associations as the National Association of Purchasing Management (NAPM). With more than 48,000 members, NAPM is a not-for-profit association based in Tempe, Arizona, that provides national and international leadership in purchasing and supply management research and education. NAPM meetings offer forums for developing important business relationships with procurement representatives of large corporations. NAPM can be contacted on the Web at http://www.napm.org or by calling 1 (800) 888-6276.

- *Consider listing your company in the Thomas Register of American Manufacturers.* Major corporations and government agencies often consult this resource for suppliers that operate nationally. Visit http://www.thomasregister.com.

- *Visit Internet sites to learn more about potential prime contractors.* For example, http://www.hoovers.com provides detailed business and financial information about major corporations.

2.4 Becoming part of a governmentwide acquisition contract (GWAC) team

Governmentwide acquisition contracts (GWACs) are multiple award task order contractual vehicles for various resources and services hosted, or "owned," by one agency of the federal government, but which other specified federal agencies can use as well. GWACs were given statutory weight through the Federal Acquisition Streaming Act (FASA) of 1994. The government can realize economies of scale by centralizing the purchase of certain types of services or products through these ID/IQ GWAC vehicles. There is a limitation on the total percentage of the GWAC that one agency can use, and each GWAC has its own specific terms and conditions. Despite a small administrative and contracting fee levied by the host federal agency, the direct benefit is that other government agencies do not need to incur major acquisition and procurement costs when using the GWAC vehicle. The General Services Administration (GSA) strongly supports this type of contracting. As part of a GWAC team, your small business can proactively sell the fiscal and efficiency benefits of GWACs to a broad group of constituencies, including government agency chief information officers (CIOs), as well as establish government-funded venture capital programs to support R&D within the IT sector.

GWACs are critical federal contractual vehicles for small, midsized, and large businesses. They are becoming the vehicle of choice for many government agencies. *Why?* Because much of the procurement paperwork, price justifications, and contractor selections can be done up front in the GWAC procurement process. Individual purchases can then be made from a small cadre of qualified vendors quickly through the use of a task order competition or purchase order. Your company should prime as many GWACs as possible and, as an alternative, be an integral part of as many GWAC teams as you can. Increasingly, these are the key pathways for doing long-term work in the federal market space. Examples of important current and upcoming GWAC vehicles include GSA STARS, GSA Alliant, NIH CIO-SP2*i*, the Department of Transportation's ITOP II, and NASA SEWP III.

The Department of Transportation's Information Technology Omnibus Procurement (ITOP) II performance-based omnibus contract provides information systems engineering, systems and facilities management and maintenance, and information systems security support services. As one of the most successful GWACs, DOT's ITOP program was saluted for its stellar contribution to the National Partnership for Reinventing Government (NPR, formerly the National Performance Review) in making the federal government a service-oriented, cost-conscious source of value for IT customers.

GSA's nationwide *8(a) Streamlined Technology Acquisition Resources for Services (STARS) GWAC* is a small business set-aside contract for technology solutions. This new contract vehicle is the replacement contract for the 8(a) FAST small business contract. STARS is a multiple-award, ID/IQ contract vehicle with a 3-year base and two 2-year option periods. The contract spans eight distinct functional areas (FAs) designated by NAICS codes, including Customer Computer Programming Services; Computer Systems Design Services; Computer Facilities Management Services; Data Processing, Hosting and Related Services; Other Computer Related Services; Internet Publishing and Broadcasting; All Other Information Services; and Wired Telecommunications Carriers. Directed task orders up to $3 million each for federal civilian agencies are allowable pursuant to 41 U.S.C. 253(c)(5). Directed task orders up to $3 million each for DoD activities are compliant with Section 803 of the 2002 National Defense Authorization Act. Importantly, 8(a) credit can be transferred back to the client agency via the Federal Procurement Data System (FPDS) Program ceiling.

Specific benefits of this new GWAC vehicle are (1) hardware, software, and services can be procured as part of an integrated solution; (2) it supports small businesses and meets procurement preference goals; (3) it provides availability and access to proven small business and 8(a) technology providers; (4) procurement lead times are short; and (5) it has low user fees (0.75%).

The GSA Small Business Government Wide Acquisition Contracts Center has offered the first contract vehicle set-aside exclusively for HUB-Zone firms. These contracts enable federal agencies to fulfill their information technology (IT) requirements through the utilization of small businesses that are located in HUBZones. Agencies have a statutory goal to spend 3% of their federal procurement dollars with HUBZone small business firms.

The GSA HUBZone IT GWAC was designed to provide federal agencies with access to qualified HUBZone contractors. Industries under this contract are primarily engaged in providing electronic data processing services including the processing and preparation of reports and automated data entry. Contract awards have been made to 36 technology firms who are certified by the SBA as HUBZone firms. These firms represent seven FAs, including On-Line Information Services; All Other Information Services; Data Processing Services; Custom Computer Programming Services; Computer Systems Design Services; Computer Facilities Management Services; and Other Computer Related Services. Keylogic Systems,

Inc., of Morgantown, West Virginia, was issued a task order worth $4 million over 3 years by the EPA, Washington, D.C. This is the largest HUB-Zone GWAC award made to date.

NASA's Scientific & Engineering Workstation Procurement (SEWP) III GWAC vehicle is geared to provide the latest in IT products for federal agencies. SEWP consists of 22 contracts, including 11 small business contracts, which offer a vast selection and wide range of advanced technology to all federal agencies including the Department of Defense. SEWP offers low prices (generally below GSA Schedule prices) and the lowest surcharge (0.6%).

The National Institutes of Health Chief Information Officer—Solutions & Partners 2 (NIH CIO-SP2*i*) GWAC contract provides information technology hardware, software, systems, and services in support of IT solutions within NIH and other government agencies. This vehicle was let by the National Information Technology Acquisitions and Assessment Center (NITAAC).

In 1999, the U.S. Department of Commerce created the COMMerce Information Technology Solutions (COMMITS) program, which is a governmentwide acquisition contract that provides information technology services and solutions (http://www.commits.doc.gov/). The COMMITS program was designed to accomplish three objectives: (1) Deliver top-quality IT services and solutions to meet government organizations' missions; (2) deliver IT services and solutions utilizing a streamlined acquisition methodology; and (3) provide a talented pool of small, small disadvantaged, 8(a), and women-owned small business contractors capable of delivering the government's IT requirements. The program provides performance-based information technology services and solutions in three major functional areas: (1) Information Systems Engineering (ISE) Support Solutions, (2) Information System Security (ISS) Support Solutions, and (3) Systems Operations and Maintenance (SOM) Support Solutions. Among the top COMMITS contractors in terms of contract awards were RS Information Systems, Inc. (RSIS), of McLean, Virginia, with $549 million, the QSS Group, Inc., of Lanham, Maryland, with $160 million, and STG, Inc., of Reston, Virginia, with $109 million.

COMMITS Next Generation (COMMITS NexGen) is the follow-on to the extremely successful COMMITS GWAC contract awarded in 1999. This GWAC, which was awarded in late 2004, has an $8 billion ceiling and a 10-year period of performance. It will be available for use governmentwide.

Finally, Alliant is a newly proposed GWAC by GSA Federal Supply Service (FSS) to consolidate and replace Millennia, ANSWER, ITOP II, Safeguard, Virtual Data Center, and other GWACs. GSA will make about 20 awards (15 to large business and 5 to small business consortia).

Functional areas will include Information Planning and Architecture; Cyber Security; Systems Integration; Facility Operations; Applications Systems Design, Development, and Implementation; Systems and Network Engineering; and Operations.

The contract ceiling is set at $150 billion, and awards are expected in May 2006. The GSA Millennia GWAC has done more than $6 billion worth of business, ANSWER more than $2 billion, and ITOP II about $2 billion.

2.5 How mentor-protégé programs can help your business

Created by Congress in 1990 (P.L. 101-510), the DoD Mentor-Protégé (MP) (http://www.acq.osd.mil/sadbu/mentor_protege/) Program is a socio-economic initiative that encourages major DoD prime contractors (mentors) to develop the technical and business capabilities of small disadvantaged businesses, qualified organizations that employ people with severe diabilities, and women-owned small businesses (protégés) in the context of long-term business relationships. Through credit toward subcontracting goals or some direct reimbursement of costs, the MP provides tangible incentives for the major prime contractors to establish and implement a developmental assistance plan designed to enable the small business to compete more successfully for DoD prime contract and subcontract awards. Mentors represented in the MP encompass a broad range of industries: environmental remediation, engineering services and information technology, manufacturing, telecommunications, and health care. These include AT&T, Northrop Grumman, Lockhead Martin, IBM, and Booz Allen Hamilton. Protégés receive assistance in marketing and business planning, management, training, needs assessment, bid and proposal preparation, human resources practices, quality assurance (including ISO certification), environmental protection, emerging technologies, and health and safety procedures.

Protégé firms participating in the program benefit by receiving training in the latest technology pertinent to their business niche. Like their mentor, protégé firms also enjoy the benefit of a long-term stable relationship, which helps create a stable business base. Also included in the benefits are expanded subcontracting opportunities throughout the mentor corporation for the protégé and the protégé's ability to pursue the larger, sometimes bundled requirements by bringing along the mentor as its partner. Another benefit whose value is not as easily quantifiable is the guidance and vision a mentor firm provides to its protégé on the particular industry's direction and trends.

The Air Force Mentor-Protégé Program (http://www.selltoairforce.org/Programs/MentorProtege/mp_program.asp) involves competition rather than application. Separate Air Force contracts are awarded for the mentoring efforts versus adding contract line items to existing contracts being performed by the mentor. The Air Force requires meaningful participation by Historically Black Colleges and Universities and/or Minority Institutions (HBCU/MIs). Program oversight and monitoring is provided through the Mentor-Protégé Center of Excellence (AFMPCOE) located at Brooks City-Base, San Antonio, Texas.

The Mentor-Protégé Program as set forth in legislative and regulatory coverage has three major goals:

1. *Provide incentives to DoD contractors.* The program seeks to provide incentives to prime DoD contractors (mentors) to develop subcontracting relationships with SDBs (protégés) and to assist the SDBs in developing relevant capabilities to compete for DoD contracts.

2. *Increase overall participation of SDBs.* A major stumbling block to the achievement of the 5% SDB subcontracting goals mandated by Congress is the assertion by large DoD prime contractors that there are not enough qualified SDBs to perform as subcontractors on DoD work. Therefore, a primary goal of the Mentor-Protégé Program is to increase the capabilities of SDBs to the point where more of them can perform significant work on DoD contracts, which should result in an increase in the overall subcontracting levels.

3. *Foster long-term business relationships.* Major prime contractors historically developed long-term business relationships, generally with other large businesses, to permit effective competition for contracts that could not be performed entirely "in house." The long-term relationships benefit both prime contractors and their "team members." It is the intent of the Mentor-Protégé Program to foster this type of relationship between mentors and protégés in order to develop a stable business base for protégés and a stable SDB vendor base for mentors.

The Department of Homeland Security's (DHS) Mentor-Protégé Program (http://www.dhs.gov/interweb/assetlibrary/mentorprotegedhsfinal3-29-04.doc) is designed to motivate and encourage large business prime contractor firms to provide mutually beneficial developmental assistance to small business, veteran-owned small business, service-disabled veteran-owned small business, HUBZone small business, small disadvantaged business, and women-owned small business concerns. The program is also designed to improve the performance of DHS contracts and subcontracts, foster the establishment of long-term business relationships between DHS large prime contractors and small business subcontractors, and strengthen subcontracting opportunities and accomplishments at DHS. Through self-identification, DHS small business outreach events, networking, and marketing, the mentor-protégé team jointly submits an agreement to the DHS OSDBU for review and approval demonstrating the mutually beneficial relationship of the two parties.

The Federal Aviation Administration (FAA) also maintains an active mentor-protégé program (http://www.sbo.faa.gov/sbo/mentor.asp). The FAA's program is designed to motivate and encourage larger businesses to assist Small Socially and Economically Disadvantaged Businesses (SEDB), HBCUs, MIs, and Women-Owned (WO) Small Businesses. The goal is to enhancing the capabilities of these protégé organizations to perform FAA prime contracts and subcontracts, foster the establishment of long-term business relationships between these entities and mentor corporations, and increase the overall number of protégés receiving FAA prime contract and subcontract awards.

Other federal government agencies that sponsor a Mentor-Protégé Program include the Department of Energy (DOE) (http://www. energy.gov/engine/content.do?PUBLIC_ID=12606&BT_CODE= BUSINESSDEVELOPMENT&TT_CODE=INFORSPOTLIGHTDOC UMENT), DOE's Oak Ridge National Laboratory (ORNL) (http://www. ornl.gov/adm/smallbusiness/mentor.shtml), the SBA (http://www.sba.gov/ 8abd/mentoroverview.html), and the EPA (http://www.epa.gov/osdbu/ pdfs/mppparticipants.pdf).

Through its annual conference, NASA's Mentor-Protégé Program (http://www.sba.gov/8abd/indexmentorprogram.html) provides small business attendees with information regarding important and recurring NASA small business programs and initiatives. These include Training and Development for Small Business in Advanced Technologies (TADSBAT), the quarterly Aero-Space Technology Small Business Forum, and the semiannual Science Forum for Small Businesses.

The U.S. Small Business Administration's Mentor-Protégé Program (http://www.sba.gov/8abd/indexmentorprogram.html) enhances the capability of 8(a) participants to compete more successfully for federal government contracts. The program encourages private-sector relationships and expands SBA's efforts to identify and respond to the developmental needs of 8(a) clients. Mentors provide technical and management assistance, financial assistance in the form of equity investments or loans, subcontract support, and assistance in performing prime contracts through joint venture arrangements with 8(a) firms.

Recent data suggest that small firms participating in the mentor-protégé programs have experienced significant growth. Specifically, among 226 agreements reviewed there was a net gain of 3,342 jobs within protégé firms; there was a net revenue gain in excess of $276 million within the protégé firms; and mentors reported an additional $695 million in subcontract awards to small disadvantaged business firms. Mentor firms reported value accrued to them as the direct result of developing a technically qualified and more competitively priced supplier base for DoD requirements. As mentor firms restructured and right-sized, they often formed strategic alliances with protégé firms for the specific purpose of outsourcing functions previously performed in-house. Similarly, the DoD gained by having an increased number of cost-effective, technically qualified small business sources for defense prime contracting as well as subcontracting requirements.

Your reasons to enter and participate in mentor-protégé programs should extend far beyond subcontracting dollars alone. Your company should be willing to invest its nonreimbursable funds and your employees' billable time for training in order to develop a long-term teaming relationship premised upon mutual trust, respect, communication, and flexibility. The process of being selected by a mentor corporation can take up to 1 year. The mentor-protégé programs present opportunities for realizing diverse technical capabilities, efficient contract management, exposure to new technologies and business practices, expanded work potential, and exposure to new federal clients. The CEO of one protégé firm stated "We measured success by what we learned rather than what was earned."

Minco Technology Labs, Precise Hydraulics and Machining Center, TN&A Associates, Kuchera Defense Systems, SUMMA Technology, Omega Environmental Systems, Fuentez Systems Concepts, and Vista Technologies are several of the 174 small disadvantaged businesses that have participated in the mentor-protégé programs. The 102 mentors span corporate giants such as SAIC, Boeing, Raytheon, Lockheed Martin, Texas Instruments Defense Systems & Electronics, IT Corporation, Bell Helicopter Textron, Earth Tech, and Hughes Aircraft as well as Booz Allen Hamilton, EDS, and Northrop Grumman Mission Systems.

ENDNOTES

1. Peterson, Ralph R., "Plotting a Safe Passage to the Millennium," *Civil Engineering News,* September 1997.

2. Mulhern, Charlotte,"Round 'Em Up," *Entrepreneur*, Vol. 28, No. 8, August 1998, p. 120.

3. "Lockheed Martin Division, Nova Group Earn SBA's Top Awards for Subcontracting," SBA Number 98-43, June 4, 1998; http://www.sba. gov/news/current/98-43.html.

4. U.S. Small Business Administration, "Vice President and SBA Administrator Announce Pact with Big Three Automakers," Press Release Number 98-09, February 19, 1998; http://www.sba.gov/ news/current/98-09.html.

Public–private partnership initiatives of value and interest to small businesses also include the comprehensive joint effort of IBM, the U.S. Chamber of Commerce, and the SBA called the Small Office Solutions (SOS). Announced in 1998, this nationwide initiative is designed to promote new opportunities for small business enterprises by educating them about the benefits of information technology. Expert guidance will be delivered free of charge to small businesses through the SBA's Business Information Centers (BICs) in Albany, New York, El Paso, Texas, and Spokane, Washington; the U.S. Chamber of Commerce and local chambers; and IBM. Small businesses seeking support on technology-related issues can contact IBM at http://www.ibm.com/ businesscenter or by calling 1 (800) 426-5800. See "IBM, U.S. Chamber of Commerce Announce Results of New Study on U.S Small Business and Technology," http://216.239.39.104/search?q= cache:v86sqMvVWmQJ:americanbusinessplans.com/1.pdf+%22IBM, + U.S.+Chamber+of+Commerce+Announce+Results%22hl=en.

Chapter 3

Marketing to and with your clients

[G]et in the path of their judgment ... to make your product part of an already existing code.
—Adcult USA (1997)

3.1 More than just selling

Demonstrate how you can help meet your client's requirements and performance thresholds, provide solutions, and minimize associated technical, contractual, and fiscal risks.

Marketing involves far more than just short-term or one-time selling in response to immediate issues and needs. It is a long-term process of dedicated commitment to and cultivation of your client's success as well as your specific business and product lines as defined in your strategic plan. Your client wants to know that you'll be around next year, and 5 years down the road, to offer him high-quality products and superior services at competitive costs. Marketing means learning about your client's technical requirements, contractual and fiscal constraints, and programmatic as well as regulatory concerns. It involves developing an in-depth understanding of your client's business strategy and culture, organizational structure and dynamics, and procurement decision-making processes and buying influences. You will want to help your client appreciate the value of not just buying a product or service, but rather buying a solution [1]. Effective marketing takes time.

Functionally, marketing is about understanding your client and his stakeholders, your competition, your potential teaming partners, and the specific technical and programmatic benefits that your company can provide (see Figure 3.1).

Figure 3.1
The point of intersection.

- What other companies have provided goods and services to your client?

- How are those companies perceived by your client?

- Because potential teaming partners are part of the knowledge equation as well, does your client favor the formation of teams? How are your potential teaming partners perceived by your client?

- How can your company contribute in positive ways to your client's bottom line (private sector), program budget (federal and state), or hard currency deficit (international clients)?

Fundamentally, marketing means championing the customer [2]. It begins with target identification and proceeds through the qualification phase. The goals of qualification are validating that you can win the opportunity, building a business case to demonstrate that the opportunity warrants bidding, gathering customer intelligence through relationship building, and jump-starting the capture team.

Within the federal government market space, the business developer or marketer's role is to gain robust insight into the customer and the customer's stakeholder community in an ethical and professional manner. The overall business development life cycle is depicted in Figures 3.2 and 3.3. This insight, or marketing intelligence, then becomes the basis on which a superior proposal solution is built. Figure 3.4 shows the key categories of marketing intelligence that will ultimately become part of the proposal solution. Notice that these categories mirror sections of most proposals.

Major areas of focus for the business developer span target identification and qualification. These include:

- Identifying marketing leads and maintaining the leads list for assigned market segments (e.g., science and engineering);
- Marketing the customer—helping the customer understand "Why you?" or "Why your team and not your competitors' teams?";

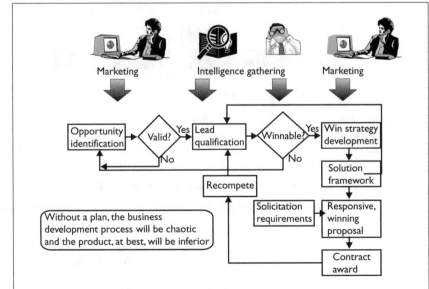

Figure 3.2
Overall business development life cycle.

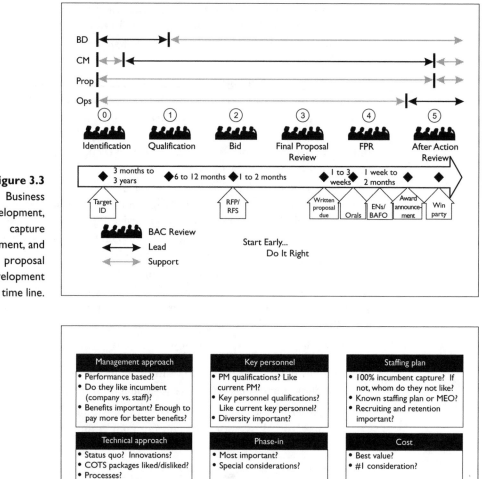

Figure 3.3
Business development, capture management, and proposal development time line.

Figure 3.4
Marketing intelligence categories.

- Knowing the customer's buying habits and sell points;
- Legitimately influencing the acquisition strategy and RFP (e.g., educating the customer about the potential GWAC vehicles under which a procurement might be released);
- Developing and executing the call plan (see Section 3.6);
- Drafting the initial win strategy and solution framework;

- Building the *business case* for pursuing the specific procurement (i.e., business case for investing B&P dollars and human resources for developing and submitting a proposal and oral presentation, as required);

- Serving as an integral member of the capture team through phase-in of the contract and after-action review of the proposal.

Specific factors that must be considered when building the business case for pursuing a given procurement are presented in Figure 3.5. Ensuring, for example, that the procurement aligns with your company's core competencies and the annual business plan are two of the most critical parameters to consider when generating a defensible business case—one that can be presented to and approved by executive management.

In the case of federal procurements, your company must also determine the following when building your business case:

- That the particular target program is real and will be funded;

- That the RFP or request for solution (RFS) will indeed be issued;

- That political and procurement conditions within the specific government agency will allow for genuine competition in the case of a recompetition;

- That you understand who the customer is;

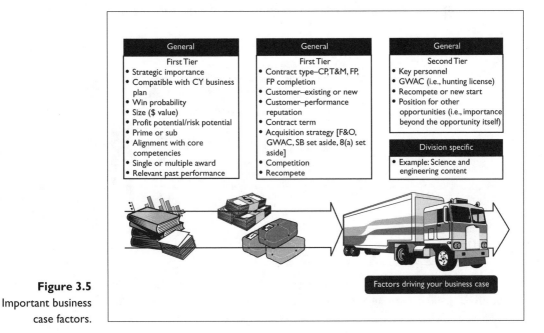

Figure 3.5
Important business
case factors.

- That you have identified the customer's *unstated* criteria for the award (*stated* criteria are given in the RFP).

Figure 3.6 defines winning proposals in the context of stated and unstated criteria. It is vital to remember that your company is not competing against the RFP, but rather against stated and unstated criteria.

Learn which civil servants will serve on the evaluation or selection board for your procurement. Does the client expect to get the most cost-effective solution? In effect, is price the real driver of the acquisition? Or is best value important to your client [3]?

Over time and in a variety of ways, inform the client about your company's distinctive human talents, technical capabilities and knowledge base, past and present contractual experience, management and subcontractor approaches, and financial stability and approved financial and accounting systems in ways that demonstrate how you can help meet his or her requirements and minimize, as well as share[1] associated technical, contractual, and fiscal risks. Be sure to meet your client's emotional or intangible needs as well—help them feel secure knowing you are looking out for them. Provide them with peace of mind [4].

If your potential client does not know your firm, he or she is less likely to buy from you. You must differentiate your company from every other company in a positive, credible way. Marketing is the process of persistently helping your client understand and believe that your brand of product or service will contribute substantively to his or her success. Build your differentiated suite of solutions into the decision pathway of your client's

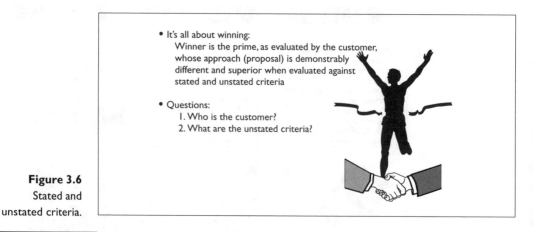

Figure 3.6
Stated and unstated criteria.

---

1   Increasingly, risk sharing is a stipulated expectation of federal clients. The Department of Veterans' Affairs' (VA) $650 million Austin Automation Enterprise System contract was a major case in point. See Appendix D for additional details.

infrastructure and buying influences. Know and cultivate ongoing relationships with the "sell points" within your customer's organization, that is, the people in the decision-making chain.

3.2 Transactions are personal—people buy from people

Face time with your client is critical. Why? Because people buy from people, and they buy emotionally. Interaction with your client should be conducted in a concerted manner by a variety of levels within your company—from marketing and senior management staff to mid-level technical personnel and project managers. Client visits by your marketing staff are necessary to open doors and establish new relationships with a potential client. These visits should be ongoing and augmented by additional information-sharing visits by members of your senior management as well as appropriate, select staff from among your technical and programmatic ranks. For your company's president to invest energy and time in focused direct client contact is essential to demonstrate and authenticate your corporate commitment to a given program or project. The only way to develop a robust, contextual understanding of your client and his mission, overall program goals, funding processes, and technical concerns is to personally interact with him and his staff as well as cognizant contracting, procurement, and management decision-making personnel (sell points).

Given the decentralized marketing activities of many firms—both small and large, technical and line managers are often in a position to make their own decisions about pursuing marketing opportunities. In my experience, this has resulted frequently in the pursuit of marginal targets that may have been appropriate from a technical standpoint but were not considered thoroughly from programmatic, cost, strategic planning, or proposaling perspectives. In fact, many times no marketing has taken place—the announcement on FedBizOpps (http://www.fedbizopps.gov) or government agency procurement Web sites is the first time that the project has been identified and considered for a bid–no bid decision. If your company does not market the client, you cannot expect to win contracts. There are, at minimum, several firms that qualify technically for most any government opportunity. Unless you can set your company apart in a positive way from your competition, you will not acquire new business.

Paul Lombardi, former chief operating officer of DynCorp (now CSC), has noted that "[u]ltimately, the government is buying people. The transactions are personal" [5]. A sales representative for a mid-Atlantic analytical chemistry laboratory remarked to me that a proposal that he had submitted received extremely high marks from the prime contractor and the ultimate government client. Why? Because this rep had marketed the client and was able to answer all of his relevant questions and concerns in the context of responding to the RFP. And in competitions between experienced contracting firms, relationships assume an even larger role in the final decision-making process. Plan on attending professional and trade association gatherings; meeting and talking with your clients at conferences, trade shows, and expositions; participating in sporting events; and going out to lunch (with each party paying his or her own check), for example.

3.3 Listen to your client

First and foremost, listen to your client—and listen to understand, not to reply [6]. Take every opportunity to learn about his or her requirements, budget concerns, technical problems, staffing goals, scheduling issues, experiences with your competitors, professional and personal likes and dislikes, and career goals. Importantly, learn who are your client's mentors, champions, advocates, and adversaries. Too many times, a company attempts to force-fit its clients into the company's own technical or programmatic compartments. Give your prospective client a chance to participate in your marketing presentations. Get your technical people in front of your client to have meaningful discussions on a variety of important technical, programmatic, and financial issues well in advance of the release of the RFP, RFS, or SF330 synopsis. Your company should be marketing an existing federal opportunity for the life of the current contract. Introduce and reinforce innovative, solution-oriented technical and management approaches that will appear later in your own proposal. Indeed, a compelling case can be made that the sale is made pre-RFP and that the proposal only closes the deal. A reasonably well-constructed proposal will not make up for inferior technical homework, poor management plans, or high costs [7]. And inferior and outdated marketing intelligence about your client, teaming partners, and competitors will most certainly result in a proposal loss.

3.4 Infuse marketing intelligence into your proposal

To be useful at all, the marketing information that your company collects must be analyzed, distributed, and archived so that it can be retrieved and updated easily. There is no point in marketing your client if you do not apply what has been learned into conceptualizing, designing, and producing your proposal. All information gathered that is relevant to a given procurement should be provided to the specific proposal manager. Appropriate marketing information must shape and also be fitted directly into your proposal at several critical stages, as illustrated in Figure 3.7. During the proposal response life cycle, there is a minimum of three critical inflection points at which marketing information should be infused into the process. These include (1) prekickoff activities, (2) the formal kickoff meeting, and (3) the formal review processes such as the Red Team review. These topics are discussed in Chapters 9 and 7, respectively. Current marketing information should shape postproposal activities as well. Procurement-specific, client-specific, and competition-specific information collected and analyzed by your company's external sales staff must be shared (as appropriate from a competition sensitivity and need-to-know standpoint) with the proposal manager and the proposal team and internal sales staff at these times. During the prekickoff phase, accurate, current marketing intelligence helps drive (in addition to the RFP requirements found in Sections L, M, C, H, and others) the architecture of the proposal outline, the benefits-oriented thematic structure of the proposal narrative, the project management approach, and the costing and staffing strategies. Without this marketing intelligence, your company would be like any other firm that was responding to the RFP only, with no further guidance from

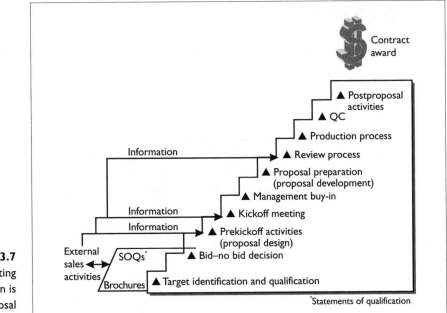

Figure 3.7 Marketing information is critical to proposal success.

the client than the words on the pages (or bytes in the electronic file) of the RFP.

At the formal kickoff meeting, in-depth marketing intelligence as shared by a member of your company's external sales group will assist members of the proposal team—particularly those charged with writing significant sections of the proposal narrative—with a reasonably full-life big picture of the procurement opportunity, the client's sentiments and concerns and hot buttons, and the competition.

Again, during the formal review cycles, marketing intelligence should be used as a benchmark against which to measure the thematic success of the proposal documents at that stage in the response life cycle. In the terminology of satellite remote sensing, marketing should supply the "ground truth" that authenticates your company's proposal for the particular client and project at hand.

As depicted in Figure 3.7, the flow of information and support moves in both directions between the internal sales staff to the external sales staff. For example, the internal sales staff will assist their external sales counterparts in the development of effective marketing brochures, briefings, and statements of qualification (SOQs) that are valuable tools in helping your client to better understand and remember your company's capabilities and talents.

3.5 Intelligence gathering and analysis techniques

One potentially effective technique of intelligence gathering and analysis is for your company's external sales support staff and division managers to visit the contracting offices of your primary federal client agencies in order to obtain specific information on existing contracts. Learn the contract title and numbers of existing contracts, the incumbent contractor(s), the projected contract renewal date, and the nature of the work, as well as the staffing size, contract dollar amount, contract duration, and contract type and vehicle (e.g., GWAC). Ask for the names and telephone numbers of the contracting officer's technical representative (COTR) and contracting officer (CO) for each contract. Supplemental information can be gathered by contacting specific federal agencies and requesting copies of their acquisition forecast lists. Public Law 100-656, the Business Opportunity Development Reform Act of 1988, amended the Small Business Act to emphasize acquisition planning. The law requires federal agencies to compile and make available projections of contracting opportunities that small businesses (including minority, women-owned, HUBZone, veteran-owned, and service-disabled veteran-owned) may be able to perform. These lists are called different names by various agencies, for example, procurement forecast, active contracts list, advanced procurement plan, and active contracts register-contractor sequence.

Many federal agencies now maintain their acquisition forecasts online on the Web. For example, the Department of Transportation (DOT) procurement forecast includes contracting opportunities from DOT's participating operating administrations, available sorted by operating administration and procurement category (http://osdbuweb. dot.gov/business/procurement/forecast.html). The FAA posts its procurement forecast at http://www.sbo.faa.gov/sbo/prc_frc_ lst.asp?rgn=1. The Air Force lists a Long Range Acquisition Estimate (LRAE) that consolidates anticipated procurements from major operational commands and activities plus the logistics, product, and test centers of the Air Force Materiel Command (AFMC). It is designed to help firms identify procurements that offer opportunities for their participation early in the acquisition process and to help them improve the timeliness and responsiveness of their proposals (http://www.selltoairforce.org/ Opportunities/lrae.asp).

Other federal agencies that provide procurement or acquisition forecasts on line include the following:

- Department of Veterans Affairs (http://www.osdbu.va.gov/cgi-bin/ WebObjects/FcoPublic.woa);
- U.S. Department of Agriculture (http://www.pforecast.net/);
- NASA (http://procurement.nasa.gov/cgi-bin/nais/forecast.cgi);
- Department of Homeland Security (http://www.dhs.gov/interweb/ assetlibrary/forecastrev0204.doc);
- Department of Energy (http://hqlnc.doe.gov/Forecast) (procurement searches are available by such parameters as NAICS code, solicitation method, release quarter, award quarter, and incumbent contractor);
- Department of the Interior (http://ideasec.nbc.gov/forecast).

When you obtain this marketing intelligence from in-person contacts, agency procurement briefings, and budget documents, Web sites, and published sources, record it carefully in a table, such as the one presented in Table 3.1. Someone within your organization must be responsible for maintaining the tracking tables and keeping them secure. It is critical that the information is accurate, current, and relevant. Try to collect information on both short- and long-term opportunities. One to two years is not too long a lead time for advanced business planning (see Figure 3.8). Once

#	Project Title	Acquisition Manager	Incumbent(s)	Contract #	Anticipated RFP Release Date	Duration of Contract	Prime/ Sub	Estimated Level of Effort per Year	Potential Contract Dollar Value
1	A-E Services to perform Hazardous Toxic & Radioactive Waste (HTRW) & environmental compliance services USACE-Mobile District	Ms. Jennifer Brothers	AABC Corp.	DACA01-99-R-0115	7/05	1 base year + 4 option years	Prime	35 full-time equivalents (FTEs)	
2	Investigation and design for remediation of misc. HTRW sites USACE-Omaha District	Mr. Tim Grayson	AAAC Corp.	DACA45-99-R-0011	10/05	Indefinite delivery	Sub	8 FTEs	
3	Environmental contract for Ft. George G. Meade, Maryland	Mr. Greg Yost	AACB Corp.	DACA31-99-R-0039	4/06	Indefinite delivery	Sub	5 FTEs	
4									
5									

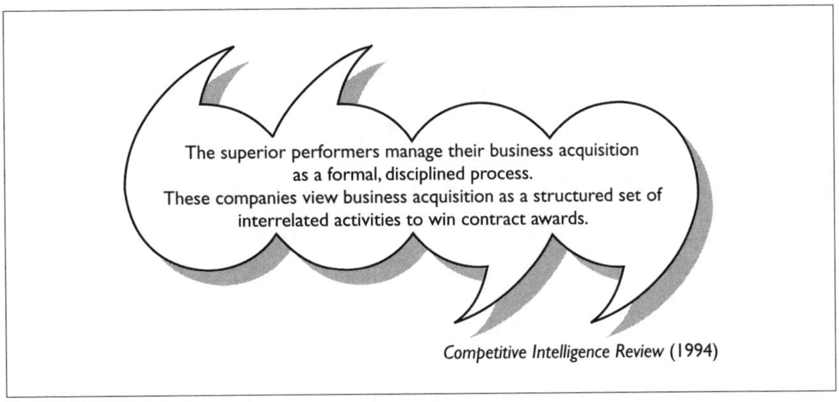

Figure 3.8
Formalized
business
acquisition
processes produce
tangible results.

The superior performers manage their business acquisition as a formal, disciplined process.
These companies view business acquisition as a structured set of interrelated activities to win contract awards.

Competitive Intelligence Review (1994)

client target tracking tables are compiled for a given federal agency, preliminary bid–no bid decisions, proposal resource allocation, and revenue stream projections can be made in a planned and rational manner. These potential marketing targets comprise your company's opportunity pipeline. This initial downselecting process, if done in a structured way, will allow your company to focus scarce resources in the direction of viable targets and prevent your internal sales group and publications staff from becoming totally overburdened. There is nothing more debilitating to morale companywide than to produce losing proposal after losing proposal because of too few resources chasing too many targets. Be selective, and then produce a stellar proposal every time you submit one to your client.

Automated tools are available to support business opportunity tracking and decision making. The WinAward marketing management system from Bayesian Systems (http://bayes.com/products.htm) in Rockville, Maryland, is one example of a Windows-based, robust software product for bid life-cycle management. WinAward retains and manages data characterizing specific marketing leads, retains costs associated with the bid process, determines award probabilities, and supports a rigorous bid decision process.

Once a preliminary downselecting procedure has been completed, capture plans and associated call plans (direct contact plans) can then be generated to pursue the qualified list of targets. Your company will want to collect and continually validate in its capture plans the following type of information:

- Project name (according to government sources, that is, use the client's name for this project);

- Client name (to the level of the line organization) and organizational chart;

- Client mission and vision statement and recent developments in the client's operational environment;
- Incumbent contractor(s), if applicable;
- Potential competitors;
- Specific services or products being procured;
- Key evaluation factors for award and client success criteria;
- Anticipated RFP or RFS release date (include anticipated dates for the preproposal conference, release of any prequalification questions, and oral presentations);
- Anticipated contract award date;
- Projected phase-in period;
- Anticipated contract dollar value to your company;
- Source evaluation board membership;
- Relevant contractual experience (for your company);
- Proposed project manager (for your company);
- Other proposed key staff (from your company);
- Potential teaming partners.

Take the time to know your teaming partners, their technical expertise, human talent, cost strategies, contractual performance, financial stability, business processes, contract-performance liabilities and conflicts of interest (COI), and any pending litigation or government inspector general (IG) investigations. Ask your client what they think of your prospective teaming partner. Consider the benefits and disadvantages of exclusive versus non-exclusive teaming arrangements. (See Chapter 18 for more discussion regarding teaming agreements.)

3.6 Call plans

A call plan should be designed to clearly indicate whom within your company will be visiting the client, as well as when, why, and at what level of the client organization. An overview of the call plan process is presented in Figure 3.9. Note the importance of electronically archiving call plan results.

Information gathered from direct client contact should include the following.

Figure 3.9
Call plan process.

Government or commercial buying organization and key personnel.

Obtain the official organizational charts for the client organization, and learn the full names and functions of the important technical and contractual decision makers (sell points).

Currently, what is the client doing technically and programmatically?

Is the client planning to migrate their automated process from one computer platform or operating system to another? Have regulatory changes imposed new demands on the client's programs? Will service-to-the-customer-style e-government necessitate a telecommunications center capable of handling significantly more queries per unit time? How does the migration toward the President's Management Agenda (PMA) and e-business drive your client's technical requirements? Bottom line: Learn the technical and managerial issues that are important to your client now and over the life of the upcoming contract.

What is planned for the future?

Through interaction with the client's staff, understand the evolutionary changes anticipated and planned for the period of performance of the contract.

What relevant documents or articles have the client and incumbent contractor personnel published?

Has the client established a public-access reading room, library, or Web site that archives documents and articles relevant to the particular contract? If so, plan to visit or browse and copy, download, or take notes from appropriate materials. Sometimes photocopying of hardcopy materials has to be scheduled in advance, so call ahead.

Strengths and weaknesses of the incumbent contractor.

Incumbent contractor refers to the company performing on the current contract.

Can your company hire incumbent personnel?

Incumbent personnel refer to those staff members employed by the incumbent contractor on the specific project that is being recompeted. Placing these people under contingent hiring agreements and including their résumés in your proposal can be highly beneficial to your prospects of winning the contract. Incumbent personnel are known quantities to the client. Those staff are knowledgeable about and familiar with such elements as the client's current and future technical directions, organizational and reporting structures, and procedures and protocol. There is a reduced learning curve associated with contract phase-in and start-up given a staffing mix that includes a significant number of key incumbent staff. *Contingent hiring* refers to the establishment of a signed agreement between an individual and a company that stipulates, upon contract award, the person will be hired to perform certain duties on the particular contract. Ensure that the client holds any incumbent staff to whom your company extends a contingent hiring agreement in high regard professionally and personally. Very importantly, interact with incumbent staff during noncore business hours away from their job site, if at all possible. You might also host an open house at a local hotel or conference center to introduce your firm to the incumbent staff. Publicize the open house and plan to have members of business development, executive management, recruiting, and human resources participate. Develop a take-away flier for attendees and deploy a Web site for prospective new hires.

One proven approach to incumbent capture is to identify one or several key incumbent staff professionals, build a rapport with them to win their hearts and minds to your company's cause, and then leverage their strong working relationships and leadership skills to bring other incumbent staff into the fold—in effect, into a formal, written commitment with your company. This Pied Piper technique requires absolute confidentiality and honoring of all agreements, commitments, and promises that were made during the incumbent capture and proposal phases. And it must be done in full recognition that the incumbent staff is, in many cases, putting their careers and reputations on the line for your company.

Which client personnel will sit on the source evaluation board?

Attempt to learn who the probable key candidates are for the SEB, including technical, contracting, legal, and administrative staff. Learning their sentiments, concerns, goals, and professional backgrounds can be very advantageous when composing your proposal narrative and developing benefits-oriented graphics.

Who is the competition? What does your company know about them?

The client's contracts office will tell you the name of the incumbent contractor, as will published and Web-based agency acquisition forecast lists. Advance knowledge of the identity and technical competence of additional competitors can be gleaned from conversations with other companies in the same industry; watching newspaper, professional-journal, and Web-based job-site classified ads,[2] preproposal conference attendance lists, and Freedom of Information Act (FOIA) requests for previously submitted proposals to the client and monthly technical progress reports for the specific program or project.

Is your company qualified to bid alone, or must you team as the prime contractor or as a subcontractor?

There are circumstances under which teaming to win a contract is very appropriate. Talk with the client about his or her predispositions about and experiences with teaming arrangements. And sometimes your company may be too small to prepare a credible management plan as the prime contractor. For example, the contract staffing requirement may be 35 full-time equivalents (FTEs) per year. If yours is a 60-person company, you may want to consider establishing a teaming relationship with a larger qualified prime.

What are your company's technical, programmatic, costing, and staffing strengths and weaknesses regarding the particular opportunity? Can you overcome your shortcomings in time?

After preparing a realistic inventory of your strengths and weaknesses, your Acquisition Team will have to determine if ongoing marketing efforts combined with hiring success rate (to address staffing deficits), cost

---

2   This practice can be very beneficial to learn what projects your competition is pursuing as well as their staffing deficits. Regularly check monster.com, careerbuilder.com, dice.com, and other job-searching sites.

restructuring [in accordance with Defense Contract Audit Agency (DCAA) stipulations], and exploring specialty subcontracting arrangements (to mitigate technical deficits and programmatic risks) will result in a favorable bid position for your company.

It is critical that the information collected during interaction with your client be recorded electronically and then distributed on a need-to-know basis within your company. You may want to develop client contact reports, which include such information as the following:

- Client;
- Contact name and title;
- Contact date;
- Company representatives;
- Contact type (such as visit, call, or conference);
- Mission (your company's objectives for the contact);
- Summary of results;
- Next action.

These reports might be made accessible to your staff via a password-protected, firewalled company intranet.

Support services contracts are often characterized by very diverse, sometimes even conflicting, requirements, because they frequently take the form of umbrella task-order, level-of-effort (LOE) contracts serving multiple government representatives. The challenges associated with marketing and preparing the proposal for such bids center on creating as complete and detailed a picture as possible of various decision makers' needs within the client's infrastructure. To this end, it is very useful to build a detailed technical-marketing activity into the proposal development life cycle, using domain experts to interview as many of the cognizant government staff as possible. The purpose of this activity is to bring your company's understanding of the requirements to the next level of technical detail after the initial round of marketing visits. The preproposal interview form presented in Figure 3.10 is very useful for this purpose. Armed with it, your company's technical marketers—usually working-level support services staff—can hold technical discussions with many government task leaders in a manner that yields fruitful management approaches and proposal themes. Note that the questions center on three areas: (1) hot buttons, that is, critical issues and their likely evolution; (2) ghosts, that is, possible problems with the current support (if there is an incumbent); and (3) discriminators, that is, approaches that allow you to propose a superior, solution well-tailored to your client's requirements that your competitors cannot.

Preproposal Interview Form

Date: _____

Interviewer(s): _____   _____

Interviewee(s): _____   _____

1. Please provide a brief description of the work that you expect contractors to perform under this contract.

   Note to interviewer: This should be a ½- to 1-page description. Attach a separate write-up to accommodate it. It is very helpful to get important reprints and other documents from the interviewee, though not to the extent that it becomes impossible to assimilate them. Concentrate on the main points! Also, be on the lookout for material for graphics.

2. What are the most critical areas of the work (i.e., the technical or management issues that are driving the contractor)?

   Provide some key sentences or bullets. This is very critical as these points will drive the proposal themes and technical response.

3. How many contractors are currently supporting this work? Do you expect this level to change?

4. Are there any particularly important contractor personnel whose names you can give us, who are key to your effort?

5. What kind of skill mix do you require for this work?

   Does the client want more high-level people ("partners") or lower-level support? On technical jobs, what kind of proportion of senior programmers, analysts, junior programmers, scientists, and engineers is desired?

6. What kind of management structure best supports this work?

   How does the client think management would best support him or her (e.g., what kind of organizational structure and processes)?

7. Are there any areas in which you think substantial change or innovation is required? "Innovation" could be either technical or managerial.

   Probe for areas where we can suggest innovation! Examples: Does the client feel that he or she needs short-term support with highly specialized skills? Is the client looking for some particular kind of management structure?

8. If you had the chance, are there any changes you would make in the nature of your contractor support?

   Tie this to the previous question. Get the client to tell you how you can look like a hero!

9. How do you see your contractor support requirements evolving in the future?

   Get information about future skill mix and staffing levels. What are the client's hopes, fears, or biases about budgeting, and how does he or she see these affecting the need for support? For example, does the client see a larger or smaller role for the support contractor in the areas of science, operations, and instrument engineering?

10. What do you think your future critical issues are likely to be?

    This is an extremely critical question! Get as much information as possible regarding projected future problems and drivers, and the directions from which the client thinks solutions might emerge. This will be a dominant part of the proposal!

11. What would you be looking for in a winning proposal for this effort? In particular, what do you see as the highest priorities that would be given the most attention in a winning proposal?

    Try to draw the client out about his or her hot buttons and likely selection discriminators. That is, what could distinguish you from your competitors in a strong, positive way?

12. Can you suggest anyone else to whom we might talk to get more information on these topics?

Figure 3.10 Preproposal interview form. (Adapted with permission from Dr. Richard Isaacman.)

As a practical matter, it is not a good idea to walk into a client interview and start reading from a list of questions; the form should be completed by the interviewer as soon as possible after the interview. When feasible, it is useful to send two company staff members to the interview; they can alternate asking questions and taking notes, and address any gaps in each other's recollections when completing the interview form later. The completed interview write-ups should be reviewed by the capture manager, proposal manager, and key proposal writers before actual writing begins in order to identify common issues out of which proposal themes can emerge. These write-ups also represent a wealth of technical knowledge that can be mined by the proposal writers in order to address client concerns in depth. Figures 3.11 and 3.12 provide additional potential interviewing tools when interacting with your private-sector client or potential federal government client.

3.7 Maintain management visibility on your contracts

A very practical and relatively easy marketing technique applicable to contracts on which your company is the incumbent (contractor performing on the current contract) is for members of your company's executive and division-level management to visit the client staff as well as your on-site personnel on a regular basis. Executives from one highly successful small business visit their clients nationwide at least once every business quarter. Their actions validate commitment! Build and continue to cultivate communication networks with client staff at both professional and appropriate social levels, including professional association meetings and government-industry council meetings. Let your client know that his contract is important to your company by showing a real physical presence at his job site during the entire contract life cycle and not merely at contract renewal. A discussion of your management's commitment to a particular project in the proposal narrative is far more persuasive when it has been substantiated by regular site visits by your senior managers over time.

Senior management can communicate on an ongoing basis with, for example, on-site company staff through e-mail, company newsletter and e-newsletter, intranet, informal briefings, friendly on-site luncheons (brown bag, bring-your-own lunches are fine), and performance recognition meetings and award ceremonies on-site. There is an enhanced sense of loyalty among your on-site staff to your company that can follow from ongoing management presence and genuine interest. Too often, on-site staff, due to sheer proximity to the client and isolation from contact with

Client:_____ **Address:**_____ **Telephone No.: (__)_____**

1. Are you meeting with a representative of corporate management or facility management?

2. What are the specific types of needs or activities of the client?

3. Are these needs managed through an in-house technical organization, professional service contracts, or both?

4. What is the total company or facility budget for all goods and services for the current and upcoming year? (Is the budget published and can we have a copy?)

5. How much of this budget will be spent on professional services or consulting?

6. Who is responsible for management and procurement of services in each specific requirement area?

7. Is there an organizational chart available that can help us understand the client? Can we have a copy?

8. Which of our competitors are under contract or have recently performed work for the client? Are any of the competitors under term or national service contracts? When are these contracts up for rebid?

9. Are there any opportunities evident from the discussions for our company to benefit the client? What is the time line for these opportunities?

10. Is there a requirement that our company be listed as an approved firm on a procurement list? Are we on it? Who in the client's organization is responsible for maintaining the list?

11. Does the client have Minority Business Enterprise/Woman-Owned Business Enterprise (MBE/WBE) goals for their procurements? Are there specific MBE/WBE firms that the client recommends?

12. How does the client conduct the procurement of service contracts? Who is (are) the decision maker(s)?

Figure 3.11
Private-sector
client information
interviewing tool.

your company's management, develop greater loyalty to the project and to their government counterparts than to your company. When recompetition comes around, your staff may not be so concerned as to which company they work for but rather that they stay associated with the project. And your competition may find them easy targets to lure away with contingent hire agreements.

It is critical that senior management take an active role in client inter-action, relationship building, and other marketing activities. "A senior executive's instinctive capacity to empathize with and gain insights from customers is the single most important skill he or she can use to direct a

Questions

Day-to-Day Operations

1. What are your specific responsibilities and organizational requirements?

2. What are the most significant challenges to meeting your performance goals today?

3. How are effectiveness and efficiency measured in your organization?

4. How is success measured within your organization?

5. What performance metrics have been established and documented? How well do they support your organizations measures for effectiveness, efficiency, and success? How would you improve these?

6. Please describe the level of importance that cost control and financial management occupy within your day-to-day operations.

7. Currently, how is your organization working to retain and leverage intellectual capital?

8. Can you suggest anyone else to whom we might talk to get more information on these topics?

Management for Moving Forward

1. Where do you believe there is substantial room for change or innovation? Innovation could either be technical or managerial.

2. Are there any sacred cows (protected structures or functions) within the organization?

3. Explain how important organizational change management—changing culture, attitudes, and policies—is for your organization.

4. What are the critical things that need to be fixed or enhanced within your area of operation? What changes in agency policies will impact your work directly and why? What do you consider to be your primary future challenges? (What are the things that keep you up at night?)

5. What kinds of management structure, staffing plan, and skill mix best support your work?

6. What are the key challenges perceived from a federal management standpoint)? What would you recommend as a path forward?

Customer and Contractor Assessment

1. Who are your current support contractors? How many contractor staff from each company support your operations?

2. How are your current contractors performing from both a technical and a management perspective?

3. What do you see as their specific strengths and weaknesses?

4. Please describe your contractors' level and type of support.

5. Are there contractors that possess a great deal of institutional knowledge and would be considered key personnel?

Figure 3.12
Client information interviewing tool.

6. Help me to understand your specific concerns regarding contract phase-in periods.

7. Have service-level baselines been established for your area of support?

8. How are your business processes now aligned with your IT portfolio? How will that alignment change over time?

9. If you had the opportunity, what changes would you make in terms of contractor support in your area?

Figure 3.12
(Continued.)

company's strategic posture" [8]. Know when to have your senior management visit your client, and make sure that your management knows what is necessary to convey as well as to learn.

3.8 Project managers as client managers

Your company's project managers (PMs) should develop self-recognition as client managers and first-tier marketing generalists. These managers are in the enviable position of being in day-to-day contact with your clients. PMs should be taught and mentored on how to cross-sell your company's services and products. They can help your client better understand and appreciate the direct benefits of your firm's core competencies. PMs can appropriately, yet persistently, inquire as to what additional services your company can provide to this client. Expanding services to existing clients is the most cost-effective form of marketing activity. You need to recognize your PMs as an integral part of your public relations, or reputation management, initiatives. PMs put a face on your company; they need to be your local advocate. Your project managers can also be instrumental in obtaining letters of commendation from your clients, which in turn can be used in proposals to authenticate your technical and programmatic expertise. Letters of commendation should be filed in your company's proposal library, and scanned into electronic format in your automated knowledge base for rapid retrieval and easy placement into a proposal document.

Your company should also regularly assess performance levels on each of your current contracts. That assessment must necessarily include an evaluation of your performance by your client. Figure 3.13 offers one such performance scorecard that might be used. Figure 3.14 presents an automated performance questionnaire.

Please rate how we are doing for the following activities:

	(Not doing well at all)				(Doing very well)	
	1	2	3	4	5	N/A
Clear and Effective Planning (schedules, updates)						
Support for Client Decision Making						
Regular Communication and Feedback						
Innovative Ideas						
Appropriate Experienced Staffing Applied						
Appropriate Technical Support Provided						
Cost-Effective Project Performance						
Issues Resolved Appropriately						
General Client Satisfaction						
Meeting Project or Task Objectives						
On-Time Deliverables						

Other comments:

Thank you.

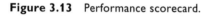

Figure 3.13   Performance scorecard.

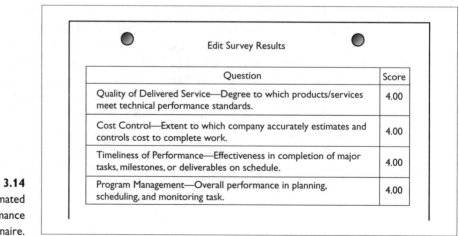

Figure 3.14
Automated
performance
questionnaire.

Question	Score
Quality of Delivered Service—Degree to which products/services meet technical performance standards.	4.00
Cost Control—Extent to which company accurately estimates and controls cost to complete work.	4.00
Timeliness of Performance—Effectiveness in completion of major tasks, milestones, or deliverables on schedule.	4.00
Program Management—Overall performance in planning, scheduling, and monitoring task.	4.00

Edit Survey Results

3.9 Commercial off-the-shelf acquisition

There is a clearly discernible migration toward commercial off-the-shelf (COTS) procurement within the federal arena, particularly in the area of software products. COTS products represent industry's best—they are tested and piloted before deployment in the marketplace. They are also readily available off of the GSA schedule. Congress and the Office of Management and Budget (OMB) within the Executive Office of the President have indicated a preference for COTS solutions, which constitutes an important consideration in obtaining funding for a federal agency.

Significantly, the goal of the E-Government Act of 2002 is to maximize return on investment (ROI) of government IT assets and portfolios. Tomorrow's enterprise success factors have changed from technology-centric ones to those premised on "business + security," "business + enterprise architecture," and "business + capital planning." COTS tools play a key role in addressing enterprise IT asset management issues as well as developing and deploying integrated enterprise technical solutions. Importantly, COTS products tend to drive down the total cost of ownership (TCO) of IT portfolios.

On March 26, 2004, the Department of Energy (DOE) deputy secretary issued a memorandum for developing a single, integrated IT infrastructure within that federal department. One of the key elements specified by the deputy secretary is to reduce operational costs and increase security by moving to a single, integrated IT infrastructure. Specific future challenges for DOE span providing a common operating environment (COE) of secure and easily accessible core applications, enhancing service performance levels across a broad range of diverse government customers; implementing a practice to build once, and then use many times; implementing an enterprisewide information cybersecurity governance, management, and architecture in accordance with the Federal Information Security Management Act (FISMA); and aligning IT direction with DOE's business strategies and mission.

Also, as the Department of Defense (DoD) pushes toward more COTS solutions, it is facing a greater need to integrate those solutions with each other and with existing systems.

This extensive use of COTS software components in system construction presents new hurdles to system architects and designers. Some federal agencies are discovering, however, that a practice that was intended to streamline acquisition—and that does so in many cases—also raises some challenging issues associated with data conversion and systems integration. For example, government requirements may have to be shoehorned to fit the features of COTS products. In specific cases, software bridges and interfaces can be designed and developed to address this issue. There are also the matters of intellectual property and licensing that must be addressed. When COTS products are modified, ownership of the new product becomes open to question. And "[f]ormerly, government officials who were in a hurry to buy off-the-shelf equipment such as personal computers could sidestep the normal, time-consuming bidding practices by awarding a sole-source contract directly to a minority-owned firm that could fill the order" [5]. According to the National Federation of 8(a) Companies based in Arlington, Virginia, the 1996 federal acquisition reform legislation allows "agency officials to make purchases of selected equipment from a large list of preapproved contractors, large and small" [5].

A broad cross-section of federal agencies has recently examined their high-end computing (supercomputing) requirements. These agencies include the Defense Advanced Research Projects Agency (DARPA), DOE, NASA, NIH, the National Oceanic and Atmospheric Administration (NOAA), the National Science Foundation (NSF), and the National Security Agency (NSA). In general, these agencies are using COTS-based high-end computing solutions. Examples of high-end computing applications areas with federal government involvement and interest include nuclear stockpile stewardship, global climate modeling, ocean state prediction, geophysical phenomena such as earthquakes, aeronautics, weapon systems design, signal processing, and materials modeling.

The Federal Acquisition Streamlining Act of 1994 stipulated that a commercial item can still be a commercial item if it has one of two modifications. The first is a modification, any kind of modification, that is customarily available in the marketplace. This is a type of modification that the contractor has done in the commercial marketplace. But the other kind of modification allowed under FASA is a modification for government use. Even if it has a modification for government use, it is still considered to be a commercial product. Commercial item services are services that are customarily sold in the marketplace for a set price for a particular task. This is a great opportunity for small business.

3.10 Pursuing firm-fixed-price and invitation-for-bid opportunities

In negotiated services contracts, relationship building through marketing is paramount to success. However, selling computers or reprographic paper—in effect, products—under firm-fixed-price (FFP) contracts often involves less in the way of personal interaction. FFP contracts require that "the contractor delivers the product or performs the services at a specific time at a price that is not subject to any adjustment of actual costs incurred during contract performance. It places 100% of the risk on the contractor and places the least amount of administrative burden on the government" [9]. Maximum profit can be earned by managing FFP contracts aggressively and identifying, understanding, and controlling risk. During business planning and when marketing, your company will want to establish an appropriate mix of targets between cost-reimbursement and FFP types of opportunities.

Invitations for bid (IFBs), or sealed bids, result in an FFP contract. Bids are opened publicly, and the award is made immediately to the lowest priced, responsible, responsive bidder. No discussions are allowed, and no exceptions may be taken by the contractor. IFBs are suitable for COTS purchases as well as for small support services contracts. Understanding your company's cost structure and your vendors' costs as well as being able to operate at maximum efficiency during contract performance are critical to making profit on IFB/FFP opportunities. Profit on cost-reimbursement type contracts is termed *fee*.

3.11 Using the request for information and the request for comment as valuable marketing tools

In circumstances under which the government is seeking to establish that adequate competition exists within industry to release a solicitation or to determine if industry has developed specific technologies that the government requires (assess project feasibility), an agency will issue a request for information (RFI), screening information request (SIR), or request for comment (RFC) on a statement of work (SOW). In fact, the Federal Acquisition Regulations encourage government agencies to promote early exchanges of information about future acquisitions (see FAR, Part 15.201). This practice is intended to identify and resolve concerns regarding such things as proposed contract type and vehicle, requirement feasibility, suitability of evaluation criteria, business size standards [determined by North American Industry Classification System (NAICS) codes], and availability of reference documents.

Your company's response to an RFI, SIR, or RFC should be prepared in the same professional manner as a proposal. Too many times in the case of an RFI, companies will submit marketing brochures or other prepared, canned materials rather than develop a customized, client-oriented package that provides specific benefits you can offer to your client. Responding to an RFI presents a significant opportunity for your company to demonstrate in writing that you are interested in supporting the client and in developing solutions for your client's issues and challenges. Also, you will learn about the client and possibly the competition. As with an RFP, respond specifically to the client's requests. Authenticate your technical capabilities with references to specific contractual experience. Help your client to better understand your company's particular strengths. Submit the RFI on time and to the correct location. Hand delivery can also be beneficial, although, increasingly, electronic submittal is required and hard copies are expressly disallowed.

3.12 Standard Form 129s and contractor prequalification statements

Certain client organizations require that a company be prequalified or preapproved prior to the company being able to receive RFPs, RFSs, RFQs, RFIs, and other solicitation documents. An example of such a prequalification vehicle is the federal Standard Form 129 (Revised 12-96), which is a "Solicitation Mailing List Application." This form can be found on the Web in PDF and other formats at http://www.gsa.gov/Portal/gsa/ep/ formslibrary.do?viewType=DETAIL&formId=BEED4B6F1221961 885256A1F005CE51C. The U.S. Army Corps of Engineers, Forest Service, General Services Administration, and various port administrations are among the many organizations that use this application. The SF129 contains 21 blocks for information ranging from the type of organization your company is and the names of officers, owners, or partners to tax identification numbers, applicable NAICS codes,[3] and net worth. The application must be signed by the principal of the company. Certain federal agencies attach a Supplemental Commodity list along with instructions to the SF129. The Standard Army Automated Contracting System (SAACONS) Vendor Information Program pamphlet (revised April 16, 1989) is one such Supplemental Commodity list. Your company will have the opportunity to identify up to 12 Federal Supply Class (FSC) categories of supplies and services on this particular list. Accurate and timely completion of the SF129 application and supplemental documents will result in your company's name being added to a particular agency's automated bidder's mailing list.

State and local governments as well as private-sector clients may issue contractor prequalification applications or contractor qualification questionnaires[4] in order to prequalify a listing of potential contractors. Examples of the wide-ranging list of clients that request such information

---

3   SIC codes have been changed to NAICS codes.

are the State of Georgia, MCI, Southwestern Bell, Philadelphia Electric Company, Shell, and Exxon Mobil. The information required can be broad-ranging in scope but often includes such information as the names of principals of the company, financial information, years of contractual experience, debarment status (has your company currently been debarred from submitting bids?), minority business status, professional registrations by state, and equipment lists. Again, timely and accurate completion is important.

3.13 Ethics in marketing and business development

Ethical conduct in all business activities, whether domestic or international, should be your company's corporate standard at all times [10]. The development and implementation of a formal set of ethics rules that prohibits any type of procurement fraud by employees, agents, consultants, or subcontractors would be a prudent step. Indeed, DoD FAR Supplement (DFARS) 203.7000, titled *Contractor Responsibility to Avoid Improper Business Practices*, notes that a contractor's management system should include a written code of business ethics and conduct as well as an ethics training program for all employees. According to an article published in the *Harvard Business Review*, new federal guidelines now recognize the organizational and managerial roots of unlawful conduct and base fines partly on the extent to which companies have taken steps to prevent misconduct [11].

In their important work titled *Formation of Government Contracts*, George Washington University law professors John Cibinic, Jr., and Ralph Nash, Jr., discuss standards of conduct in two broad categories, namely, "those dealing with improper influence on Government decisions and those requiring honesty and disclosure of relevant facts in dealing with the Government" [12].

FAR subpart 9.5 addresses the organizational conflict of interest. Part 3 of the FAR presents guidance regarding "Improper Business Practices and Personal Conflicts of Interest." One example, the Anti-Kickback Act of 1986 [codified at 41 U.S.C. §§ 51–58 (1988)] discussed in FAR 3.502-2, is legislation designed to deter subcontractors from making payments and contractors from accepting payments for the purpose of improperly obtaining or rewarding favorable treatment [13] in connection

---

4   These questionnaires can be called bidder's list applications, general subcontractor surveys, potential supplier profile questionnaires (PSPQs), preliminary qualification questionnaires, qualified suppliers lists (QSLs), and vendor forms.

with a prime contract or a subcontract relating to a prime contract. The term kickback means any money, fee, commission, credit, gift, gratuity, thing of value (for example, promise of employment), or compensation of any kind. Unlike its precursor in 1946, the 1986 act includes all types of government contracts. Finally, the 1986 act places new reporting requirements on prime contractors and subcontractors. Possible violations of kickback laws must be reported in writing to the IG of the appropriate federal agency, the head of the contracting agency if the agency has no IG, or the U.S. Department of Justice.

The Procurement Integrity Act, about which there is considerable confusion on the part of both civil servants and contractors, provides that the government cannot impart information to one bidder or proposer without making that information public to all potential bidders or proposers. And the False Claims Act Amendment of 1986 facilitated prosecution of complex defense procurement fraud cases.

Many private-sector corporations—your potential prime contractors or clients—have enacted written policy codes that govern gifts and entertainment. One major East Coast manufacturing firm stipulates that its employees are forbidden to accept any incentives in exchange for increased purchases through a given supplier or subcontractor. Gifts are defined very strictly, to include dinners (unless incidental to a business meeting), money, discounts on personal purchases, holiday presents, tickets to theaters and sporting events, airline tickets, hotel accommodations, and repairs or improvements at prices other than for fair market value.

Finally, the Defense Industry Initiative on Business Ethics and Conduct (DII) (http://www.dii.org/) was established in June 1986 by 32 major defense contractors. These contractors pledged to adopt and implement a set of principles of business ethics and conduct that acknowledge and express their federal procurement-related corporate responsibilities to the Department of Defense as well as to the public and to each other. The defense industry signatories are united in their commitment to adopt and implement principles of business ethics and conduct that acknowledge and address their organizational responsibilities. Further, they each accept the responsibility to create an organizational culture in which ethics is paramount and compliance with federal procurement laws is a strict obligation. The DII's essential strength lies in sharing best practices to maintain the highest ethical standards, encouraging employees to ethical conduct, and requiring compliance in the course of its business activities.

3.14 Advertising, trade shows, and high-impact public relations

As with brand name recognition in consumer advertising and marketing, increasing the baseline level of name recognition for your company is important to your overall business development success. Well-targeted public relations (PR), trade show, and advertising efforts will probably not result directly in sales or contracts but will help to build an image of legitimacy and solid foundation for your company that can and should be leveraged by your external and internal sales teams and public relations or reputation management efforts (see Figure 3.15).

There are many freelance graphics artists, for example, who can develop high-impact ads on a cost-per-ad basis for placement in key business and trade publications without the significant fees associated with retaining an advertising and design agency. Figure 3.16 provides an example of an ad developed for a small company. Watch for special issues of publications important to your lines of business that may focus directly on market segments that are part of your strategic plan. Cases in point are *Washington Technology; World Dredging, Mining, and Construction;* and *Chemical Week.* The latter publication, for instance, ran a special supplement devoted to the state of New Jersey. Companies that did chemical or environmental business or were positioning themselves to do such business in New Jersey could then place ads in that particular supplement.

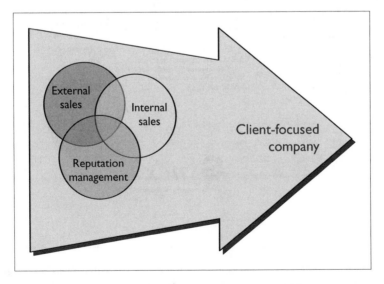

Figure 3.15
One common
vision.

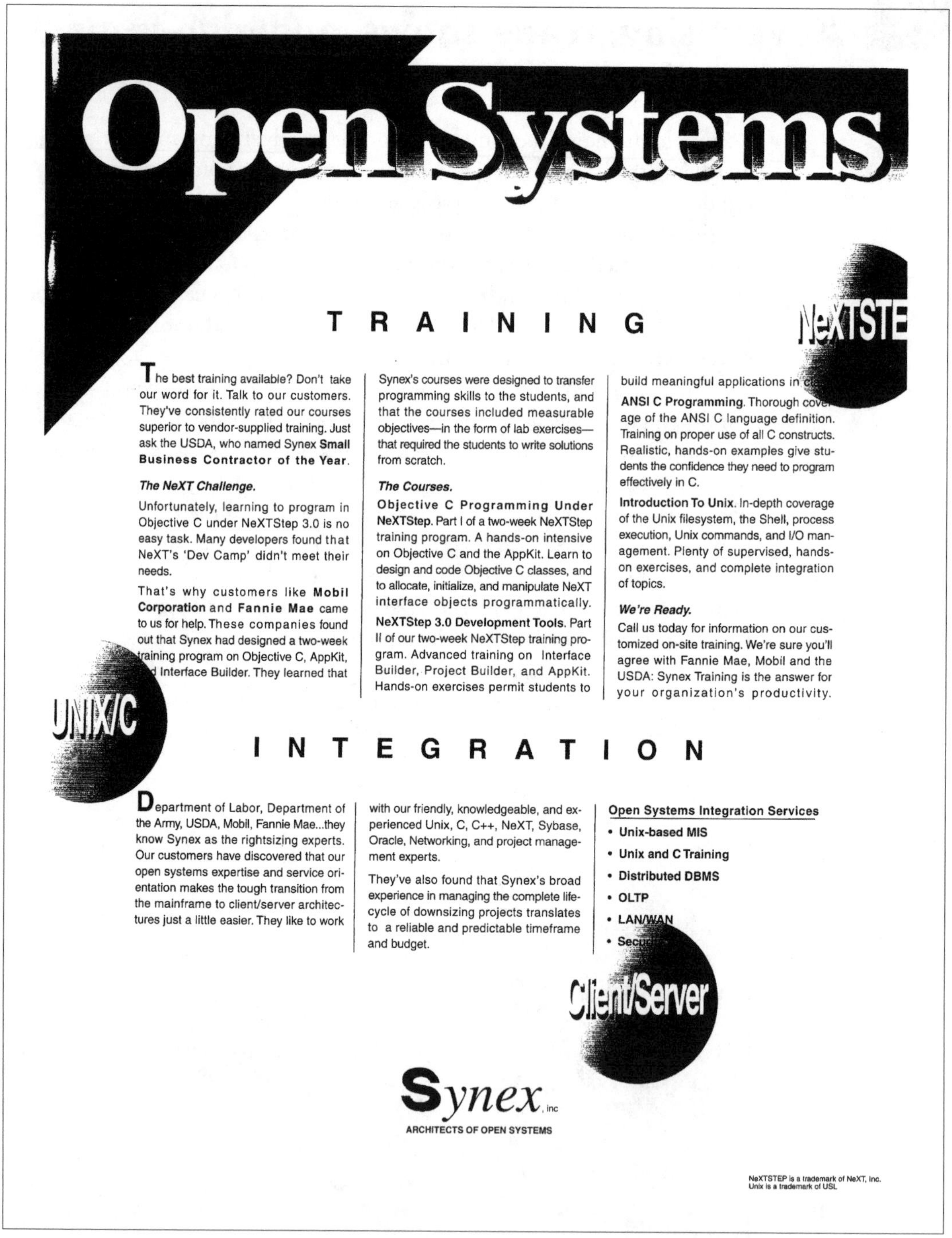

Figure 3.16 Advertisement for a small company that appeared in *Washington Technology's 8(a) Report.* (Ad developed by Robert S. Frey for Synex, Columbia, Maryland. Reprinted with permission from Robert S. Conner, president, Synex.)

Increasing the baseline of name recognition for your company is important to your overall business development success.

Appearance at selective trade shows will increase your company's visibility in the specific business community you serve. Professional-looking table-top trade show booths can be designed and populated with attractive, powerful photographs and other imagery for only a few thousand dollars total. If your company does work for the Department of Defense, for example, that agency maintains public domain photographs that are available for nominal fees at its Anacostia Naval Air Station near Washington, D.C. NOAA has a Web-based photo library at http://www.photolib.noaa.gov/search.html. NASA also maintains an extensive Web-based image library at http://www.nasa.gov/multimedia/image/gallery.htm. An outstanding Web site that points to a host of public domain image galleries spanning the USDA, Air Force, Navy, and U.S. Geological Survey (USGS) is located at http://mciunix.mciu.k12.pa.us/~spjvweb/cfimages.html. Professional stock imagery may be procured from photo houses or from the Internet (e.g., http://www.fotosearch.com/photodisc) for limited usage instead of taking your own photographs.

Other public relations efforts can take the form of news releases to appropriate business, financial, and professional and trade journals. News releases will, of course, need some kind of news hook; that is, you will need to provide a financial or business editor with something around which to build a headline and weave a story line.

Your company's credibility depends in part upon the efficient distribution of news about your success. You might consider employing the services of a media relations firm, such as BusinessWire (http://www.bizwire.com). BusinessWire, a San Francisco–based "bricks-and-clicks" company founded in 1961, is used by public relations and marketing professionals to disseminate news releases, photos, multimedia, regulatory filings, and other information. News release subjects include breaking news, earnings results, product announcements, mergers and acquisitions, public policy, legal issues, webcasts, and press conference advisories (http://home.businesswire.com/portal/site/home/index.jsp?epi_menuItemID=2773f616199c2cdf3126a365c6908a0c).

BusinessWire also offers a suite of automated PR support tools and technologies. It is critical for an organization to monitor and measure the effectiveness of its PR functions. For this reason, BusinessWire NewsTrak provides a suite of measurement and monitoring services for an organization's public relations and communications efforts. These presentation-ready reports include NewsTrak Reach, with detailed statistics on the BusinessWire distribution of an organization's news release. NewsTrak Posting provides navigable links to a news release as it appears on key news and information Web sites. Another tool, NewsTrak Access, provides ongoing aggregate viewership data of a news release. Finally, NewsTrak Clips provides ongoing media monitoring, and NewsTrak Compass offers ongoing media analysis of an organization's public relations and communications initiatives (http://home.businesswire.com/portal/site/home/?epi_menuItemID=78214c987aca550475299010e6908a0c&epi_menuID=10c2d33c773a4256f8d35d87a0908a0c&epi_baseMenuID=384979e8cc48c441ef0130f5c6908a0c).

Importantly, BusinessWire's news wire circuits enable organizations to target their news to select audiences based on specific geographic and industry market needs. BusinessWire's delivery network is comprised of 60 news agencies, financial and information providers, and Web-based news services around the world (http://home.businesswire.com/portal/site/home/index.jsp?epi_menuItemID=6fe6ec23611b3703ac4b05e6c6908a0c). Your news releases will also be delivered automatically to 150 on-line services and databases worldwide, including Reuters Business Alert, *Wall Street Journal Interactive, New York Times,* Dow Jones News, MSNBC News, Bloomberg Financial Markets, *USA Today, Washington Post,* and Yahoo! Finance and AOL.

Another effective PR technique related to trade and professional journals is to generate and place articles in these publications [14]. This process is highly cost-effective, and reprints of articles that have been published by your company's staff can be used in proposals to authenticate the firm's and the proposed staff's expertise. It is advantageous to contact the editorial staff of appropriate trade journals to obtain the agendas and themes of upcoming issues as well as submission guidelines. In addition, your company can create and maintain its own Web page on the Internet, replete with logo, easy-to-use icons, and engaging content, and accessible for data mining with well-known search engines, such as http://www.google.com (metasearch engine), http://www.dogpile.com (metasearch engine), http://www.hotbot.com, http://www.metacrawler.com, http://www.yahoo.com, http://www.altavista.com, http://www.webcrawler.com, http://www.lycos.com, and http://go.com, as well as http:// www.excite.com. Your Web page should clearly identify your company's lines of business up front, and provide user-friendly, quick-loading access to information about how your firm can support clients successfully. Include an electronic fill-in form that Web site visitors can complete and e-mail or mail to your marketing staff. You might build in the text from recent external news releases, information about your office or project locations and client base, corporate awards, core competencies, GSA Schedule, qualifying NAICS codes, and upcoming trade show appearances. Add your Web site address to business cards, marketing brochures and fliers, and promotional items such as golf balls, key rings, or calculators.

Finally, include electronic point-and-click hyperlinks on your Web pages to other pertinent Web sites that provide additional regulatory, technological, business, or contractual information. Register your Web site with appropriate search engine providers on an ongoing basis.

In developing your company's Web site, three key elements will require significant attention: (1) design, (2) functionality, and (3) audience appeal (see Figure 3.17). Under the design element, important factors include readability, clear purpose of the site, and site architecture. Within the element of functionality, considerations include navigability, intuitive search capacity, order management/e-commerce, actionability [15], and download speed. Under audience appeal, you need to consider such key parameters as visitor retention, consistent Web site mood, and brand name recognition.

Remember that whatever the level of funding that can be allocated for these types of important activities, the critical thing is to do them on a consistent basis. Clients are more likely to buy goods and services from companies they know or have heard about in the press or trade literature. Public relations helps bring your company and its capabilities to life and keeps your company's name embedded in decision makers' minds.

For very select major-impact proposal efforts, your company might consider a full-scale corporate communication campaign that uses the following media and tools:

- Newspaper ads;
- Selective radio ads for key events (e.g., open houses and trade-shows);
- Attendance at industry or customer events;
- Billboards;
- Web sites;
- Signage within appropriate offices and conference centers.

The key to this type of expanded campaign is advanced planning and *consistency of message*—both in the written text and in the graphical images.

We have now introduced the federal competitive proposal environment and the strengths and constraints of small businesses operating in that arena. A company organizational scenario was offered, along with a presentation of strategic and mission planning. Specific, effective marketing techniques, as well as ethical conduct for business development, were discussed.

We will now examine the RFP, the solicitation document issued by both government and private-sector client organizations. The RFP should not be a call to action, as it many times is, but rather a continuation of the ongoing process of marketing your client. As a former fighter pilot and

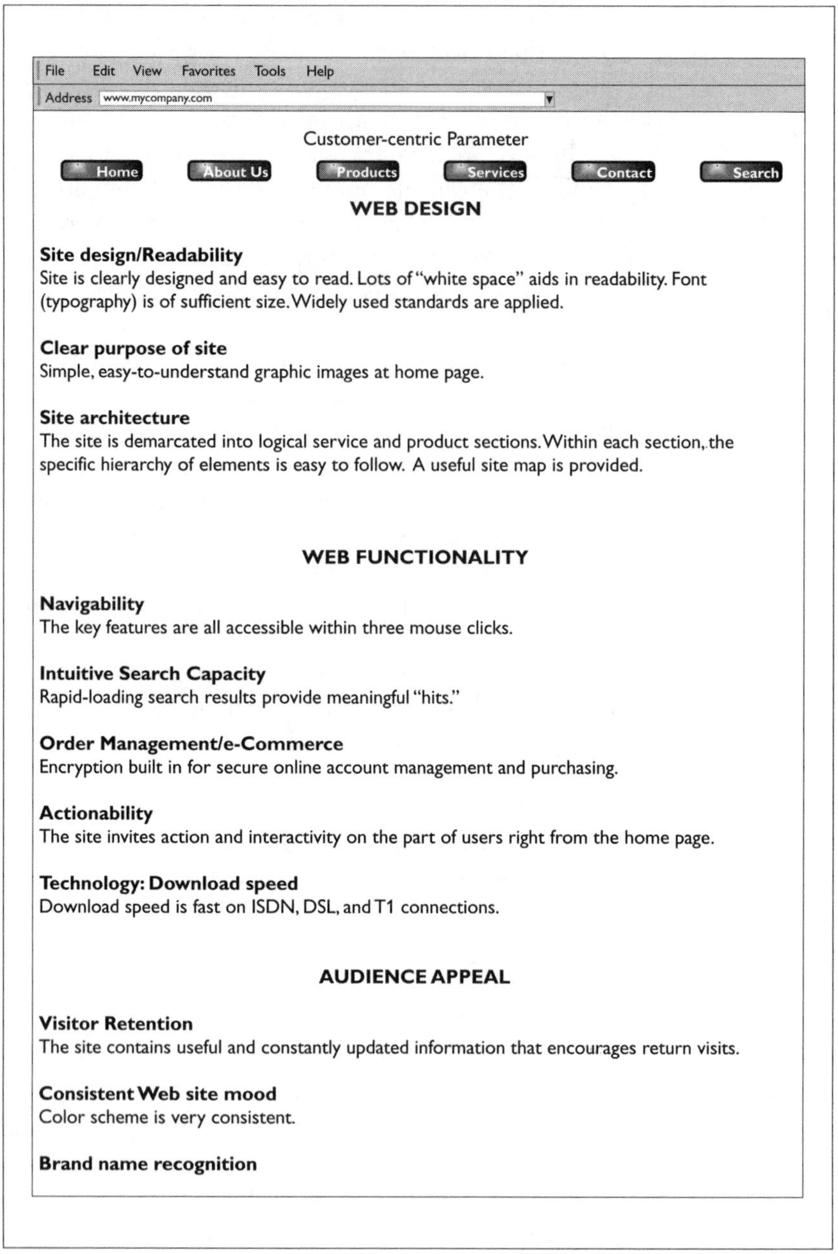

Figure 3.17
Important Web
site parameters.

business executive observed during a major Air Force Space Warfare Center procurement, "The proposal is a terrible place to educate the client."

ENDNOTES

1. Graham, John R., "Getting to the Top, And Staying There," *The New Daily Record,* Baltimore, MD, July 26, 1997, p. 7B.

2. This metaphor was borrowed from Marshall, Colin, "Competing on Customer Service: An Interview with British Airways' Sir Colin Marshall," *Harvard Business Review*, November/December 1995.

3. "Any time the government makes an award to a contractor who is neither the highest technical offeror nor the lowest price offeror, then the government has made a best-value determination; it has made a trade-off between technical and cost features." See Hackeman, Calvin L., "Best Value Procurements: Hitting the Moving Target," McLean, VA: Grant Thornton, 1993, p. 3, (a white paper prepared in connection with Grant Thornton's Government Contractor Industry Roundtables).

4. Ray, Dana, "Filling Your Funnel: Six Steps to Effective Prospecting and Customer Retention," *Selling Power,* July/August 1998, p. 44.

5. As quoted by Behr, Peter, "Just Say Know," *Washington Post*, April 1, 1996.

6. Goldberg, Mim, "Listen to Me," *Selling Power,* July/August 1998, p. 58.

7. "Selected Viewgraphs from Judson LaFlash Seminars on Government Marketing and Proposals," *Government Marketing Consultants,* July 1980.

8. Gouillart, Francis J., and Frederick D. Sturdivant, "Spend a Day in the Life of Your Customers," *Harvard Business Review*, January/February 1994.

9. McVay, Barry L., *Proposals That Win Federal Contracts: How to Plan, Price, Write, and Negotiate to Get Your Fair Share of Government Business*, Woodbridge, VA: Panoptic Enterprises, 1989, p. 29.

10. For a variety of articles that all focus on business and professional ethics (including international marketing ethics), see *BRIDGES: An Interdisciplinary Journal of Theology, Philosophy, History, and Science*, Vol. 2, 1990.

11. See Paine, Lynn Sharp, "Managing for Organizational Integrity," *Harvard Business Review*, March/April 1994.

12. Cibinic, John, and Ralph C. Nash, *Formation of Government Contracts*, 2nd ed., Washington, D.C.: George Washington University, 1986, p. 107.

13. According to Ropes and Gray, examples of favorable treatment that are illegal include receiving confidential information on competitors' bids, obtaining placement on a bidders' list without meeting requisite qualifications, obtaining unwarranted waivers of deadlines, obtaining unwarranted price increases, and recovering improper expenses. Contractors should also be aware of the Fraud Awareness Letter issued in September 1987 by the DoD Council on Integrity and Management Improvements, which identified "indicators of potential subcontractor kickbacks." See "Complying with the Anti-Kickback Act: Guidelines and Procedures," *Developments in Government Contract Law,* No. 10, September 1990.

14. Schillaci, William C., "A Management Approach to Placing Articles in Engineering Trade Journals," *Journal of Management in Engineering*, September/October 1995, pp. 17–20.

15. Smith, P. R., and D. Chaffey, *eMarketing eXcellence: The Heart of eBusiness*, Oxford, England: Butterworth-Heinemann, 2003, p. 164.

Chapter 4

Requests for proposals

4.1 Overview

In addition to federal government RFPs, there are RFPs released by state, municipality, and city governments, as well as by private-sector companies and international governments.

The federal government RFP or performance-based Request for Solution (RFS) that your company requests and receives or downloads from the Web is the culmination of a lengthy planning, budgeting, and approval process. It is a solicitation document issued to obtain offers from contractors who propose to provide products or services under a contract to be awarded using the process of negotiation. Federal procurement is governed by strict statutory and regulatory frameworks that are intended to reduce the risk of fraud, favoritism, and undue influence. In addition, and very importantly, these frameworks are designed to reassure taxpayers that their tax dollars are being spent appropriately [1]. The federal procurement process is highly structured to ensure that the award of a contract is made fairly (see Figure 4.1).

The RFP is a complex document, often prepared under proposal-like conditions. That is to say, it is written and reviewed by a variety of civil servants under tight schedule constraints and is subject to delay caused by, for example, late inputs, protracted legal reviews, program modifications, funding issues, or changes in contractor support of the SOW development.[1] RFPs often contain conflicting or ambiguous requirements, as well as incorrect cross-references, particularly in Section L, "Instructions, Conditions, and Notices to Offerors." (An offeror is any company or organization that responds with a proposal to the RFP.) This may result in part because RFP documents are often assembled using boilerplate materials from previous or

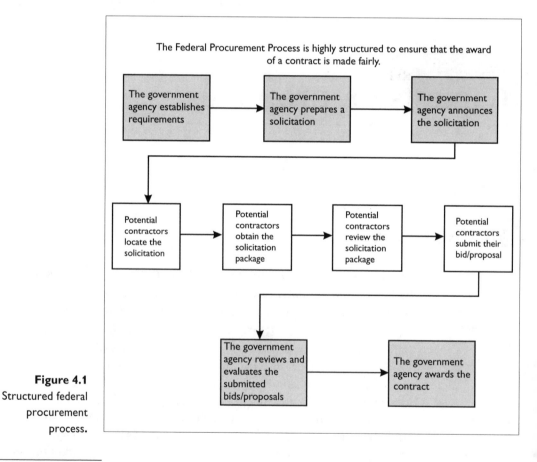

The Federal Procurement Process is highly structured to ensure that the award of a contract is made fairly.

- The government agency establishes requirements
- The government agency prepares a solicitation
- The government agency announces the solicitation

- Potential contractors locate the solicitation
- Potential contractors obtain the solicitation package
- Potential contractors review the solicitation package
- Potential contractors submit their bid/proposal

- The government agency reviews and evaluates the submitted bids/proposals
- The government agency awards the contract

Figure 4.1
Structured federal procurement process.

---

1 Many SOWs and specifications are researched, written, and prepared by contractor personnel under separate contract to the government procuring agency. Because of conflict-of-interest issues, companies that have contracts to prepare SOWs cannot legally compete on particular procurements for which they have prepared the SOW.

similar RFPs. When there are conflicting guidelines, submit questions for clarification in writing to the appropriate government point of contact by the date specified in the RFP. Your questions,[2] along with those from other firms, may result in the government issuing an amendment to the RFP. Within the DoD, Integrated Product Teams (IPTs) have been employed to ensure the proper integration of the various parts of the RFP.

Contractors should not assume that the RFP reveals the full intent of the client's preferences, sentiments, or requirements. There are the *unstated* criteria to address, as discussed in Chapter 3. By the same token, care must be exercised to respond precisely to the RFP requirements in the order in which they are provided. For example, if the RFP calls for a discussion of technical understanding and approach, past performance, and key personnel in that order, your proposal should be structured to respond in exactly that order. Do not build in additional levels of quality or unrequested services that will inflate your company's costs when compared with your competitors. Do not propose a Mercedes when the client wants a Buick! A general rule is to take no deviations or exceptions to the stated RFP requirements, and do not submit alternate proposals, even though they may be allowed. The government has spent considerable time, energy, and resources in performing functional and data requirements analyses; conceptualizing measurable evaluation factors (e.g., past performance), subfactors (e.g., project profiles), elements (e.g., management of complex contracts), and standards (e.g., meeting specific performance metrics); and developing the RFP and its SOW[3] to reflect its concept of a technical solution for a given project or program. It can, therefore, be perceived as presumptuous on a company's part to propose an alternative. Your pre-RFP marketing efforts should have provided intelligence as to the feasibility and acceptability of alternate proposals.

RFPs can range from a few pages to thousands of pages, replete with attachments, for major aerospace and defense procurements. Many RFPs are now being distributed via electronic media (Internet, CD-ROM, agency electronic BBSs) instead of in hardcopy form.

Importantly, when an RFP is posted on the Web, assign a member of the proposal team to check and recheck that Web site several times each day during the entire proposal life cycle. Amendments or other modifications can appear there only minutes before proposal due date and time. This was, in fact, the case with a $180 million DOE–National Nuclear

---

2   Because they will become public record, be certain that your questions do not reveal a weakness or lack of understanding about the project on the part of your company.

3   A statement of objectives (SOO) is sometimes used instead of a government-written SOW to maximize the flexibility afforded to offerors to propose innovative and cost-effective approaches and solutions.

Security Administration (NNSA) solicitation. With some government Web sites, such as Wright-Patterson Air Force Base (AFB), you can register to be notified of changes to the specific procurement via e-mail message. Other Web sites do not have or allow this function.

When hardcopy RFP documents are mailed or picked up at the client site, some contractors elect to scan the RFP into electronic files using some type of optical character recognition (OCR) or intelligent character recognition (ICR) technology. Scanning can facilitate electronic searches for RFP requirements, which in turn can be transferred to electronically stored storyboard templates. It also facilitates searches for hard-to-locate requirements or duplications. A word of caution regarding scanning—time must be allotted to spell check electronically, review, and correct the scanned file. Scanning accuracy levels vary widely depending upon the scanner technology itself and the physical quality of the hardcopy document.

In accordance with the Uniform Contract Format established at FAR 15.406-1, government COs must prepare and assemble RFPs in a specific manner as enumerated in the following sections.

4.2 Part I—the schedule

Section A: Solicitation/Contract Form (Standard Form 33);

Section B: Supplies or Services and Prices/Costs;

Section C: Description/Specifications/Work Statement (The SOW may include system specifications, contractor tasks and services, products, contract end items, data requirements, and schedules, for example. It is an essential part of the RFP.)

Section D: Packaging and Marking;

Section E: Inspection and Acceptance;

Section F: Deliveries or Performance;

Section G: Contract Administration Data;

Section H: Special Contract Requirements.

4.3 Part II—contract clauses

Section I: Contract Clauses.

4.4 Part III—list of documents, exhibits, and other attachments

Section J: List of Attachments.

4.5 Part IV—representations and certifications

Section K: Representations, Certifications, and Other Statements of Offerors ("Reps and Certs");

Section L: Instructions, Conditions, and Notices to Offerors;

Section M: Evaluation Factors for Award (used to determine proposal page allocations, writing emphases, and thematic structure).

Presented a slightly different way, the RFP does the following:

- Describes the requirement in Sections B, C, D, E, F, and J;
- States the government agency's terms in Sections A, B, G, H, I, K, and J;
- Describes the evaluation criteria in Section M;
- Prescribes the proposal format and content in Section L;
- Provides process information in Sections A and L.

Figure 4.2 presents the table of contents (Section 11) from an RFP issued by the Bureau of Census. This form is a Standard Form 33, "Solicitation, Offer and Award," which appears in federal RFP documents.

The contractor's response to the government's RFP is called the proposal, the focus of Chapters 7 and 8 of this book.

4.6 The importance of Section L (instructions to offerors)

Section L of the RFP or RFS provides specific instructions for preparing and structuring the proposal document. (Note, however, that on occasion these instructions are found in Section M. RFPs from the Sacramento District of the U.S. Army Corps of Engineers' are cases in point.)

Turn to Section L first when evaluating a new RFP. Section L drives the structure and form of your company's proposal response. Together

SOLICITATION, OFFER AND AWARD	1. This Contract is a Rated Order Under DPAS (15 CFR 350)	Rating		Page 1	of pages 1

2. Contract No.	3. Solicitation No. YA1323-04-RP-0001	4. Solicitation Type ☐ Sealed Bid (IFB) ☒ Negotiated (RFP)	5. Date Issued 05/18/2004	6. Requisition/Purchase No. 12345-0-0

7. Issued By   Code COACOSU   8. Address Offer To (If other than item 7)   Code

BUREAU OF CENSUS
ACQUISITION BRANCH, ROOM G-312-3
4700 SILVER HILL ROAD
WASHINGTON DC 20233

NOTE: In sealed bid solicitations "offer" and "offeror" mean "bid" and "bidder".

SOLICITATION

9. Sealed offers in original and 6 copies for furnishing the supplies or services in the Schedule will be received at the place specified in Item 8, or if handcarried, in the depository located in Suitland Federal Center, Bldg 3, Rm G312 until 03:00 PM (hour) local time Jun 21, 2004 (date)

CAUTION – LATE Submissions, Modifications, and Withdrawals: See Section L, Provision No. 52.214-7 or 52.215-1. All offers are subject to all terms and conditions contained in this solicitation.

10. For Information Call:	A. Name PHILIP M. HODGES PMH	B. Telephone No. (include area code) (NO COLLECT CALLS) 301-763-2835 or 301-763-4016

11. TABLE OF CONTENTS

OFFER (Must be fully completed by offeror)

NOTE: Item 12 does not apply if the solicitation includes the provisions at 52.214-16, Minimum Bid Acceptance Period.

12. In compliance with the above, the undersigned agrees, if this offer is accepted within _____ calendar days (60 calendar days unless a different period is inserted by the offeror) from the data for receipt of offers specified above, to furnish any or all items upon which prices are offered at the price set opposite each item, delivered at the designated point(s), within the time specified in the schedule.

13. Discount for Prompt Payment (See Section I, Clause No. 52.232-8)	10 Calendar Days %	20 Calendar Days %	30 Calendar Days %	Calendar Days %

14. Acknowledgment of Amendments The offeror acknowledges receipt of amendments to the SOLICITATION for offerors and related documents numbered and dated.	Amendment No.	Date	Amendment No.	Date

15A. Name and Address of Offeror	Code Facility	16. Name and Title of Person Authorized to Sign Offer (Type or print)
15B. Telephone No. (Include area code)	**15C. Check if Remittance Address is difference from above. Enter such address in Schedule.**	**17. Signature** **18. Offer Date**

AWARD (To be completed by Government)

19. Accepted as to Items Numbered	20. Amount	21. Accounting and Appropriation

22. Authority for Using Other Than Full and Open Competition: ☐ 10 U.S.C. 2304 (c)( ) ☐ 41 U.S.C. 253 (c)( )	23. Submit Invoices to Address Shown in (4 copies unless otherwise specified)	Item

24. Administered By (If other than Item 7) Code	Payment Will be Made By Code

26. Name of Contracting Officer (Type or print)	27. United States of America (Signature of Contracting Officer)	28. Award Date

IMPORTANT – Award will be made on this form, or on Standard Form 26, or by other authorized official written notice.

NSN 7540-01-152-8064
PREVIOUS EDITION NOT USABLE

33-134

STANDARD FORM 33 (Rev. 4-85)
Prescribed by GSA – FAR (48 CFR) 52.214 (c)

Figure 4.2 Table of contents from an RFP.

with Sections C (SOW), M (Evaluation Criteria), and H (Special Contracts Requirements), Section L provides the basis for the architecture of the proposal documents. Of note is that recent versions of Section L within the U.S. Department of Transportation and National Institutes of Health (NIH) include an element titled "Why Should We Contract with You?"

Margin requirements, font family and size, number of foldout (11 × 17–inch) pages permitted, page count, double-sided photocopying, binding and packaging specifications, résumé formats, project experience formats, number of copies of each volume, and oral presentation slide specifications are among the publication parameters that may be addressed in this important section of the RFP. In addition, software application, version, and compatibility for electronic upload or CD-ROM submittal may be stipulated. Here, you may also find specific guidance as to how your proposal should be structured in terms of the outline and numbering conventions. For example, you may see that your technical volume should consist of six major sections, each numbered according to the convention I.A, I.B, I.C. Outlining the proposal volumes should take into account guidance from Section L, as well as Sections M and C and other parts of the RFP or RFS as appropriate. Remember to use the verbiage from the RFP itself, particularly Section C, to build your proposal outline. Evaluators, their administrative support staffs, and government-hired consultants or evaluators[4] will be looking for those same words to ensure that you have a compliant submittal.

The proposal manager for a given RFP should ensure that your internal publication group receives a copy of all of the instructions from Section L during the prekickoff planning activities. Your publication group can then develop RFP-specific page templates and provide the technical writers and subject matter experts (SMEs) with estimated word counts per page. Words-per-page guidelines are important so that the writers adhere to the page allocations established either within the RFP itself or by the proposal manager. Vague instructions, such as, Write 10 pages of text on systems engineering, could result in 2,500 words or 7,500 words, depending upon font, margin, and linespacing settings. Instead, provide written instructions to each writer, indicating that one should provide, for example, 1,500 words for Section 2.2, "Effective Software Engineering Approaches."

---

4   Certain government agencies use consultants from industry (e.g., MITRE) or academia to verify past performance client references and validate technical discussions or proposed solutions.

4.7 Section M (evaluation criteria): toward maximizing your score

The general criteria by which the government will evaluate your proposal are presented in Section M of each RFP. The Competition in Contracting Act (CICA), as implemented in the FAR, requires that price or cost to the government be included as an evaluation factor in every source selection. In some cases [such as National Aeronautics and Space Administration (NASA) RFPs], points are clearly allocated for each scored portion of the proposal, such as mission suitability and personnel management; however, many times the evaluation criteria must be derived from a somewhat vague narrative, as in Table 4.1.

Table 4.1
Example of
Nebulous
Evaluation
Weightings

Criteria	Weight
1. Offeror's understanding the problem and proposed technical approach	Most important
2. Offeror's applications-related experience	Less important
3. Experience and training of individuals who will work under the contract	Same importance as 2
4. Offeror's general experience in developing software of comparable size, complexity, and content to this project	Same importance as 2
5. Offeror's general experience in maintaining and operating (M&O) a computer-based system of comparable size, complexity, and content to this project	Less important than 2
6. Offeror's proposed management plan	Less important than 5

4.8 Greatest or best-value approach

Although it can increase the cost of acquisitions, federal agencies have been encouraged proactively by the U.S. General Services Administration (GSA) to develop and issue source selection procedures that include the greatest or best-value approach, especially as it applies to the acquisition of Federal Information Processing (FIP) resources. Approximately half of the states employ some type of best-value procurement [1]. In states that permit best-value procurement, the specific method may be called "performance-based procurement," "competitive best value bidding," or "innovative procurements," among other terms. Among the strengths of the best-value approach from the federal government's perspective are that it (1) allows greater flexibility to balance technical and total cost factors on the basis of the proposals actually received rather than on the basis of a predefined formula that combines technical and total cost ratings or rankings; (2) leverages the experience and personal judgment of senior governmental decision makers; and (3) permits the evaluation to take place just before the start of actual contract performance [2]. Ultimately, the goal of best-value procurement at both the federal and state levels is to combine the twin goals of efficient contracting and enhanced taxpayer trust in the procurement process [1].

4.9 Emphasis on performance-based acquisition (PBA)

Performance-based acquisition is a major trend within federal procurement.[5] In FY2002, for example, the Department of Defense awarded more than 20% of its service requirements using performance-based specifications [3]. According to an Office of Management and Budget (OMB) report on "Bush Administration Priorities" issued in April 2004, the federal government as a whole has as a goal of achieving 40% performance-based contracts for 2005 [4]. The Bush Administration has reenergized PBA as a key initiative by emphasizing the importance of *contracting for results*. This emphasis is also an integral message in the President's Management Agenda (PMA), which calls for creating a better government that is citizen-centered, results-oriented, and market-based. As a direct result of the Bush Administration's re-energizing of the PBA approach, the OMB in 2001 issued a mandate (OMB Memorandum M-01-11 [2/14/01 memo] and M-01-15 [3/19/01 memo]) that explicitly directs agencies to write performance-based techniques on a specific percentage of the total eligible service contracting dollars worth more than $25,000. The Services Acquisition Reform Act (SARA), which was included in the National Defense Authorization Act of 2004 (P.L. 108-136), provides incentives for performance-based services acquisitions in a manner similar to what the Federal Acquisition Streamlining Act (FASA) and the Clinger-Cohen Act did with regard to federal hardware procurement.

The principal objective of PBA is to express government needs in terms of required and measurable performance objectives, outcomes,

---

[5]   Note that the term *performance-based acquisition* (PBA) has replaced *performance-based contracting* (PBC).

and/or results, rather than the method of performance. This approach is to encourage industry-driven, competitive solutions and commercial best practices. The overarching focus is on the *results,* not the processes used to deliver those results. Either a performance work statement (PWS) or statement of objectives (SOO) may be used for performance-based service acquisition (PBSA). Performance-based RFSs describe how the contractor's performance will be evaluated in a quality assurance surveillance plan (QASP). Positive and negative incentives are identified as appropriate.

The PWS is a statement within the solicitation that identifies the technical, functional, and performance characteristics of the agency's requirements. The SOO is an alternative to the PWS. It is a summary of key agency goals and outcomes. The QASP is a plan for assessing contractor performance in order to ensure accomplishment of the government's objectives. The level of surveillance is linked to contract dollar amount, risk, and complexity of the requirement.

Executive Order 12931, issued on October 13, 1994, directed executive agencies of the U.S. government to "place more emphasis on past performance and promote best value rather than simply low cost in selecting sources for supplies and services." Additionally, in accordance with the Office of Federal Procurement Policy (OFPP) Best Practices Guide of May 1995, "the use of past performance as an evaluation factor in the contract award process makes the award (de facto) 'best value' selections." The U.S. Navy led the way in this acquisition innovation in 1989 when it employed a methodology for "greatest value source selection" of firm-fixed-price supplies in which cost and past performance were the only factors for award.

It is a central principle of acquisition reform in the federal arena that source selections must place greater weight on past performance empirical data and commensurably less on the technical and management proposal writing skills in determining best value. A performance-based SOW focuses on contractor accountability for required outcomes and results rather than on the details of how the contractor is expected to accomplish the tasks. Performance-based SOWs include methods and standards that will be used for performance measurement, clearly established deliverables and other reporting requirements, and well-defined task completion criteria. Performance metrics may take the form of automated system availability (uptime), resolution times for user issues, and time to hire and place qualified, cleared, technical personnel.

Contract data requirements lists (CDRLs) have also migrated toward a performance-based structure within Department of Defense procurements. Performance-based specification of contract data requirements defines the government's need for data (in hardcopy, electronic, e-mail, or other format) and gives the contractor latitude to propose the content and format of data to be provided to meet the requirement.

Because severe competitive pressure has encouraged some contractors to promise more than they can deliver, the government has found that it is particularly useful to compare actual performance with promised or bid performance on earlier or similar contracts. Did the contractor deliver the quality of work promised? Was the contract executed in a timely manner with major milestones met consistently? Was cost adequately controlled? With respect to the proposed effort, are proposed costs realistically estimated? Are salaries consistent with experience for the area? Do they raise questions about the ability of the contractor to perform as proposed and, thus, put the government at risk in terms of continuity of operations? Using a best value approach, these criteria can be compared and contrasted for each offeror to determine the overall greatest value for the government.

Your company's strategy for responding to an RFP should definitely take into account the evaluation criteria. For example, if "Key and Other Résuméd Personnel" are stipulated to count 50 out of 100 total points, then carefully tailored résumés and biographies should receive significant emphasis in your proposal response. Keep in mind, however, that every evaluation point is critical. Do not overlook sections that count 100 points, and only focus on those that count 250 points. A point is a point is a point. Evaluation criteria should also serve as a guide for page allocations. Heavily weighted items should have an appropriately high number of pages allocated to them. Section M should also serve as a guide in your company's final bid–no bid decision for the particular procurement. Let us suppose that the evaluation criteria place significant scoring weight (50%) on past performance with the particular agency. Your firm, although it has marketed the client well, has only performed on one or two small contracts for this client. When you evaluate Section M and determine this, you may want to rethink the decision to expend the time, resources, and money to prepare a credible proposal as the prime contractor. Subcontracting or teaming may be a more appropriate and cost-effective solution.

4.10 Influencing the content of an RFP—legitimately

RFPs are generated by client staff who have problems to solve and needs to meet within certain budgetary and time constraints. Your company's marketing efforts should be directed toward understanding those problems and constraints and introducing the government staff to innovative solutions that can be applied better, faster, more cost-effectively, and with less performance risk than your competition. By talking with your client's staff at a variety of technical, programmatic, and contractual levels well in advance of the release of an RFP, your company may be successful in helping to shape the requirements of that RFP. Help your client understand his or her problem or issue in terms that your company can meet. Consider the following: a government agency has a need to archive important hard-copy documents. Your company supplies storage space, file boxes, and warehousing staff. Your goal is to help the government staff understand their requirements in terms of physical warehousing and rapid, just-in-time access and retrieval. Now let's say that your company specializes in microfilming services. Your goal is to help that same government agency understand their requirement in terms of microfiche and microfilm processing. Finally, let us consider that your company's focus is on high-resolution document-scanning technologies and long-term electronic storage media. Your marketing goal is to define the requirement in terms of leading-edge image processing and mass storage. Then when the RFP is written, the requirements stipulated will reflect the industry or technology that the government believes will provide the best solution. Influencing the RFP legitimately is a skillful craft, but it can be and is accomplished all the time. You will often hear of RFPs being "wired" for this or that company. A case could be made that with effective marketing, every RFP is predisposed in terms of technical requirements and position descriptions, for example, for one vendor or another, or at least one class of vendor over another. It is your company's task to predispose RFPs in your direction!

Influencing the RFP legitimately is the product of relationship building, careful listening, and idea sharing. Let us suppose that your company employs Ph.D.-level scientists with combined specialties in water resources management and advanced software design. Convincing your client that his problem requires the analytical depth that sophisticated computer models can provide might very well result in RFP position descriptions that require dual training in scientific disciplines and computer modeling for key staff. And this capability may be one that your competition does not have and cannot get easily.

In the course of marketing your client, you may determine that rapid, on-site response to task orders is perceived to be critical to the upcoming contract's success. You may be able to convince the client to stipulate a geographic distance requirement in the RFP. That stipulation might require all offerors to have offices within 15 miles of the client's site—something which you have but which will cost your competition time and money to establish.

If a draft RFP is released, provide a meaningful critique that includes ideas for enhancement. In a respectful manner, identify any inconsistencies with industry best practices.

4.11 Other types of solicitation documents

In addition to federal government RFPs, there are RFPs released by state, municipality, and city governments as well as by private-sector companies and international governments (see Chapter 11).[6] Many of the marketing and proposal planning, designing, scheduling, writing, and publication guidelines and processes, as well as KM approaches, that apply to federal procurements can be used effectively with these other important types of proposals. Central to any proposal effort is developing a response that carefully, concisely, and completely meets your client's needs—whether that client is the U.S. Air Force Space Command, Princeton University, PEMEX in Mexico (petroleum organization), the Environmental Protection Administration in Taiwan, Ford Motor Company, or the Port Authority of New York and New Jersey.

Because private-sector companies are not obligated to adhere to the federal Uniform Contract Format, their solicitation documents can vary substantially. General Electric's or Public Service Electric and Gas's RFPs, for example, can be quite different than DuPont's or British Telecom's RFPs. Many private-sector solicitations, however, address the following major elements:

- Understanding the requirements;
- Technical approach;
- Project organization and management approach;
- Project schedule;
- Health and safety;
- Qualifications and experience: résumés, key competencies, and project summaries;
- Terms and conditions;
- Subcontractors;
- Cost and other financial terms (supplier financing, countertrade, and letters of credit).

In addition, some companies require the completion of a complex set of forms as an addendum to the proposal itself. These forms can request a variety of institutional, financial, and contractual information and can be very time-consuming to prepare.

---

6   International RFPs may be called tenders or terms of reference.

With international proposals, efforts must be made to remove references to U.S.-centric references and regulations from project summaries and the main proposal narrative. In many cases, time must be allocated in the proposal response milestone schedule to have the proposal and its associated documents (such as corporate certifications and articles of incorporation) translated into another language, such as Spanish or Mandarin Chinese. Your company's ranking by prestigious business journals or organizations is often important to include in international proposals.

Let's take a closer look now at private-sector solicitations.

ENDNOTES

1. Heisse, John R., " 'Best Value' Procurement: How Federal and State Governments Are Changing the Bidding Process," April 29, 2002; http://www.constructionweblinks.com/Resources/Industry_Reports_ _Newsletters/April_29_2002/best_value_procurement.htm.

2. See *Source Selection: Greatest Value Approach*, Document #KMP-92-5-P, U.S. General Services Administration, May 1995.

3. *BMRA Procurement Alert Notice*, Fairfax, VA: Business Management Research Associates, Inc. Vol. 8, No. 3, Fall 2003.

4. Muzio, David, "Bush Administration Priorities," Office of Management and Budget, April 8, 2004; http://www.amciweb.com/ con301_oe/ppt_pres/HHS/2004%20CON%20301%20-%20 April%208.ppt.

Chapter 5

Private-sector solicitation requests

IN ADDITION TO FAR-driven RFP- or RFS-based proposals submitted in response to federal government procurement solicitations, companies should consider other important types of proposal documents such as private-sector proposals, product proposals, R&D internal proposals, grant proposals for education, international proposals, and health care proposals. Many times, nongovernment proposals are most analogous to bids in that price is the primary determinant for award. In the commercial sector, communication with the client organization during the proposal process is not precluded by FAR-type regulations, so your company's marketing efforts can be more aggressive and protracted. In private-sector solicitation documents, there may be no instructions to offerors (i.e., Section L) or published evaluation criteria (i.e., Section M).

A product proposal tends to be built upon a point-by-point format. There is no management plan or other response required wherein you have to take into consideration client preferences that go well beyond what is explicitly in the RFP. Health care proposals are similar to product proposals, and lend themselves to boilerplate response far more easily than solution-driven RFP-based proposal responses.

Private-sector solicitation requests may be issued in hardcopy form as RFPs or Inquiries documents, and might also be presented as an informal telephone call or FAX from the client organization. Whatever the form or format of the request from the client, the response on the part of your company or organization must be customized and targeted, reflect the relevant marketing intelligence you have obtained, and concisely convey the benefits of doing business with your firm. Private-sector proposal documents that you will prepare can range from a very focused letter (1 to 5 pages) to volumes that are well in excess of 100 pages. In many instances, private-sector clients will include specific vendor qualification and data questionnaires, bid forms, and professional services agreement (PSAs) that must be completed fully and accurately and included with your proposal submittal package.

Most private-sector proposals must be submitted to a named location and point of contact by a specific date and time. Although corporations and other private entities are not bound by the stringent proposal acceptance parameters of the Federal Acquisition Regulations (FAR 52.215-1) when conducting business with other companies for private projects, timely submittals are becoming increasingly important. Plan to have your private-sector proposal arrive on time and with the appropriate number of copies in your submittal package. And be sensitive to your client: Refrain from sending a proposal to United Parcel Service via Federal Express, for example.

The following listing represents the variety of elements that may be required in a private-sector proposal. Note the similarity to federal proposal responses. Boldface items are among the ones encountered most frequently. Some elements may be presented as appendixes in your proposal document.

- **Introduction or Executive Summary**
- **Scope of Work**
- Understanding of the Project
- **Technical Approach**
  Alternative approaches
  Bibliographic references
- **Experience and Qualifications**
  - Corporate contractual experience
  - Experience of key personnel
- **Team Organizational Structure**

- **Project Management Approach**
  - Roles and responsibilities
  - Project control
  - Ongoing communications with client
  - Availability of staff
- **Proposed Project and Task Milestone Schedules**
- **Task Subcontracting Approach**
- **Project Deliverables**

  Quality control and quality assurance programs

  Future action plans
- **Client References**
- **Résumés**
- **Project Summaries**
- **Sample Work Products and Deliverables**
- **Annual Reports, 10Ks, Financial Statements, and Dun & Bradstreet (D&B) Reports**

  Equal Opportunity Employer Program and goals

  Training certificates

  Licenses, registrations, and certifications
- **Certificates of Liability Insurance**
- Affirmative Action Plan

  Existing contract disclosure

  Complete list of contractor affiliates and their locations
- **Representations and Certifications**
  - Type of business organization
  - Taxpayer identification number (TIN)
  - Corporate status (sole proprietorship, partnership)
  - Certification of Independent Price Determination (Noncollusive Bidding and Code of Ethics Certification)
  - Preference for labor surplus area (LSA)
  - Certification of nonsegregated facilities
  - Certification regarding a drug-free workplace
  - Cost accounting practices and certifications
  - Certification of no investigation (criminal or civil antitrust)

- **Indemnification** (protection for the client against the risk of legal claims)
  - Damage to persons or property
  - Negligence on the part of the contractor
- **Nonconflict of Interest Statement**
- **Confidentiality Agreement**
  Health and safety plan
  Commercial terms and conditions
  Equipment billing rate schedules
- **Project Costs**
  - Names, titles, and hourly rates for specific job classifications
- Sample Invoice
- Bid Bond
- Comprehensive General Liability Insurance and Worker's Compensation Insurance

As part of your proposal library or automated knowledge base, your company should have a collection (in both hardcopy and electronic form as applicable) of various licenses, certifications, insurance coverage, TIN numbers, training certificates, and so forth. Ready, albeit secure and controlled, access to this information will make building your responses to private-sector solicitations much easier and quicker.

As with federal government proposals, the focus of the grant proposal is on selling. Grant proposals are, fundamentally, sales documents. Nevada-based grants and fund-raising expert David G. Bauer speaks of two distinct types of grantsmanship. One is *reactive* and the other is *proactive*. Reactive grantsmanship involves developing the project first, looking for funders, and then scrambling to meet a deadline. On the other hand, proactive grantsmanship begins with researching funding sources and then expands to cultivating professional relationships with relevant grantmakers. Your organization should definitely practice *proactive* grants development.

5.1 Grant proposals—winning what you bid

As with federal government proposals, the focus of the grant proposal is on selling. Grant proposals are, fundamentally, sales documents. Nevada-based grants and fund-raising expert David G. Bauer speaks of two distinct types of grantsmanship. One is *reactive* and the other is *proactive*. Reactive grantsmanship involves developing the project first, looking for funders, and then scrambling to meet a deadline. On the other hand, proactive grantsmanship begins with researching funding sources and then expands to cultivating professional relationships with relevant grantmakers. Your organization should definitely practice *proactive* grants development.

5.1.1 Letters of inquiry

In many cases, organizations must first prepare a brief letter of inquiry before receiving an invitation to submit a full grant proposal. The letter of inquiry provides a valuable forum for vetting project ideas with potential funding sources. It allows you to build a connection between your project's goals and the grantmaking foundation's or agency's mission and interests. Be concise in this letter of inquiry. The first paragraph should provide a synopsis of the project and the requested funding amount. Focus on the tangible and intangible benefits of your proposal, and convey a passionate dedication to your project.

Before you start writing your grant proposal, invest the time to plan your approach and develop a realistic schedule for completing the grant application (see Figure 5.1). Importantly, build in time for internal reviews by principal investigators (PIs) and other colleagues.

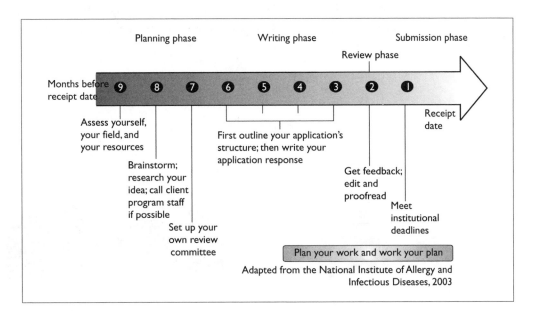

Figure 5.1 Forward-looking preparation time line.

5.1.2 **Balancing the technical and the nontechnical**

Place yourself in the peer reviewers' shoes and, thus, balance the technical and the nontechnical throughout your grant proposal (see Figure 5.2). To begin, develop a title for your project that invites reviewers to read your proposal. Most grantmaking staffs are generalists. Therefore, you should include both technical and nontechnical information throughout your proposal. Begin each paragraph simply, and then progress to more complex information. You might use *Scientific American* as a model for the level of writing to use in the nontechnical paragraphs.

5.1.3 **Standard grant proposal components**

The 11 components of most grant proposal applications are presented in Figure 5.3. Tell your overall story in nontechnical terms up front in 1—Proposal Summary. Grab positive attention and build trust quickly. Convey what you want to do and what preliminary work you have accomplished. Clearly identify why your project is important. What is the intellectual merit of the proposed activity? What are the broader impacts of the proposed activity? Articulate the tangible benefits to industry and government as well as to the grantmaking organization. Explain why you will succeed. Identify how much your project will cost, and how your budget is cost-effective. Show how your project is related to other research in the field. Keep the proposal summary to one page in length. It serves as the foundation of your grant proposal. Importantly, it may be the only section reviewed by all evaluators.

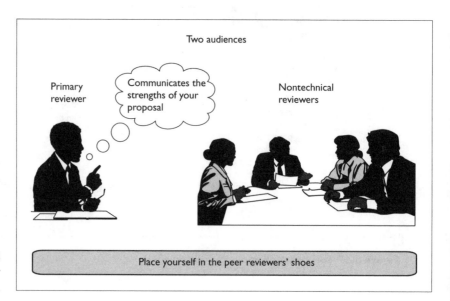

Figure 5.2
Writing to your
audiences.

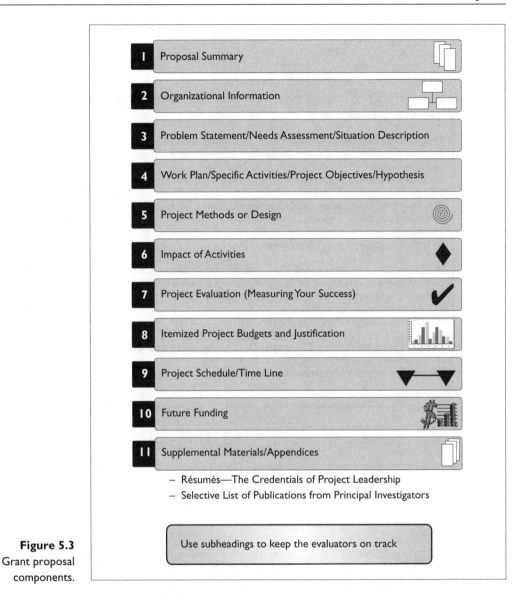

Figure 5.3
Grant proposal
components.

Web-based grants development resources are provided in Figure 5.4.

5.2 Nongovernmental organizations (NGOs)

Nongovernmental organizations (NGOs), international nongovernmental organizations (INGOs), and local nongovernmental organizations (LNGOs) are nonprofit, voluntary associations of citizens organized on a local, national, or international level that operate independently of

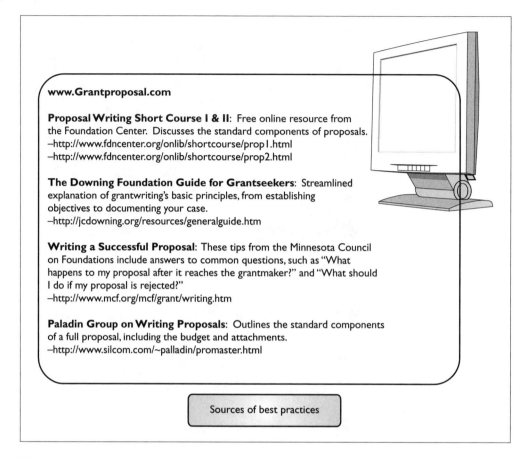

www.Grantproposal.com

Proposal Writing Short Course I & II: Free online resource from the Foundation Center. Discusses the standard components of proposals.
–http://www.fdncenter.org/onlib/shortcourse/prop1.html
–http://www.fdncenter.org/onlib/shortcourse/prop2.html

The Downing Foundation Guide for Grantseekers: Streamlined explanation of grantwriting's basic principles, from establishing objectives to documenting your case.
–http://jcdowning.org/resources/generalguide.htm

Writing a Successful Proposal: These tips from the Minnesota Council on Foundations include answers to common questions, such as "What happens to my proposal after it reaches the grantmaker?" and "What should I do if my proposal is rejected?"
–http://www.mcf.org/mcf/grant/writing.htm

Paladin Group on Writing Proposals: Outlines the standard components of a full proposal, including the budget and attachments.
–http://www.silcom.com/~palladin/promaster.html

Sources of best practices

Figure 5.4 Grant proposal components.

government or business structures and have noncommercial objectives. NGOs include the Canadian Red Cross, Cooperative for Assistance and Relief Everywhere, Inc. (CARE), Doctors Without Borders [Medecins Sans Frontieres (MSF)], International Medical Corps (IMC), and Catholic Relief Services (CRS). Task-oriented and driven by people with a common interest, NGOs perform a variety of services and humanitarian functions, bring citizens' concerns to the attention of governments, monitor policies, and encourage political participation at the community level. They provide analysis and expertise, serve as early-warning mechanisms, and help monitor and implement international agreements. Some are organized around specific issues, such as human rights, the environment, or health care.

NGOs must often compete for funding sources via a proposal process. An example of an RFP open to select NGOs was issued through Border Partners in Action (Border PACT), an extensive network of

U.S.–Mexico-border higher education institutions dedicated to building human capacity through education and training. Based at the University of Arizona, Border PACT is coconvened by the Consortium for North American Higher Education Collaboration (CONAHEC), the American Council on Education (ACE), and the Asociación Nacional de Universidades e Instituciones de Educación Superior (ANUIES). Working in smaller regional groups, as well as within the larger Border PACT network, 65 member institutions have joined forces across the border to serve as agents of change in solving the many challenges they face as borderlands communities. Border PACT's goal is to provide a means for higher education institutions, researchers, student groups, NGOs, and American-Indian and Mexican indigenous border communities on both sides of the U.S.–Mexico border to develop joint programs in conjunction with their communities intended to derive shared solutions to common problems.

Recently, Border PACT issued an RFP for its grants program. In order to apply, a brief proposal of five pages maximum, 1.5-spaced, in no smaller than 10-point font size (12-point preferred) was required that addressed the following points:

- Names of the two lead binational participating institutions, including the primary contacts at each institution, their positions or titles, physical addresses, post office box, phone numbers, faxes, and e-mail addresses.

- Description of the proposed activity, including purpose, persons involved, and a timetable of activities. (Note: it would be advisable to include a schedule with major milestones in Gantt chart format.)

- Description of how the proposed project will benefit the borderland community or communities involved.

- Discussion of how the project will be publicized and promoted, along with identification of the targeted audience for the project and the number of people impacted.

- Resources, capabilities, and infrastructure to implement the project.

- Description of how the activity will be sustained beyond the life of the seed grant.

- Proposed budget, including matching or in-kind resources.

The required format for the proposal included a hardcopy of the proposal and supporting documents, as well as a copy of the proposal on a 3.5-inch floppy diskette, IBM Windows version, in document Microsoft Word 97 format, or rich text format (RTF). The proposal was stipulated to include both a time line and work scope for the proposed projects spanning August to December. All proposals required discussion of concrete outcomes, evaluation mechanisms, and funding strategies for continuing the project beyond its initial stages.

The specific submission process for this particular RFP involved the following:

- Submissions of proposals must include a letter of endorsement by the president of each of the institutions, or the president of the board of directors for NGOs, or chief executive officer of an agency.
- A hardcopy of the proposal and supporting documents.
- A copy of the proposal on 3.5-inch floppy diskette, IBM Windows version, in Microsoft Word 97 or 2000 document or RTF.
- Submission of all materials by 5:00 P.M. Tucson time.

Next we will examine the dynamic arena of federal government acquisition in an era of e-government. There are many new dimensions to consider. We will begin by focusing on how RFPs emerge from the federal acquisition process.

Chapter 6

The federal acquisition process: emerging directions

6.1 Overview

How do RFPs and RFSs originate? The answer is that they are extraordinarily complex contractual documents prepared by a variety of government employees in accordance with an intricate, formalized sequence of regulated procedures. It was during World War II that the federal government began to carefully regulate its procurement processes. From that time until the mid-1990s, U.S. procurement law evolved toward increased complexity (multiphased contracts and elaborate evaluation methods) and increased competition, even in subcontracting. The focus was on accountability, that is, objective and defensible acquisition decisions [1]. Contracting officials and technical staff alike were extremely concerned that the competitive efforts for which they had oversight responsibility were conducted in full accordance with appropriate agency-specific procurement regulations. However, federal acquisition streamlining and reform legislation enacted in 1994 and 1996 along with the President's Management Agenda (PMA) of 2001 has placed more decision-making and discretionary authority back in the hands of government contracting officials.

6.2 Statutory and regulatory requirements for competition

The FAR system is part of the Code of Federal Regulations (CFR) [2]. It is the FAR that is the primary source of procurement regulations used by all federal agencies in their acquisition of supplies [3], construction, services [4], and research and development with appropriated funds. All of the provisions and clauses that are used in government contracting are found in the FAR, which is interpreted and applied in areas of dispute through a complex process of litigation and court and special governmental board decisions [5]. The DoD FAR Supplement, NASA FAR Supplement, Air Force Supplement, and Department of Energy Supplement all augment and amplify the FAR and should be used in conjunction with the FAR when determining acquisition regulations relevant to DoD and Air Force contracts, and NASA and DOE policies and procedures. Your company can order copies of the FAR and its supplements directly through the Superintendent of Documents, U.S. Government Printing Office, Washington, D.C. 20402, fax (202) 275-0019. Agency supplements to the FAR and services regulations complement the FAR but do not contradict it. You can also review and download the FAR in HTML, PDF, and other formats from the Web at http://www.arnet.gov/far/.

The FAR is the primary source of procurement regulations used by all federal agencies in their acquisition of goods and services.

The FAR system, which became effective on April 1, 1984, replaced the Federal Procurement Regulation (FPR) used by civilian agencies of the federal government, the Defense Acquisition Regulation (DAR) used by DoD, and the National Aeronautics and Space Administration Procurement Regulation (NASAPR) used by NASA. The Commission on Government Procurement, established on November 26, 1969, by P.L. 91-129, was given the charter of studying the statutes, policies, regulations, and practices affecting government procurement and recommending improvements. Following years of effort, the FAR was announced in the *Federal Register* on September 19, 1983 [6]. You can locate the important SFs included in RFPs in FAR Part 53.

The Federal Acquisition Regulations System was established to codify and publish uniform policies and procedures for the acquisition of goods and services by all executive agencies. The FAR record is divided into Subchapters a through h, and Parts 1 through 52. Give particular attention to Subchapter c, "Contracting Methods and Contract Types." The opportunity for your company to submit written comments on proposed significant revisions to the FAR is provided through notification of proposed changes in the *Federal Register*. The *Federal Register* is the official daily publication for rules, proposed rules, and notices of federal agencies and organizations, as well as executive orders and other presidential documents. The current volume of the *Federal Register* (Volume 69, 2004) can be found online at http://www.gpoaccess.gov/fr/index.html. In addition, you can search the *Federal Register* from 1994 to present at this Web site.

FAR Part 15—Contracting by Negotiation, establishes that in negotiated procurements the bidder must be responsible; evaluation criteria are flexible, that is, tailored to the procurement; and the acquisition process may be competitive or noncompetitive. Unlike sealed bidding (IFBs), negotiated procurement is not merely a series of steps. Every federal government agency has a somewhat unique pattern of procurement activity. Be sure to visit the contracting offices of your company's target client agencies and obtain copies of the handbooks that govern procurement practices for that particular agency.

6.3 The source selection process

Competitive negotiation is formally called *source selection*. This process, which is regulated by the FAR at Subpart 15.3 and designed to select the proposal that represents the best value to the government, normally involves the following steps.

First, an RFP is prepared and publicized by the government. Federal government RFPs all look essentially the same in terms of major sections (A through M) as a result of the application of the Uniform Contract Format (UCF), established at FAR Subpart 15.204.1 (see Table 15.1 in the FAR).

Second, technical and cost proposals are submitted to the client organization by offerors in the contractor community.

Third, proposals are analyzed, evaluated, and rated by a team of client staff against documented, weighted criteria and unstated standards.

An award can be made at this stage without discussions, based on the decision of the Source Selection Authority (SSA).

Fourth, potentially successful proposals are identified and included in the competitive range (shortlisted, see FAR Subpart 15.306) based on price and other factors; all others are eliminated from further competition and the offerors notified in writing of such.

Fifth, oral and written discussions are conducted with those offerors in the competitive range for the principal purpose of eliminating deficiencies in their proposals.

Sixth, those offerors are given an opportunity to submit BAFOs (see FAR 15.306).

Seventh, BAFOs are evaluated and a contract award is made to the offeror whose proposal is most advantageous to the client, as determined on the basis of stated evaluation criteria.

Eighth, unsuccessful offerors are notified promptly in writing (per FAR 15.503).

Finally, debriefings are held with unsuccessful offerors.

According to FAR 2.101, *acquisition* means the acquiring by contract with appropriated funds of supplies or services by and for the use of the federal government. There are four primary phases in the acquisition process used by the U.S. government: (1) needs assessment and determination; (2) acquisition planning; (3) solicitation release, proposal evaluation, contractor selection, and contract award (source selection phase); and (4) contract administration. Within NASA, for example, there are the following major steps in the acquisition process: (1) selection of a candidate project; (2) commitment to project planning; (3) project planning review; (4) project approval by Deputy Administrator; (5) RFPs; and (6) system design, development, and operation (see Figure 6.1). The federal government fiscal year (FY) begins on October 1 of each year.

Fundamentally, the source selection involves two processes:

1. Selection of a contractor;

2. Formation of a contract.

The federal government uses a hierarchical source selection organization, the size of which depends upon the complexity, size, and importance of the procurement [7]. The source selection process has traditionally been dominated by weapons procurement. The formal source selection procedures of the U.S. Air Force serve as a frequent model for other federal agencies as well as international governments. All formal U.S. government source selection systems, such as presented in the *NAVSEA Systems Command Source Selection Guide* (February 15, 1984), are structured for objectivity, legality, and thoroughness. Most systems use successively weighted levels of evaluation, which allow the government to assign relative importance to each evaluation criterion. In the Air Force system, an SSA is

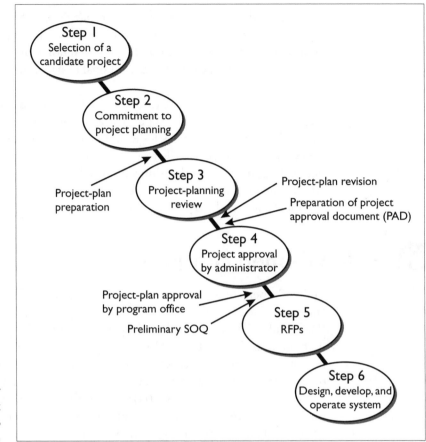

Figure 6.1
NASA acquisition
process. (*Source:*
Statements of Work
Handbook NHB
5600.2., p. 3.)

supported by a Source Selection Advisory Council (SSAC), which in turn
is supported by a Source Selection Evaluation Board (SSEB).[1] The SSEB
is actually groups of teams that evaluate the proposal. The SSEB teams are
further divided into subpanels, areas, items, factors, and subfactors. It is at
the level of subfactor that the actual scoring (evaluation) occurs. The overall evaluation is then a compilation of subscores. Scoring can take the form
of numbers; a plus, check, minus scheme; colors; or adjectives (pass/fail;
outstanding to unsatisfactory). Scoring of proposal responses is weighed
against the prescriptions (standards) [8] set forth in the government's
highly proprietary *Source Evaluation Guide* (or *Handbook*). The NASA

---

1   The U.S. Army Corps of Engineers' (USACE) source selection organization is the SSEB. In this agency, the SSEB
     is comprised of separate technical-evaluation and cost-evaluation teams. This organization is designed to ensure
     active, ongoing involvement of appropriate contracting, technical, logistical, legal, cost analysis, small business,
     and other functional staff management experience in the procurement process.

system includes a Source Selection Official (SSO), an SEB, committees, panels, and subpanels. Again, evaluation occurs at the subpanel level. Of note is the fact that evaluation scores are not binding on source selection officials "as long as the official has a rational basis for the differing evaluation" [9].

In accordance with FAR 1.602-1(a), COs have the authority to enter into, administer, or terminate contracts and make related determinations and findings (D&F). The D&F detail the proposals that will be included in the competitive range and the reasons for those decisions. The determination is the conclusion or decision supported by the findings. The findings are statements of fact or rationales essential to support the determination and must cover each requirement of the statute or regulation (FAR 1.701).

6.4 Full-and-open competition

Full-and-open competition means all responsible sources are permitted to compete. The major requirement of the Competition in Contracting Act (CICA) of 1984[2] was that full-and-open competition is the required standard for awarding contracts. CICA '84 established the civil service position of competition advocate to promote and ensure the use of full and open competition whenever feasible [10]. "The Office of Federal Procurement Policy Act (41 U.S.C. § 404) requires that each executive agency of the Federal Government designate an advocate for competition for the agency and for each procuring activity within the agency" [11]. The fact that cost and technical volumes are evaluated separately is not a statute or part of the FAR but is simply part of traditional practice. In full-and-open competitively negotiated bids, the CO is not bound by the Source Selection Board's decision. He or she can, in fact, override the conclusions and recommendations of the selection board. And no Final Proposal Revision is required.

No matter how objectively the RFP and the proposal evaluation process are structured, or how much the client's contracts office enforces competition in contracting protocol, in the final analysis the evaluation is one of human judgment. Remember that the proposal must address both *stated* and *unstated* criteria. Human judgment certainly pertains to the unstated criteria.

---

2   The CICA '84 was implemented in the FAR through Federal Acquisition Circulars (FACs) 84-5 and 84-6.

6.5 Major contract types

There are two basic types of contracts that result from the negotiated bid process: fixed price and cost reimbursement. A firm-fixed-price contract provides for a price that is not subject to any adjustment on the basis of your company's cost experience in performing the contract. This contract type places full responsibility and maximum risk on your company for all costs and resulting loss or profit. The latter calls for paying the contractor all incurred direct and indirect costs, as defined and specified in the contract, plus some profit [12]. When bidding fixed-price opportunities, your company must ensure that it fully understands the scope of work and terms and conditions of the contract in order to ensure an acceptable profit.

In addition, there are several unusual combinations of fixed-price and incentive-type contracts. These include Fixed-Price Incentive Firm (FPIF) (FAR 16.403, http://www.arnet.gov/far/) and Fixed-Price Award Fee (FPAF) (FAR 16.404, http://www.arnet.gov/far/).

The fixed-price incentive (FPI) contract provides for an adjustment of profit and the establishment of the final contract price by means of a formula based on the relationship of final costs to a negotiated target cost. Under this type of contract, the following elements are negotiated at the outset: a target cost, a target profit, a ceiling price, and a formula for establishing final price and profit.

When costs are less than the target cost, the contractor's profit is increased in accordance with the formula negotiated. When costs exceed the target, the contractor's profit is reduced. Therefore, both the government and the contractor share in the risk. However, the contractor still shoulders significant risk, because if actual costs far exceed the target, the formula for adjustment of profit may yield a negative figure or a net loss.

The government's assumption of risk is limited also by the ceiling price negotiated, as that is the maximum amount the contractor can be paid. To provide an incentive consistent with the circumstances, the formula should reflect the relative risks involved in contract performance.

Under FPAF contract vehicles, award-fee provisions may be used in fixed-price contracts when the government wishes to motivate a contractor and other incentives cannot be used because contractor performance cannot be measured objectively. Such contracts shall establish a fixed price, including normal profit, for the effort. This price will be paid for satisfactory contract performance. Award fees earned (if any) will be paid in addition to that fixed price. FPAF contracts provide for periodic evaluation of the contractor's performance against an award-fee plan.

6.6 Significant recent paradigm shifts in federal government acquisition

Change has been *the* constant in the federal government acquisition arena since the mid-1980s. Two decades ago, a gradual shift began in *what* the federal government was buying [13]. Specifically, in FY1985, supplies and equipment accounted for 56% or $145 billion (in 1999 constant dollars) of the contracting dollars, compared with services, construction, and R&D [14]. By FY1999, however, the largest acquisition category was services, standing at $78 billion, or 43% of total spending [14]. Among the top items in the services category was professional, administrative, and management support [13].

In addition to the changes in *what* the federal government was buying, there have also been significant changes in *how* the government has been buying goods and services [13]. The decade of the 1990s saw such landmark legislation as the Federal Acquisition Streamlining Act (FASA) of 1994 and the Federal Acquisition Reform Act (FARA) of 1996. Congress had enacted these important pieces of procurement reform legislation to enable the government to take full advantage of the commercial marketplace and to allow contracting officers to exercise sound business judgment, initiative, and creativity in satisfying the needs of their agency customers [15].

This paradigm of procurement change continued into our new century with the President's Management Agenda (PMA) of 2001. And the trajectory of change continues to this day, with the goals of *increased accountability* from both the federal government and the contractor communities; *streamlined acquisition processes;* procurement that focuses on *measurable results,* service quality, and customer satisfaction (performance-based contracting and acquisition); and *expanded confidence* on the part of the American people in their federal government and the disposition of their tax dollars.

In addition, federal agencies are making increased use of contracts awarded by other agencies as well as Federal Supply Schedule (FSS) contracts awarded by the General Services Administration (GSA) [16]. Use of the FSA FSS grew from $4.5 billion in 1993 to $10.5 billion in 1999 [17]. Most of this growth was in the area of information technology. Of note is that GSA studies have shown that acquisition time is reduced significantly under schedule buys [17].

Collectively, these acquisition changes tell us that the federal acquisition environment is now characterized by a greater reliance on services and information technology [17]. Additionally, electronic commerce (EC) is now being used as the preferred approach to accomplish a variety of procurement tasks and to streamline and improve federal buying practices [17]. Of note, however, is that electronic procurement carries with it associated cybersecurity, personal privacy, reliability, and data integrity issues. Because the U.S. government is one of many players in the global services- and information-driven economy, it will have to continue to become a more savvy, commercial-oriented buyer [17].

Figure 6.2 presents in sequential order some of the critical legislation, executive orders, and other federal guidance that have been instrumental in shaping U.S. government procurement decisions and actions as well as contractor proposal responses during the past two decades.

Competition in Contracting Act (CICA) of 1984 (P.L. 98-369): This act amended the Armed Services Procurement Act and the Federal Property

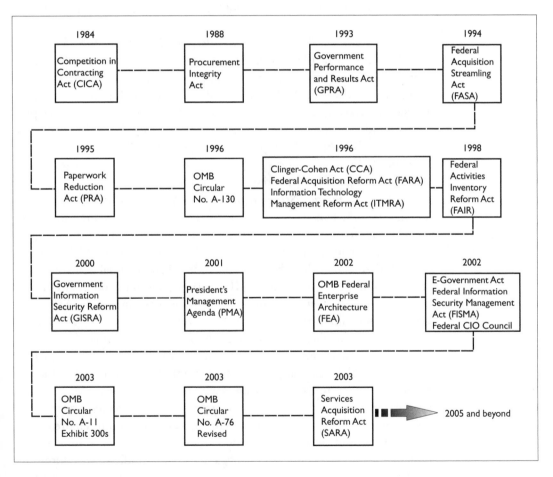

Figure 6.2   Critical legislation, regulations, and policy guidance that have driven federal procurement actions and contractor proposal responses.

and Administrative Services Act of 1949 to enhance competition in federal contracting by requiring a justification for any procurement in which full-and-open competition would not be obtained. This public law was enacted for the purpose of increasing the number of government procurements conducted under the principles of full-and-fair competition, as opposed to contracts that are issued under noncompetitive arrangements such as "sole source" or "set-aside" awards. Contracting officers are required to promote and provide for full-and-open competition in soliciting offers and awarding U.S. government contracts over and above the simplified acquisition procedures (SAP) threshold. Maximum competition is desirable from a public citizen perspective because it results in the timely delivery to the U.S. government of quantity products and services at reasonable cost.

Procurement Integrity Act (1988): This act prohibits the disclosure of "contractor bid or proposal information" and "source selection information."

Government Performance and Results Act (GPRA) of 1993: This act established strategic planning and performance measurement in the federal government to improve the efficiency and effectiveness of federal programs. Under the GPRA, the director of the Office of Management and Budget (OMB) now requires each federal agency to prepare an annual performance plan covering each program activity set forth in the budget of the agency. OMB must establish performance goals to define the level of performance to be achieved by a program activity, and express such goals in an objective, quantifiable, and measurable form.

Federal Acquisition Streamlining Act (FASA) of 1994 (Title V, FASA VP.L. 103-355; signed into law October 13, 1994): This act required federal agencies to establish cost, schedule, and measurable performance goals for all major acquisition programs, and to achieve on average 90% of those goals. FASA focused largely on the purchase of commercial items and smaller dollar buys (those under $100,000). Importantly, the act exempted commercial items from many unique government requirements. Acquisitions over the micro-purchase limit ($2,500) but not exceeding $100,000 were reserved for small businesses. Under FASA, contracting officers were encouraged to use approaches in awarding contracts that leveraged SAP. In addition, FASA gave agencies statutory authority to access computer records of contractors doing business with the federal government. The act also placed a greater emphasis on the use of past performance when selecting a contractor [18]. FASA "repeals or substantially modifies more than 225 provisions of law to reduce paperwork burdens, facilitate the acquisition of commercial products, enhance the use of simplified procedures for small purchases, transform the acquisition process to electronic commerce, and improve the efficiency of the laws governing the procurement of goods and services" [19]. "The bill strongly encourages the acquisition by federal agencies of commercial end-items and components, including the acquisition of commercial products that are modified to meet government needs " [20]. Instrumental in the generation of this specific legislation was Executive Order 12862, the Government Performance and Results Act (1993), and the Chief Financial Officers Act of 1990.

Finally, FASA encouraged the use of electronic commerce, and established the statutory framework for task and delivery order contracting [21].

Paperwork Reduction Act (PRA) of 1995: The Paperwork Reduction Act established a broad mandate for agencies to perform their information resources management (IRM) activities in an efficient, effective, and economical manner.

OMB Circular No. A-130, "Management of Federal Information Resources": This circular provides uniform governmentwide information resources management policies as required by the Paperwork Reduction Act of 1980, as amended by the Paperwork Reduction Act of 1995.

Clinger-Cohen Act of 1996 (P.L. 104-106, signed into law on February 10, 1996): Enactment of the Clinger-Cohen Act was driven by the federal government's growing reliance on information technology (IT) and the resulting increased attention and oversight on its acquisition, management, and use. Building on FASA, Clinger-Cohen provided the statutory foundation to streamline IT acquisitions and minimize layered approvals. The act emphasized accountability, outcomes-based performance, and results-based IT management. It promoted the improved performance of the civilian agency acquisition workforce, and it allowed contracting officers to select competitive contractors more efficiently. In addition, Clinger-Cohen required federal agencies to use a disciplined capital planning and investment control (CPIC) process to acquire, use, maintain, and dispose of information technology. The use of commercial-off-the-shelf (COTS) products was to be maximized.

The Clinger-Cohen Act rescinded the *Brooks Act* (P.L. 92-582), also known as Qualifications Based Selection (QBS), which was enacted on October 18, 1972. This 1972 act had established the procurement process by which architects and engineers (AEs) were selected for design contracts with federal design and construction agencies. The Brooks Act had also established a qualifications-based selection process, in which contracts for AEs were negotiated on the basis of demonstrated competence and qualification for the type of professional services required at a fair and reasonable price.

Federal Acquisition Reform Act (FARA) of 1996 (Division D of P.L. 104-106; 40 U.S.C. 1401; signed into law on February 10, 1996): This act enables the federal procurement system to emulate many of the most successful buying practices used in the commercial marketplace. Focused on reforming how the federal government makes larger dollar purchases and acquires IT, FARA increased the discretion of federal contracting officers in making competitive range determinations and lowered the approval levels for justification and approvals resulting in efficient competition. FARA also permitted the use of SAP in the acquisition of commercial items up to $5 million.

Information Technology Management Reform Act (ITMRA) (Division E of P.L. 104-106, signed into law on February 10, 1996): This act enables the federal procurement system to emulate many of the most successful buying practices used in the commercial marketplace. It is focused on reforming how the federal government makes larger dollar purchases and acquires IT. Establishes the role, duties, and qualifications of the CIO within federal agencies.

Federal Activities Inventory Reform Act (FAIR) of 1998: This is legislation that requires agencies to identify functions that could be performed by the private sector [17].

Government Information Security Reform Act (GISRA) of 2000: This act brought together existing IT security requirements from the Paperwork Reduction Act of 1995 and the Clinger-Cohen Act of 1996. GISRA also codified existing OMB IT security policies found in OMB Circular A-130 and IT security budget guidance in OMB Circular A-11, "Preparation, Submission, and Execution of the Budget." Specifically, GISRA directed agency CIOs to conduct annual IT security reviews of their systems and programs. Review results are to be reported to OMB. After GISRA expired in November 2002, the *Federal Information Security Management Act* (FISMA) was signed into law as Title III by President George W. Bush as part of the Electronic Government Act of 2002. FISMA permanently reauthorized the framework established by GISRA.

President's Management Agenda (PMA) (launched in August 2001): President George W. Bush's vision for government reform is guided by three important principles: (1) citizen centered, (2) results oriented, and (3) market based. In the PMA, Mr. Bush identified five governmentwide initiatives and nine program initiatives. Governmentwide initiatives include Strategic Management of Human Capital, Competitive Sourcing, Improved Financial Performance, Expanded Electronic Government, and Budget and Performance Integration. The PMA was launched as a strategy for improving the management and the performance of the U.S. federal government. Importantly, federal agencies have been held publicly accountable for adopting the disciplined approaches of the PMA through a governmentwide colorimetric scorecard system (GREEN—YELLOW—RED). GREEN indicates that a given agency has met all of the established standards of success under the PMA, or that the agency's implementation is proceeding according to plan.

Federal Enterprise Architecture (FEA): To facilitate efforts to transform the federal government into one that is citizen centered, results oriented, and market based, the OMB is developing the Federal Enterprise Architecture (FEA), a business-driven and performance-based framework to support cross-agency collaboration, transformation, and governmentwide improvement. Begun on February 6, 2002, the FEA is being constructed through a collection of interrelated "reference models" designed to facilitate cross-agency analysis and the identification of duplicative investments, gaps, and opportunities for collaboration within and across federal agencies. FEA provides OMB and other federal agencies with a new way of describing, analyzing, and improving the federal government and its ability to serve the American citizen. The outcome of this effort will be a more citizen-centered, customer-focused government that maximizes technology investments to better achieve mission outcomes.

Electronic Government (E-Government) Act of 2002 (P.L. 107-347, 44 U.S.C. Ch. 36; signed into law December 17, 2002): This act, which advocates a more citizen-focused approach to current governmentwide IT policies and programs, was designed in part to institutionalize the PMA. This law requires federal agencies to develop performance measures for implementing e-government. This act also requires agencies to conduct governmentwide e-government initiatives and to leverage cross-agency opportunities to leverage e-government. The federal CIO Council, originally established in 1996 by Executive Order 13011, was codified under the E-Government Act. This Council is comprised of the CIOs of 29 federal agencies as well as representatives from OMB. Its charter is to implement elements of GPRA, FISMA, and ITMRA. The federal CIO Council is the principal interagency forum to assist CIOs in meeting the goals of the PMA.

In addition, the E-Government Act established an Office of Electronic Government within the Office of Management and Budget to improve federal management of information resources.

OMB Circular No. A-11: The Office of Management and Budget uses the information reported in Circular A-11, Exhibits 53 and 300, to assist with budget decisions. Federal agencies must map their IT investments to the FEA reference models to help OMB and other federal organizations identify potential opportunities to collaborate and eliminate redundant spending.

OMB Circular A-76 (revised May 29, 2003): Circular A-76 is a set of policies and procedures to help determine whether public or private sources will undertake the federal government's commercial activities and services, ranging from software consulting and research and lab work to facilities management. The roots of Circular A-76 can be traced from the former Bureau of the Budget's Bulletin 55-4 (issued January 15, 1955), which stated that the federal government would "not start or carry on any commercial activity" that the private sector could do. Revisions have been made periodically ever since. A-76 is a federal government management tool. The A-76 competition provides an opportunity for government managers to streamline organization, implement best business practices, increase productivity, enhance quality, increase efficiency of operations, lower operational costs, and adjust IT initiatives to new regulatory drivers such as the PMA. The A-76 study process focuses on:

- Government/customer requirements;
- Organizational structure;
- Work processes;
- Defined outcomes;
- Competition.

Historically, the government wins 50% to 84% of competitions. But regardless of who wins, 30% to more than 40% savings are achieved.

Services Acquisition Reform Act (SARA) of 2003 (P.L. 108-136; Title 14 of the FY2004 National Defense Authorization Act): This legislation provides additional incentive for use of performance-based contracting for services.

The federal government spent $2.1 trillion in FY2003. Managing and modifying the complex set of processes by which the government procures goods and services has proven to be daunting, but progress is definitely underway.

6.7 Understanding the Federal Acquisition Streamlining Act (FASA)

The trend in federal acquisition is toward electronic commerce, purchase of commercial products, and streamlining the procedural framework.

As illustrated in Figure 6.3, the federal procurement landscape is very dynamic and vectored in part toward faster, better, cheaper (FBC).[3] Emerging trends span fast-track e-business procurement, electronic proposal submittal and evaluation, best-value source selection, and

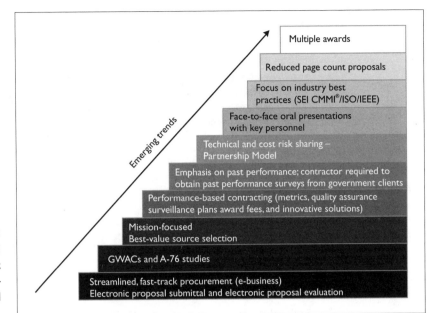

Figure 6.3
New federal
procurement
environment—
market-based and
results-oriented.

performance-based acquisition with associated technical, schedule, and cost control metrics, as well as award fees. In addition, federal procurement since 1999 has followed much more of a partnership model wherein the government expects contracting firms to not only mitigate but share both technical and cost risks. Important resources for information about e-government, electronic commerce, and federal acquisition reform are presented in Table 6.1.

The focus now is on innovative solutions, presented in proposal documents and oral presentations in concise form, that infuse industry best practices, such as the Software Engineering Institute (SEI) Capability

---

3   In 1992, then NASA Administrator Daniel S. Goldin challenged all of NASA, including the aerospace industry and academia, to use an FBC approach to project management. By initiating FBC project management, NASA intended to maximize the overall amount of scientific results obtained on a mission while minimizing the impact of a failed spacecraft. From 1996 to 2000, there were 6 mission failures out of 25 total missions flown, which, according to a March 13, 2001, audit report issued by the NASA Office of Inspector General (IG-01-009) "were clearly associated with exploring the boundaries stimulated by FBC." A Department of Defense review of five rocket failures between August 1998 and May 1999 found that government contractors were trying to do too much with too little resources and needed more training for mission-critical technical staff. Today, there seems to be two distinct management philosophies within NASA and the aerospace industry. One philosophy adheres to the value of faster, better, cheaper, if organized and configured properly. Proponents of the FBC approach point to Microsoft, which, during the past 20 years, has shortened schedules, cut costs, and enhanced performance. Conversely, the other philosophy says faster, better, cheaper—pick any two. Cheaper spacecraft are equated with lower performance, according to this approach.

FirstGov

http://www.firstgov.gov
Official portal that offers easy and comprehensive access to all online U.S. government resources. FirstGov is an interagency initiative administered by the U.S. General Services Administration.

U.S. Small Business Administration (SBA)

http://www.sba.gov

Central Contractor Registration (CCR)

http://www.ccr.gov

Federal Business Opportunities (FedBizOpps)

http://www.fedbizopps.gov
Formerly the Electronic Posting System (EPS), FedBizOpps is the single government point of entry for federal government procurement opportunities that exceed $25,000. Federal agencies had until October 1, 2001, to complete their transition to, or integration with, FedBizOpps. As of October 1, 2001, all agencies are now required to use FedBizOpps to provide the public access to notice of procurement actions over $25,000.

E-Government Web Site

http://www.whitehouse.gov/omb/egov
Official Web site of President Bush's e-government initiatives

Army Single Face to Industry (ASFI)

https://acquisition.army.mil/asfi
Portal to U.S. Army business opportunities. Includes an acquisition tool set and procurement notification mechanism.

Minority Business Development Agency (MBDA)

http://www.mbda.gov
The Minority Business Development Agency (MBDA) is part of the U.S. Department

Table 6.1
EC, e-Gov, eB, and Acquisition Reform Resources

of Commerce. MBDA is the only federal agency created specifically to foster the creation, growth and expansion of minority-owned businesses in America. Headquartered in Washington, D.C., with regional offices in Atlanta, Chicago, Dallas, New York, and San Francisco, MBDA provides assistance to socially or economically disadvantaged individuals who own or wish to start a business. Such persons include: Asian-Pacific Americans, Asian Indians, black Americans, Eskimos/Aleuts, Hasidic Jews, Native Americans, Spanish-speaking Americans, and Puerto Ricans. In addition, MBDA provides funding for Minority Business Development Centers (MBDC), Native American Business Development Centers (NABDC), Business Resource Centers (BRC), and Minority Business Opportunity Committees (MBOC) located throughout the United States.

DoD EMALL

http://www.emall.dla.mil
The DoD EMALL is the single entry point for DoD and other federal customers to find and acquire off-the-shelf, finished good items from commercial marketplace. The DoD EMALL offers cross-store shopping for the purpose of comparison pricing and best-value decision making. All vendors meet FAR and DFAR requirements and statutory requirements. The DoD EMALL is primarily composed of three corridors: parts and supplies, information technology, and training. It also provides a one-stop visibility of order status.

Defense, Procurement, and Acquisition Policy (DPAP) Electronic Business

http://www.acq.osd.mil/dpap/ebiz

Federal Acquisition Institute (FAI)

http://www.fai.gov
Publishes the Federal Acquisition *Insight* newsletter online.

Procurement Technical Assistance Centers (PTACs)

http://www.dla.mil/db/procurem.htm
The Defense Logistics Agency, on behalf of the Secretary of Defense, administers the DoD Procurement Technical Assistance Program. PTACs are a local resource that can provide assistance to businesses in marketing products and services to federal, state, and local governments.

Table 6.1
(Continued)

WomenBiz.gov
http://www.womenbiz.gov Portal for women-owned businesses selling to the government. Established by the Interagency Committee on Women's Business Enterprises (IACWBE).
Small Business Innovation Research (SBIR) and Small Business Technology Transfer (STTR) Program Information
Department of Energy (DOE): http://sbir.er.doe.gov/sbir/ Department of Defense (DoD): http://www.acq.osd.mil/sadbu/sbir/ Navy: http://www.navysbir.com/ Air Force Research Laboratory (AFRL): http://www.afrl.af.mil/sbir/ NASA GSFC: http://sbir.gsfc.nasa.gov/SBIR/SBIR.html National Science Foundation (NSF): http://www.eng.nsf.gov/sbir/ Defense Advanced Research Projects Agency (DARPA): http://www.darpa.mil/sbir/ U.S. Small Business Administration: http://www.sba.gov/sbir/
FedWorld Information Network
http://www.fedworld.gov Established in 1992 by the Department of Commerce to serve as an online locator service for a comprehensive inventory of information disseminated by the federal government. The site is now managed by the National Technical Information Service (NTIS).
Navy Electronic Commerce Online (NECO)
http://www.neco.navy.mil This interactive Web site provides direct, online access to Navy business opportunities.
Federal Marketplace Procurement Data Warehouse (Wood River Technologies, Inc., Ketchum, Idaho)
http://www.fedmarket.com Site includes federal bid opportunities, federal procurement forecasts, and SBIR program information. Importantly, the site's Vendor Center provides a significant number of helpful, easily understood, brief articles on business development, preproposal activities, proposal management, proposal development, and proposal writing.

Table 6.1
(Continued)

NASA Acquisition Internet Service (NAIS)

http://prod.nais.nasa.gov/cgi-bin/nais/welcome.cgi
The NASA Acquisition Internet Service (NAIS) provides easy access to procurement forecasts across the NASA centers. NAIS business opportunities are posted by date and classification. One can also use the NAIS "Search" function to find business opportunities by date, classification, NASA center, or keyword. NASA's Procurement Reference Library provides tools such as a search of the FARs and the NASA FAR Supplement (NFS).

Assistant Secretary of the Air Force/Acquisition (SAF/AQ)

http://www.safaq.hq.af.mil
Provides information on Air Force-specific acquisition reform initiatives and success stories.

Department of Transportation (DOT) Office of Small and Disadvantaged Business Utilization (OSDBU)

http://osdbuweb.dot.gov
Includes the agency procurement forecast as well as tips for marketing to the DOT.

Netlizard (maintained by Panamax, St. Petersburg, Florida)

http://www.netlizard.com/acqpol.html
Provides listings of federal opportunities, federal regulations and procedures [e.g., Federal Acquisition Regulations (FAR), agency-specific FAR supplements, and Federal Acquisition Circulars (FAC)], statutes governing federal contracting and acquisitions, court and other decisions affecting federal acquisitions and procurements, selected state opportunities, and selected international opportunities.

Acquisition Reform Network (ARNet)

http://www.arnet.gov
Run by the Office of Federal Procurement Policy (OFPP), the Acquisition Reform Network was created to foster and propagate measurable breakthrough improvements in the way government obtains goods and services. Specific elements on the site include e-government initiatives, FAR, agency home pages, competitive sourcing, agency procurement forecasts, and contract administration links.

Table 6.1
(Continued)

Governmentwide Acquisition Contracts
http://contractsdirectory.gov/ The contracts available on this site are for various types of governmentwide contract vehicles [e.g., governmentwide agency contracts (GWACs), multiagency contracts (MACs), blanket purchase agreements (BPAs), and schedules].
National Women Business Owners Corporation (NWBOC; formerly the Women Business Owners Corporation), Washington, D.C.
Federal Acquisition Reform: http://www.nwboc.org/pfar.html State and Local Government: http://www.nwboc.org/stategov.html The National Women Business Owners Corporation, a national 501(c)(3) nonprofit corporation, was established to expand competition for corporate and government contracts through implementation of a pioneering economic development strategy for women business owners. NWBOC seeks to provide more corporations with the opportunity to enhance their procurement practices and to provide to women suppliers the opportunity to compete.
***Government Executive* Magazine's Procurement Links**
http://www.govexec.com/procurement GovExec.com is a government business news daily. This site has a specific focus on e-government, A-76 and outsourcing, Homeland Security, and defense.
Service Corps of Retired Executives (SCORE)
http://www.score.org Founded in 1964, SCORE is a nonprofit association of 10,500 volunteer business counselors serving 389 chapters in urban, suburban, and rural communities throughout the United States and Puerto Rico. SCORE is a resource partner with the U.S. SBA. SCORE members are trained to serve as counselors, advisers, and mentors to aspiring entrepreneurs and business owners. These services are offered at no fee, as a community service.
Federal Register
http://www.gpoaccess.gov/fr/index.html Published by the Office of the Federal Register, National Archives and Records Administration (NARA), the *Federal Register* is the official daily publication for rules, proposed rules, and notices of federal agencies and organizations, as well as executive orders and other presidential documents.

Table 6.1
(Continued)

Maturity Model Integration® (CMMI®) structured software engineering methodologies, ISO 9001:2000 and 9002:2000 standards, Institute of Electrical and Electronics Engineers (IEEE), and IT Infrastructure (ITIL) disciplined approaches. ITIL provides the foundation for quality IT service management. This series of documents was developed originally by the British government.

Many procurements require face-to-face oral presentations between your proposed program staff and government technical, programmatic, and contractual leadership. Multiple, rather than single, awards are very common. And GWAC vehicles have become critical contractual gateways to priming contracts with a variety of federal agencies.

Centralized contractor registration is required to inform the government that your company is ready to do business with the government. Register your small business with the new CCR/PRO-*Net* portal (http://www.ccr.gov) right away. As of January 1, 2004, the U.S. Small Business Administration's PRO-*Net* database had been combined with the DoD's Central Contractor Registration (CCR) database. *The result?* One Web portal for vendor registration and for searching small business sources—and an integrated, more efficient mechanism for small businesses to market their products and services to the federal government. All of the search options and company information that existed in PRO-*Net* are now found at the CCR's "Dynamic Small Business Search" site. The new CCR/PRO-*Net* portal is part of the Integrated Acquisition Environment (IAE), one of the important e-government initiatives under the President's Management Agenda (http://www.whitehouse.gov/omb/budget/fy2002/mgmt.pdf).

It was Federal Acquisition Circular (FAC) 2001-16, dated October 1, 2003, that amended the Federal Acquisition Regulations (FARs) to require contractor registration in the CCR database. This same circular also eliminated the Federal Acquisition Computer Network (FACNET) and designated a single governmentwide point of entry on the Internet called FedBizOpps (http://www.fedbizopps.gov) where federal agencies are to provide universal and convenient access to information about their procurement opportunities. In addition, FedBizOpps replaced both the paper and electronic *Commerce Business Daily.*

Now that we have an understanding of RFPs and how they originate in the federal acquisition process, we will examine the contracting community's proposal response life cycle and its critical components.

ENDNOTES

1. *Writing Winning Proposals*, Farmington, UT: Shipley Associates, 1988, p. 3-3. The phenomenon of increased competition is also noted by Cibinic, John, and Ralph C. Nash, *Formation of Government Contracts*, 2nd ed., Washington, D.C.: George Washington University, 1986, p. 522.

2. The National Archives has published the CFR annually since 1938. This compilation of executive orders, proclamations, and rules and regulations for departments and agencies does for administrative law what the U.S.C. does for statute law. Material for the CFR is drawn from the calendar year entries in the *Federal Register*, a daily publication of Executive Branch documents and notices of public applicability and legal effect. [*History at NASA* (NASA HHR-50), Washington, D.C.: NASA Headquarters, 1986, p. 11.]

3. Per FAR 2.101, *supplies* means all property except land or interest in land.

4. Per FAR 37.101, *services contract* means a contract that directly engages the time and effort of a contractor whose primary purpose is to perform an identifiable task rather than to furnish an end item of supply.

5. The Comptroller General of the United States and the General Accounting Office (GAO) that he heads "have been given a prominent role in the oversight of Government procurement and procurement related functions" (Cibinic, John, and Ralph C. Nash, *Formation of Government Contracts*, 2nd ed., Washington, D.C.: George Washington University, 1986, p. 53).

6. *Introduction to the Federal Acquisition Regulation Training Course*, Vienna, VA: Management Concepts, pp. 1–3.

7. *Writing Winning Proposals*, Farmington, UT: Shipley Associates, 1988, pp. 3–11.

8. Standards assist the government evaluators in determining how well a proposal meets the requirements. Several standards might be developed for one evaluation factor. Depending upon the evaluation factor, standards may be qualitative or quantitative. Standards are necessary because evaluators are not supposed to compare proposals. They must, therefore, have some standard against which to determine

if a proposal satisfies the RFP. See *Writing Winning Proposals*, Farmington, UT: Shipley Associates, 1988, pp. 3–23.

9. Cibinic, John, and Ralph C. Nash, *Formation of Government Contracts*, 2nd ed., Washington, D.C.: George Washington University, 1986, pp. 644–645.

10. McVay, Barry L., *Proposals That Win Federal Contracts: How to Plan, Price, Write, and Negotiate to Get Your Fair Share of Government Business*, Woodbridge, VA: Panoptic Enterprises, 1989, p. 11.

11. Cibinic, John, and Ralph C. Nash, *Formation of Government Contracts*, 2nd ed., Washington, D.C.: George Washington University, p. 311.

12. Holtz, Herman, and Terry Schmidt, *The Winning Proposal: How to Write It*, New York: McGraw-Hill, 1981.

13. Hinton, Henry L., *Federal Acquisition: Trends, Reforms, and Challenges*, Washington, D.C.: United States General Accounting Office (GAO), March 16, 2000, p. 6. Testimony before the Subcommittee on Government Management, Information, and Technology, House Committee on Government Reform, GAO/T-OCG-00-7.

14. Hinton, Henry L., *Federal Acquisition: Trends, Reforms, and Challenges*, Washington, D.C.: United States General Accounting Office (GAO), March 16, 2000, p. 4.

15. Hinton, Henry L., *Federal Acquisition: Trends, Reforms, and Challenges*, Washington, D.C.: United States General Accounting Office (GAO), March 16, 2000, p. 8.

16. Hinton, Henry L., *Federal Acquisition: Trends, Reforms, and Challenges*, Washington, D.C.: United States General Accounting Office (GAO), March 16, 2000, pp. 6–7.

17. Hinton, Henry L., *Federal Acquisition: Trends, Reforms, and Challenges*, Washington, D.C.: United States General Accounting Office (GAO), March 16, 2000, p. 7.

18. Hinton, Henry L., *Federal Acquisition: Trends, Reforms, and Challenges*, Washington, D.C.: United States General Accounting Office (GAO), March 16, 2000, p. 9.

19. Brockmeier, Dave, "Help Shape Federal Acquisition Regulations," *Business Credit*, Vol. 97, No. 3, March 1995, p. 40.

20. Brockmeier, Dave, "Help Shape Federal Acquisition Regulations," *Business Credit*, Vol. 97, No. 3, March 1995, p. 42.

21. Hinton, Henry L., *Federal Acquisition: Trends, Reforms, and Challenges*, Washington, D.C.: United States General Accounting Office (GAO), March 16, 2000, p. 18.

Chapter 7

The proposal life cycle

You can't close the sale when you have not made the sale.

L ET'S NOW TAKE A CLOSER look at what proposals really are, how they fit into a small business's total marketing life cycle, and the important planning, decisions, knowledge products, organization, and reviews required to prepare a successful proposal.

7.1 What is a proposal in the competitive federal and commercial marketplace?

Technically, a proposal is an offer prepared by a contractor (such as your company) to perform a service, supply a product, or a combination of both in response to an RFP or RFS document posted at FedBizOpps or an SF330 synopsis issued by a procuring agency. Proposals are legal documents, often incorporated by reference into the final contract. That means that they must be factual and accurate—including information provided in résumés and project summaries.

A proposal is designed to sell both technical and managerial capabilities of a company to accomplish all required activities on time and at a reasonable cost. Your company's proposal document(s) is scored, literally,[1] by the client's evaluators against formalized, specific standards. A proposal is the written, electronic, or oral (or any combination of the three) result of informed thinking, knowledge assimilation, and marketing processes, supported in turn by hard and directed work and buoyed by a positive collective attitude and commitment within your company's proposal team. A proposal is, first and foremost, a sales document. To be sure, it includes a host of technical, programmatic, institutional, pricing, and certification information, but it must remain sales oriented.

A proposal involves marketing your company to a government, private-sector, or international client. It is a closing sales presentation, supported concretely by traceable and auditable credentials and tailored to persuade your client to select your company for the award because you are best-qualified to achieve the results your client wants to achieve [1]. In the words of a senior U.S. DOT official, write your proposal from the government COTR's perspective. Help him or her to feel comfortable that your company will meet schedules, budgets, and other performance parameters. Demonstrate that you understand and can function well in the COTR's world. Make the COTR want to link his professional future with your company. A proposal is, necessarily, continued dialogue with your client. Proposals are, in fact, the primary vehicle for winning new federal government business in the competitive arena.

A proposal is a closing sales presentation, presented persuasively in story form and supported concretely by traceable and auditable credentials.

To reiterate, a proposal (as well as a bid, that is, a response to an IFB) is an offer or response [2]. A proposal is not necessarily a contractual document. But under specific circumstances, an offer can be made into a contract by the government. The client uses your proposal and those of your competitors as the primary sources of information upon which to base their selection of a winning contractor. Government evaluators are in no way obligated or encouraged to review publicly available material about your company that is not included within your proposal. That means that

---

1   Scoring can take the form of numeric scores, color-coded scores, pluses and minuses, or qualitative (adjectival) narrative comments, such as "very good" and "good."

you have to ensure that all relevant and salient materials are included along with your proposal within the parameters of any page or font limitations.

To continue, a proposal is a package of carefully orchestrated and documented arguments woven into a compelling story. Nothing should be put into your proposal by accident! Each section of a proposal should present carefully constructed discussion and meaningful evidence to convince your client that you should be awarded the contract because of the superiority of what you are proposing. To do this, you have to support the following messages in writing:

- You understand your client's project requirements, critical issues, and success criteria.

- Your innovative, solutions-oriented approach satisfies all requirements (be careful not to imply that your approach exceeds requirements unless you can confirm that such enhancements can be delivered for no additional cost; this can result in the client thinking that you are proposing and charging more than that for which they are asking).

- Your approach offers tangible and intangible benefits to the client.

- Your approach minimizes schedule, cost, and technical performance risk.

- You are better (that is, more reliable, more experienced, and less expensive) than your competitors.

Why us? Why not our competitors? These are the simple yet profound questions that internationally known and respected proposal consultant Hyman Silver[2] insists proposals must answer effectively. Through the vehicle of your proposal, your company must finish the process of convincing your client to select you over your competition.

Proposals are important deliverable products. One carelessly written and poorly presented proposal can damage your company's reputation with your client or potential client. Experience suggests that clients tend to remember contractors that have submitted inferior proposal responses. I have heard many senior managers say that they want to submit a proposal to "introduce the company to the client." If this is not a waste of B&P funds, nothing is. Introduce your company through marketing calls, leave-behind brochures, carefully tailored letters, or at

---

2 This phrase was coined in 1975 by Hyman Silver, president of H. Silver and Associates (HSA) based in Los Angeles, California. Mr. Silver held senior engineer and marketing positions with McDonnell Douglas and Rockwell International before founding HSA.

industry trade shows and conferences. But do not expend the human, time, and financial resources to respond to a solicitation document unless you intend to win.

7.2 Where does the proposal fit into the total marketing life cycle?

The proposal life cycle (in particular, for procurement efforts on which your company is the prime contractor) begins well before the actual writing and production of a response to an RFP. Figure 7.1 presents an overview of the entire marketing and proposal life cycle.

There are a number of salient elements in the overall contractor marketing and proposal process. The proposal process may include additional

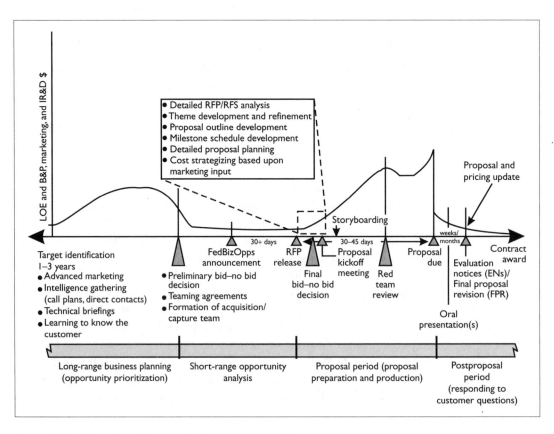

Figure 7.1 Overview of the marketing and proposal life cycle.

color team reviews (Blue, Pink, and Gold, for example). Please note that the concepts and processes introduced in the following items will be expanded in subsequent chapters of this book.

Long-term, preliminary marketing intelligence collection and direct client contact

- Presell the client on your company's qualifications and commitment to complete the job successfully. Create customer sentiment[3] favorable to your company through technical briefings and discussions at various levels of the client's organization. Learn about such concerns as your client's technical requirements, financial concerns, scheduling issues, and decision-making processes. Verify that the project is a real, funded target.

For companies that provide support services, well-managed human talent and applicable contractual expertise are precisely what they are selling.

Formation of your company's acquisition/capture team for a particular qualified procurement opportunity

- Assemble a small team of management, marketing, and technical staff that will meet on a regular basis to develop a capture plan and share information gleaned from client contacts conducted in accordance with the call plan. Themes, or sales messages, for your proposal should begin to emerge. Identify your company's competitive strengths and weaknesses. Identify staff and relevant project experience to be used in your upcoming proposal. For companies that provide support services, well-managed human talent and applicable contractual expertise are precisely what they are selling.

- Résumés and project descriptions (project citations or contract summaries) can be tailored and fine-tuned well in advance of the release of the RFP. Take the time to gather the kinds of detailed educational, certification, technical, and programmatic experience information that your clients normally require in their RFPs. Because staffing and past performance are critical evaluation

---

3   Hyman Silver coined the phrase.

factors in many RFPs, and because collecting specific information on staff and projects can require so much time, it is highly advisable to get started early. Build this information into your proposal knowledge base.

Government's announcement of the procurement online at FedBizOpps (http://www.fedbizopps.gov)

In some cases, your company must provide a written request to the appropriate government procuring agency for each particular RFP. Telephone or facsimile requests are generally not accepted. Increasingly, e-mail requests are acceptable. In addition, many RFPs are downloadable from government Web sites.

- The time to market your client for a particular project is drawing to a close. Once the RFP or RFS has hit the street, most agencies will not allow any marketing relevant to the project under competitive bid.

Release of the RFP by the specific government procuring agency

- If possible, arrange to physically pick up the RFP from the procuring agency. This will save several days delay in mailing time. Frequently, RFPs are available for downloading from client Web sites. For select procurements, your company will have to register with your client to obtain password access to the specific RFP posted on the client's Web site.

Government preproposal (bidders) conference

- Plan to attend the preproposal conference. If allowed, prepare questions for the client and submit them in writing in advance of the conference.

- Pay attention to client requirements that may be mentioned at the conference but not written in the RFP. Some conferences might include facility or site visits. Check in advance to see if photography is permitted and, if so, take relevant photos or video.

Final internal bid–no bid decision

- All such decisions should be made according to defined, structured business criteria and include (at a minimum) a member of senior management, the business development or advanced planning staff, the appropriate division manager, and the proposal manager.

- Good marketing intelligence will help to facilitate an appropriate decision.

- What is the probability of winning?

- Favorable financial return is an important consideration.

- Does your company have the people and facilities necessary to prepare a superior proposal and perform the project successfully?

- Does the project have potential strategic business advantages to your company?

- Teaming arrangements must be considered.

- Is your company the incumbent; if not, can you unseat the incumbent realistically?

Refer to Figure 7.2 for a representative bid–no bid decision matrix. Your company might use a total score of 75 to pursue a project.

Prekickoff planning meeting (conducted by the proposal manager)

It is often helpful for your company's project managers (PMs) or program managers—the people who most likely will be called upon to serve as proposal managers—to envision the up-front proposal planning process in terms of deliverable products. These are very much analogous to the types of products that are integral to every program, project, task, and delivery order in your company's contractual experience base. Just as every PM develops and adjusts staffing levels to meet changing requirements, generates project milestone schedules in accordance with the client's requirements, and tracks project budgets to ensure cost control, so too must your proposal manager work with the capture manager and other staff in your company to develop a series of discrete products early in the proposal response life cycle, as shown in Figure 7.3. Particularly in very small companies, there may be insufficient overhead resources (nonbillable staff) available to maintain a proposal library or cybrary, provide proposal coordination support, assist with résumé and project summary development, and solidify teaming agreements. Those critical tasks must be accomplished by the proposal manager, and in a very limited time frame. The specific work products should be assembled in a document called the Proposal Directive.

The Proposal Directive is a carefully structured "living document" that serves as the "encyclopedia" of a win for a given procurement opportunity. During the course of the acquisition/capture and proposal process, the Proposal Directive may undergo 10 or more iterations as new intelligence

Date: _____ GO: _____ NO GO: _____

Client: _____
Project: _____
Prepared by: _____
Approved by: _____

	10	9	8	7	6	5	4	3	2	1	0	Ranking	Comments
Strategic relevance (how well does project fit company marketing strategy?)	Fits well with marketing strategy				Fits somewhat into the marketing strategy				Does not fit into marketing strategy				
Relationship with client	Close and excellent relationship with no problem projects				Fair or good relationship with some project problems				Strained relationship; problem project; selection questionable				
Knowledge of project	Strategic project/prior knowledge and marketing				Prior knowledge; no prior marketing				No prior knowledge or marketing				
Company staffing capabilities	Staff can perform all project requirements				Staff can perform most of the project requirements				Minimal or no staff capabilities				
Project competition	Competition manageable				Face formidable competition				Project wired for competition				
Company qualifications relative to competition	Technically superior; can demonstrate experience on similar contracts				Equivalent to competition; may have slight edge				Inferior to competition				
Potential for future work	Project will lead to future work				Possible future work				One-time project; no future work				
Profitability of project	Good profit potential				Moderate profit				Little profit				
Project schedule	More than adequate				Adequate				Not adequate				
Time, B&P costs, and resources to compete effectively for project	Favorable				Reasonable				Excessive				

Figure 7.2 A formalized bid–no bid decision matrix facilitates rational, business-based decision making.

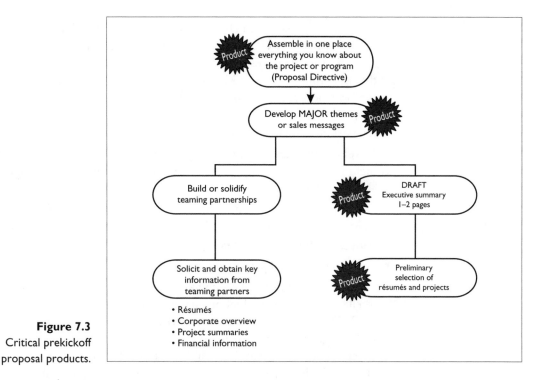

Figure 7.3
Critical prekickoff
proposal products.

about the competition, for example, is uncovered and other information
(e.g., win themes) is further refined. The Proposal Directive is the repository that contains all of the relevant information about the customer organization, competition, specific program/project, and relevant technologies. It
contains the call plan reports from customer visits; a synopsis of the hopes,
fears, biases, success criteria, and critical issues of key government decision
makers; the win strategy white paper for your company along with the proposal outlines for each volume; win themes and discriminators; proposal
response milestone schedule; and specific proposal-related action items.
Thus, the *marketing intelligence* that Business Development staff gather is
collected in the Proposal Directive. The *solution sets* that the Acquisi-
tion/Capture Team develop based on the marketing intelligence are documented in the Proposal Directive. And the *proposal development work
products*—schedules, outlines, action items, and so forth—are also
included within the Proposal Directive. It is important to note that the Proposal Directive is a highly competition-sensitive document and, therefore,
must be safeguarded very carefully.

The following critical tasks are part of the prekickoff planning process.

- Develop a detailed outline for each proposal volume, using RFP
  Sections L, M, C, and H, as well as the Contract Data Requirements

List (CDRL), Data Item Descriptions (DIDs), and RFP attachments as appropriate. Take the time to assign responsible parties to each section. Allocate pages to each section in accordance with the relative weighting of Section M evaluation criteria. For example, if past performance will count for 20% of the total score and there is a 50-page limit to the proposal, plan on approximately 10 pages for the past performance section.

- Develop a workable proposal response milestone schedule that includes adequate writing, review, recovery, production, QC, and shipping time.

- Develop and modify existing proposal themes. Themes are conclusive and compelling reasons why the client should select your company over and above your competitors.

- Select résumés and project summaries for the proposal.

- Generate a draft executive summary based upon your marketing intelligence and the evaluation factors found in Section M of the RFP. This will help to guide your writers to produce consistent, well-integrated narrative and graphics. Include graphics in your executive summary if at all possible. A full-color foldout page (11×17 inches) that captures the technical and programmatic essence of your proposal is one high-impact way to sell the benefits of doing business with your company early in the proposal.

- Develop a preliminary design concept for the proposal volume covers.

- Identify critical supplies so that they can be ordered in advance (such as three-ring notebooks, special colored paper, customized tabbed divider pages, and specialized software[4]).

Proposal kickoff meeting [with subcontractor(s), if any] directed by the proposal manager

This meeting should be attended by senior management, appropriate marketing staff, the proposal manager, the proposal coordinator, all key technical staff who will be engaged in writing major sections of the proposal, key reviewers, a representative from contracts, and if necessary, a representative from human resources and a representative from publications.

---

4   Certain RFPs require that cost and technical volumes be submitted in specific software applications (e.g., Microsoft Word, Excel, Corel WordPerfect, Lotus 1–2–3) and versions of those applications that your company may have to purchase.

- Emphasize team spirit, a winning attitude, and the support and interest of senior management.

- Distribute detailed technical, management, and cost volume outlines as part of the Proposal Directive.

- Distribute DRAFT executive summary. Having a draft of the executive summary at this stage will help everyone involved in the proposal-writing process to better understand what the client expects. It will also help to establish the major strategic themes your company wants to integrate throughout the proposal.

- Share critical marketing intelligence about your client, and discuss key win strategies.

- Technical and management writing assignments are made, along with page limits (give writers an estimate of the number of words per page) and allocations (in accordance with Sections M and L of the RFP) and writing conventions (such as capitalization, hyphenation, and punctuation). Themes, critical issues of the project, and success criteria are explained to the writers. Clear responsibilities for action items will be assigned.

- Stipulate electronic file formats, software application and version, and file configuration control procedures.

- Assignments are made to contracts and human resources staff.

- Items that require the approval of senior management should be presented and finalized.

- Address the criticality of proposal security for both electronic and hardcopy versions of the resulting proposal documents.

Storyboarding, to include the development of "elevator speeches," annotated outlines, and populated "pain tables"

Storyboarding (the associated work products are also referred to as "scenarios" and "scribble sheets") is a critical management process that focuses on building a coherent, consistent, and compelling proposal story. Building on the marketing intelligence and solution sets for a given procurement, key stakeholders in the proposal process (your capture manager, proposal manager, business development staff, subject matter experts, and so forth) meet in a series of brainstorming sessions and generate the following critical work products: (1) annotated outline for each major section of the proposal that provides an expanded architecture for each proposal section; (2) "elevator speeches" that tell the overarching story for each section; and (3) populated "pain tables" that illustrate your company's understanding of the customer's "pain points" and major issues, your approach/solution, and your company's value proposition (see Figures 7.4 through 7.7). Your value proposition includes the tangible and intangible benefits that your solution will bring to the customer organization.

Storyboarding allows your company and your proposal and capture teams to achieve the "right dive angle off the board." Just as a championship diver cannot adjust her position relative to the water when she's a mere foot above the surface, neither can a proposal team move in a fundamentally different direction when it is deep into the proposal response life cycle—at least not without a lot of depleted morale, extremely long hours, wasted resources, and interpersonal challenges across the team.

Coming out of the storyboarding process, one should be able to review the elevator speeches from end to end and grasp the entire fact-based storyline of the proposal response.

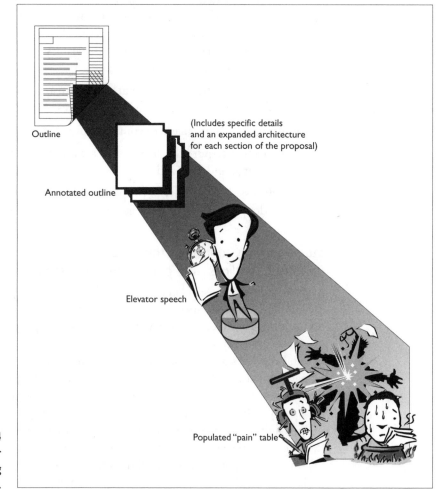

Figure 7.4
Process flow for storyboarding sessions.

Major work products
- Annotated outline
- Elevator speech
- Populated "pain table"

Sample elevator speech

> Our team will employ a systems engineering life-cycle process to ensure effective IT infrastructure management at DOE HQ. Our team has collected and documented many best practices and lessons learned from our past systems design and engineering support activities. As recognized industry leaders in the provision of engineering support services, our team will provide exceptional support for DOE's Systems Engineering Environment. We are working together today within the DOE complex—at field sites, at operations offices, and at headquarters. We have a low-risk plan to ensure continuity of operations from Day One. Our approach is based on the Global Services Methodology (GSM), which recognizes the importance of corporate business drivers as the basis for technology change and complies with Federal Enterprise Architecture guidance throughout the systems engineering life cycle. GSM delivers consistent, repeatable processes to the management of systems design, engineering, installation, and maintenance activities.

Sample pain table

Understanding the Environment			Approach/Solution		Value Proposition
Major Issues	Contractor Challenges	Critical Success Factors	Process	Schedule	Measurable Outcomes
Improved Service Delivery	• Continuously maintaining high levels of service availability	• Measurable increased customer and stakeholder satisfaction	• Enhanced tools tied to TIVOLI to deliver near real-time network performance information	• Day 1- forward	• Exceed the current availability SLA metric • Increase network infrastructure MTBF
Reducing support costs incurred today because of redundant system solutions operating at HQ	• Overcome system engineering implementation inefficiencies and security vulnerabilities due to lack of central planning and coordination	• Successful network consolidation that centralizes systems engineering planning and operations currently performed in several HQ organizations	• Systems engineering planning is centralized using consolidated resources, configuration management processes, and a consistent, structured planning methodology	• Day 1- forward	• Better leverage valuable technical resources across HQ • Reduce redundant solutions • Minimize security vulnerabilities • Provide better incident response from the Help Desk • Improve unified solution ROI
Validating DOE Enterprise Architecture compliance	• Systems design and engineering must be driven by DOE Enterprise Architecture	• Architecture guidance and CPIC governance process are applied to technology and engineering projects consistently	• Integrate Architecture guidance and CPIC governance in the FOCUS Phase of project planning	• Day 1-180	• Exhibit 300 rationale for technical and engineering projects accepted by OMB on initial submission

Figure 7.5   Major work products emerging out of the storyboarding process.

Timely submission of first draft technical and management input to proposal manager (text and graphics), which should be prepared and produced in close association with the text

- Ongoing interaction and feedback between the proposal manager and the proposal writers are critical for developing a viable draft document quickly and efficiently.

2.0	Management plan
2.1	Phase-in plan
2.2	Personnel management
2.2.1	Staffing plan
2.2.2	Cultural diversity plan
2.2.3	Training plan
2.2.4	Total compensation plan
2.3	Business management
2.3.1	Organization
2.3.1.1	Logistics
2.3.1.2	Imaging technology
2.3.1.3	Publishing
2.3.1.4	Metrology
2.3.1.5	Library
2.3.1.6	Administrative programs
2.3.1.7	Clerical support
2.3.2	Subcontract management
2.3.3	Management of government furnished property
2.3.4	Financial tracking database
2.4	Cost control plan
2.4.1	Cost-effectiveness
2.4.2	Cost fluctuation

Figure 7.6
Basic outline.

Editing and review of technical and management input

Submission of technical and management input to publication group

- Ensure that electronic files are compatible.
- A secure extranet or firewalled File Transfer Protocol (FTP) site can be used for rapid file transfer from remote sites or teaming partners in other geographic locations.

The key is to use the Internet to transact business while simultaneously employing security technologies to ensure full confidentiality. IPsec is a standard that provides a common means of authentication, integrity, and Internet Protocol (IP) encryption. Corporations can save money as IPsec virtual private networks (VPNs) solutions are implemented. Their remote users can use the Internet via an ISP instead of dial-up lines to access the corporate network. Accessing a local ISP for connection and using IPsec

2.0	**Management plan**
2.1	Phase-in plan
	Preaward activities
	• Establish transition "Tiger" team • Conduct open house • Prepare continuity of operations plan (COOP) • Prepare security plan • Prepare staffing plan • Define subcontraction plan and IPT approach • Identify transition challenges; define corrective action plan
	Postaward activities
	• Kickoff meeting • Deliver COOP, security, and staffing plans • Advise and recruit • Observe operation; collect intellectual knowledge • Define invoice format
2.2	Personnel management
	Management tools and techniques
	• Program management plan • Task order management systems • Performance measurements and metrics
	Personnel retention
	Resolution of personnel problems
2.2.1	Staffing plan
	General staffing plan and incumbent capture
	Key personnel, qualifications, certifications, security clearances, and commitment
	Other staff qualifications, certifications, clearances, and commitment
	Surge protection planning
	Staffing risks
2.2.2	Cultural diversity plan
	RSIS professional staff
	Community contributions
2.2.3	Training plan
	Comprehensive cross-training program
	Technical refresh training
	Speciality and mandatory training programs
	• Security CBT awareness training
2.2.4	Total compensation plan
	Incentize employees
	• Semiannual performance bonus program • Spot "bonus" program • Employee-of-the-month awards • President's Award

Figure 7.7
Corresponding
annotated outline.

	Company benefits plan Prepaid tuition plan PC-buy program

2.3	Business management
	Process optimization
	  • Use of industry's best practices
	  • Incorporate lessons learned
	  • Continuous communication improvements

2.3.1	Organization

2.3.1.1	Logistic
	Seat management
	Procurement management
	  • Spares depot
	  • Equipment repair and return services
	  • Hardware and software purchasing
	Vendor management

2.3.1.2	Imaging technology
	Image tools
	  • Access and viewing
	Image storage
	  • Database management
	Image analysis

2.3.1.3	Publishing
	Content management and guidelines
	  • Web development
	  • Database management
	  • Section 508 compliance
	  • Export control compliance
	  • Security

2.3.1.4	Metrology Measurement technology Calibration methods Metrology equipment and tools Methrology laboratories Standards

2.3.1.5	Library
	Organizational transformations
	  • Information database
	  • Search and browse tools
	  • Change management

2.3.1.6	Administrative programs
	Content management
	  • Program description storage
	  • Accessing and viewing program information
	  • Database security

Figure 7.7
(Continued.)

2.3.1.7	Clerical support
	Resource management
	Task order management
	Performance measurements
	Reporting and communication process
2.3.2	Subcontract management
	Finalize teaming agreement
	Define work assignments
	Establish performance monitoring, metrics, and measurements
	Define deliverables
	Implement cost control
2.3.3	Management of government-furnished property
	Property tracking and accountability
	• GFE inventory
	• Maintain software licenses
	• Hardware and software upgrades
2.3.4	Financial tracking database
2.4	Cost control plan
	Collecting and aggregating labor hours and costs
	Controlling indirect rates and costs
	Using automated program management tools
	• Deltek
2.4.1	Cost-effectiveness
2.4.2	Cost fluctuation
	Tracking and analyzing variances
	Establishing and documenting cost savings

Figure 7.7
(Continued.)

for encryption can lower telephone charges and equipment costs significantly

In addition, IPsec offers the ability to create virtual, protected links through the Internet to your customers, vendors, and other business partners. IPsec can be used to create trusted virtual work groups to help protect sensitive corporate data. For example, your business development department can be protected from other departments that do not have a need to know with respect to this department's confidential information. Or employee records residing in the human resources department can be protected from unauthorized access.

Submittal of cost and staffing information to contracts

- It is critical to conceptualize and develop the cost volume in close association with the technical and management volumes (see

Figure 7.8). Staffing selections, technical solutions, facilities and computer resources proposed, and your proposed management plan and organizational structure can all have profound cost impacts. Proposal costing is sophisticated business, not mere number crunching. The key is to have financial staff with direct program or project experience who can interact appropriately and effectively with the technical proposal team.

Preparation of complete first technical or management draft with one master copy only

Proofing and review of first draft

- Entire document frozen 24 hours prior to Red Team review.

Red Team review of technical, management, and cost volumes (1 to 2 days only)

Integration of all Red Team review comments (1 to 2 days) by proposal manager, selected technical staff, and the proposal coordinator (if needed)

Quality check of format and content by publication group

Preparation of second complete technical or management draft

Proofing and review of draft

Preparation of revised cost volume

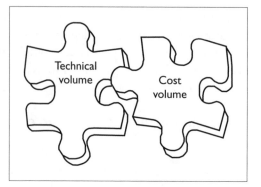

Figure 7.8
Develop the cost volume in close association with the technical volume.

Gold Team final review of technical, management, and cost volumes

- Entire document frozen 72 hours prior to final delivery to client.

Final production, integration of text and graphics, quality check, photoreproduction, collation, assembly, and preparation for delivery (3 days)

Signatures on cover letter and cost proposal forms (it is suggested that you purchase a stamp that reads ORIGINAL)

Final delivery

- Late proposals most likely will not be accepted by the government (see FAR 14.304). Even 1 minute past due is still late! And with electronic uploads, 1/100th of a second is too late. Request, obtain, and archive a time- and date-stamped receipt.

Archiving of hard and electronic copy of proposal text and graphics in your corporate library

- These materials should be stored along with all of the marketing intelligence documents, Freedom of Information Act (FOIA) requests, competitor information, the original RFP and amendments, and other materials relevant to this procurement. Being able to easily retrieve this material, even months or years later, is critical if you are required to respond to Evaluation Notices (ENs), submit a Final Proposal Revision (FPR), or deliver an oral presentation. This material should form the core of your proposal knowledge base.

Oral presentation defense, FPR, and facilities inspection

- Certain agencies require that FPR submittals or ENs be submitted as change pages to your original proposal submittal. Being able to retrieve your original submittal easily is a vital time saver.
- Take time to strategize and develop a powerful, well-rehearsed presentation.

Attend government debriefing when your company is an unsuccessful offeror and also when you win (see FAR 15.1003). Learn why you won.

Develop lessons learned and discuss with key proposal staff. Share lessons learned with other proposal managers. Archive and review these lessons, looking for patterns.

7.3 Bid–no bid decision-making process

Being very selective with the procurements your company decides to pursue is a first step toward long-term proposal and general business development success. Careful selection of bid opportunities improves your proposal win probability and maximizes the return on general and administrative (G&A) dollars your company has invested in the proposal effort (see Figure 7.9). Identifying sufficient high-dollar, high-probability opportunities for which your company is qualified is a necessary step prior to selection of particular targets. The essence of the bid–no bid downselection process relates to four primary factors.

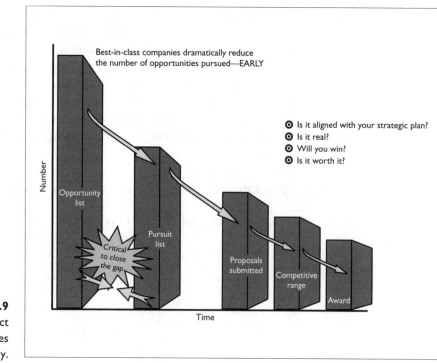

Figure 7.9
Downselect opportunities early.

- *The probability of winning.* Does the client know your company and does your company know the client? Is there an incumbent contractor? Is that contractor performing satisfactorily? What is the nature of the relationship between the incumbent and the client? Does marketing intelligence indicate that your company is in a favorable competitive position to win?

- *The reality of the procurement.* Will it receive funding and will the competition be genuine, or does the government really intend to retain the incumbent?

- *The desirability of obtaining the contract.* What are the profit expectations? Will the contract result in a strategic business advantage for your company, such as a foothold in a new geographic location or market sector? Importantly, will the procurement align with and support your company's stated objectives as presented in your strategic plan and mission statement? What are the inherent risks and liabilities? What is the add-on or follow-on potential? Does the procurement require any up-front capital expenditure or cost sharing by your company? Will new staff with specific subject matter expertise have to be recruited and hired?

- *The necessary facilities, resources, and human talent to perform the work on schedule and within budget, as well as to prepare a compliant proposal that tells a compelling, client-focused story.* Are the very best staff available to write and review the proposal documents? Are there other key competing proposal priorities that will dilute the effectiveness of the proposal development group? Is there secure floor and wall space available in your facilities to house and support a proposal team? Are there sufficient computers and printers that will be able to be dedicated to the proposal publication effort?

The bid–no bid process must balance the cost of responding; the availability of human, material, and other resources; and the magnitude of the business opportunity. The availability of in-house or consultant staff, facilities, and equipment resources to provide effective proposal support cannot alone drive the decision-making process but certainly must be considered. If your company cannot commit the best resources to prepare the proposal, then you should strongly reconsider any decision to bid the procurement. For the small business, the vice president of business development must perform this downselection process, working with a multi-disciplinary team. The company's owners should get involved as well.

The major objective of the bid–no bid process is to arrive at a rational, business-based consensus within your company to pursue or not pursue a given procurement opportunity. It is advisable to conduct both a preliminary (pre-RFP release) and final (post-RFP release) bid–no bid decision. This applies to small as well as large companies.

Poor bid–no bid decision making is arguably the most common cause of business development failure in the federal contracting marketplace. Most small businesses fail within their first 5 years of operation. Merely because your company has related technical expertise does not mean that you have a high probability of winning the business or that the business fits strategically with your goals and direction. The shotgun approach to business development and proposal development will be unsuccessful in the long term. Laser-like focus will serve you much better.

7.4 Planning and organizing

Once you receive an RFP or RFS, your acquisition/capture team for that procurement should conduct an audit of the document, carefully mapping and accounting for all requirements, including the CDRL and DIDs. Review the staffing level requirements and position descriptions in the RFP. Formulate appropriate questions to be submitted in writing to the government for clarification. Be prudent in asking questions because your competitors will be given or see posted copies of those questions along with the government's response.

Various automated tools are available to support the RFP analysis and proposal development processes. These software products include Proposal Master, RFPMaster, PresentationBuilder (Sant Corporation; http://www.santcorp.com), and *RES*TRIEVE (Applied Solutions; http://www.appliedsolutions.com).

7.4.1 Draft executive summary

Prior to your company's formal kickoff meeting for each procurement effort, an individual appointed by the proposal manager should prepare a one- to two-page executive summary that introduces high-level themes, demonstrates your company's understanding of the importance of the client's project or program, indicates that your company has the credentials and talent to do the project, provides an overview of your company's approach, as well as its direct benefits to the client, and highlights key project staff.

Your executive summary should applaud your client and your client's achievements. It should honor—and demonstrate your alignment with—your client's mission and vision. In addition, it should be structured to address in summary form the evaluation factors found in Section M of the RFP. Help your client understand exactly how your company will enhance their mission achievements, funding, congressional notoriety, technical excellence, and so forth.

7.4.2 Theme development

A theme is a substantiated sales message, a point of emphasis, an advantage, a unique or superior benefit, or a supported claim. Themes reappear, woven inextricably throughout the proposal volumes and serve to focus and unify the entire presentation (see Figure 7.10). A good theme is direct, addresses a program issue or customer concern, and is supported by concrete evidence. Your company's strengths and the competition's weaknesses can be the bases for themes. Themes are highly sensitive material and should always be marked as competition-sensitive materials when distributed to staff throughout your company and extended team for review and comment. It is suggested that your company purchase ink stamps that read Competition-Sensitive.

There are two types of themes: (1) common themes and (2) unique themes, or discriminators. Common themes, although obvious and unprofound, are still important to incorporate into your proposal. High performance, low risk, is an example of a common theme. Unique themes, or discriminators, are based upon particular advantages your company can offer and upon distinct disadvantages of your competition. Discriminators are vital, credible points that are meaningful to your client and constitute

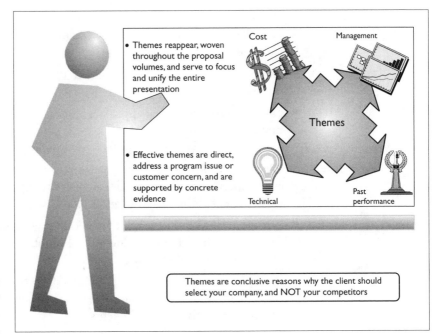

Figure 7.10
Themes are major story elements in your proposal.

the real sales hooks. You need to do your marketing intelligence-gathering homework to identify these discriminators and then locate them at appropriate points in the proposal documents. Perhaps your company is the only known proposer that has developed personal-computer-based air-traffic-controller instructional materials for a particular office of the FAA. A discriminator that you would certainly include in your proposal is a statement that highlights this unique contractual experience. It should appear, as appropriate, in section headings, the narrative, graphics, and captions. This message can also be conveyed via photographs or your proposal volume cover design. You can even reinforce this important unique theme in your compliance matrix.

In small proposals (fewer than 50 pages), your company may wish to develop 8 to 10 major themes and 15 to 20 minor (or subsection) themes. In larger proposal efforts, there may be 25 or more major themes and several hundred minor themes.

The following are select examples of themes. The particular procurement situation dictates whether these may be common or unique, major or minor.

- Best practices and lessons learned from our phase-ins with ONR, NAVAIR, and DOE will be applied to ensure a smooth transition from day one of this important contract.

- Our real-time supervision and automated performance assessment methods have proven to be responsive and timely for other USACE clients.

- Our competitive benefits have proven to be a powerful tool for motivating and retaining key personnel.

- Our exemplary past performance with the U.S. Navy is the credibility base for this procurement.

- Our innovative technical approach based upon SEI CMMI®–structured methodologies allows for cost-effective software engineering with a focus on your stakeholders' requirements.

- We offer established relationships with world-class computer science departments at universities such as George Washington University to ensure an ongoing, accessible base of talent.

- We understand and have direct experience with the Clinger-Cohen Act and the Department of Energy's IT capital planning requirements and financial planning.

- We will ensure continuity of operations (COOP) through industry-trend-setting employee retention rates made possible by superlative, yet cost-competitive fringe benefits.

- We will use specific metrics to quantify and monitor customer perceptions of and satisfaction with our level of technical proficiency and interpersonal skill.

- As extensions of the Air Combat Command's staff in Europe, much of our energy will be dedicated to developing and supporting solutions-based decision making consistent with the external and internal drivers that impact day-to-day operational activities.

- We will address problem resolution proactively and positively in the context of partners working toward common goals.

- We will conduct business at the operational level in strict accordance with FIP, commercial leasing, and software licensing procedures and protocols.

- The ABC team: superlative service built upon knowledge, common sense, and that intangible personal touch.

- We have a high level of competence demonstrated by NASA customer-approval ratings of 94.8% in the last quarterly review period.

- The accuracy of our cost and labor projections, as well as our accountability, are illustrated by costs being within 2% of budget each month.

- We will prove to be highly capable representatives as we honor the daily and big-picture needs of an agency whose work and mission are critical to the economic health of America within the global marketplace.

- Our management and procurement structures have been specifically designed to support accelerated schedules, immediate mobilization, and midcourse changes.

- We will support DOE's Strategic Action Plan and the Department's innovative e-government initiatives to help achieve alignment with the President's Management Agenda (PMA).

- Our cost-effective approach to environmental solutions meshes closely with the USACE's nationwide initiative to improve business practices and to produce quality results.

- Our well-established advocacy, mentoring, and practice of process control and checking also links with the Navy's emphasis on process improvements directed toward revolutionizing effectiveness, being more responsive to customer needs, and being more accountable to taxpayers.

- We transitioned the contract in Savannah River with 100% retention of qualified, incumbent personnel, while maintaining 99.89% system availability at all levels.
- Our networked environment, business culture of technical resource sharing, proactive administrative support, and ongoing project communication channels will ensure that staff and resources are on the job when and where they are needed to support the U.S. Postal Service.
- We are able to propose proven leadership that allows program continuity while preserving program institutional knowledge.
- ABC Company has crafted a very effective, successful, and proven recruitment process. We have cultivated a companywide business culture that attracts and retains highly talented professionals.
- This single-vendor management approach will eliminate program accountability complexities, reduce costs, and minimize overall management risks and personnel issues and concerns to the Department of Health and Human Services.
- We bring an established track record of competing and submitting deliverables to the Department of Transportation ahead of schedule.
- Our demonstrated record of careful cost controls will result in best-value support to NOAA throughout the life of the contract.
- In-house IR&D on interactive software is directly applicable to the technical solution for this important NASA/MSFC project, thereby saving you money.
- The total Washington, D.C.–based data management resources of our company will be available for application to specific needs of this important EPA program through our program manager.
- COTS hardware that we have proven in similar situations will minimize start-up time requirements, development costs, and reduce the phase-in period.
- Dedicated, hand-picked software engineer offers 18 years of documented success in developing and managing automated systems for the DoD.
- All our proposed professional staff are currently working on the existing contract and hold active secret clearances.
- We will leverage our automated trend analysis database to support risk mitigation.
- Proximity of our company's headquarters to Fort Bliss will ensure effective, real-time communications with the COTR and CO.

- We are a progressive minority-owned firm with an established $3 million line of credit to ensure financial flexibility and stability.

- Our company staffs and maintains an office within 5 miles of the IRS work site.

- Our company was the incumbent on the last modeling and simulation project, which successfully formed the basis for this one.

- Our company has served the Air Force Center for Environmental Excellence (AFCEE) successfully for the past 2 years. Our legacy of field-tested performance...

- State-of-the-art technology meets the EPA's current and projected requirements; that is, no new technology is needed.

- An excellent team of well-qualified, well-integrated subcontractors reduces program risk. We know how to work together successfully.

- Our company's organizational structure has a one-to-one correlation with the Air Force's project organization for ease of interface and communication.

- Our QA-QC program, already implemented at McClellan AFB, will ensure rapid problem isolation and resolution.

- Our integrated project team (IPT) approach will focus upon the formulation of small, flexible teams to evaluate emerging technologies and offer informed recommendations.

Your company's Acquisition/Capture Team members should convey appropriate themes to your client during the client contact and marketing phases of the proposal life cycle. Then, when those themes reappear in the proposal, the client will tend to view them as natural outcomes of previous conversations and meetings rather than unvarnished marketeering tactics. And once again, themes should be woven into the proposal body, carefully integrated into headings, text, captions, figures and tables, and even compliance matrices (under a comments section).

7.4.3 **Storyboards**

Proposals are choreographed exercises in knowledge and information management, assembly, and packaging.

Storyboards, bullet drafts, scenarios, mock-ups, and scribble sheets[5] are proposal tools designed to break the proposal planning, control, and tracking process into bite-sized pieces. Proposal choreography provides the proposal manager with a mechanism designed to develop a coherent, consistent document from multiple authors with varying degrees of writing skill in a limited time frame. Storyboards help to ensure that all RFP or RFS requirements are addressed. They also assist in linking text with graphics and help technical authors overcome any initial writer's block. Many times these tools prove vital for coordinating subcontractor input. Choreographic techniques are most effective and appropriate for the technical volume of a proposal, but can certainly be applied to other volumes as well.

Select proposal choreographic guidelines include the following:

- Identify the scope and context of the section or subsection to be choreographed.
- Identify appropriate sections of the RFP or RFS that correspond to that particular proposal section or subsection.
- Develop an introductory paragraph called an "elevator speech" (see Figure 7.11) that sets the tone and provides the emphasis and focus for the narrative of the section or subsection.
- Prepare major and minor theme statements for the section or subsection.
- Prepare five to eight statements that clarify and direct your writing approach for the section or subsection. These short statements are to include the major and minor points to address in the narrative. The bullet statements are to build upon and resonate with the thematic statements.
- Generate at least one graphic concept and compose a thematically meaningful caption.
- Prepare a concluding or summation sentence for the section or subsection.

The number of storyboards, bullet drafts, or scenarios to be developed for any given proposal depends upon the complexity of the RFP. Small procurement efforts, for example, might only require 10 to 15 storyboard sheets. A sample bullet draft worksheet is shown in Figure 7.12 and a representative storyboard form is presented in Figure 7.13.

---

5   Proposal choreography, that is, the manner by which a proposal response is planned, controlled, and tracked section-by-section, can be referred to by a variety of names: storyboards (a term borrowed from Hollywood moviemaking), bullet drafts, scenarios (the method of Hyman Silver and Associates), mock-ups (a process used by Shipley Associates), and scribble sheets (a method employed by TRW). Choreography will be used herein to refer interchangeably to storyboards, bullet drafts, scenarios, mock-ups, and scribble sheets.

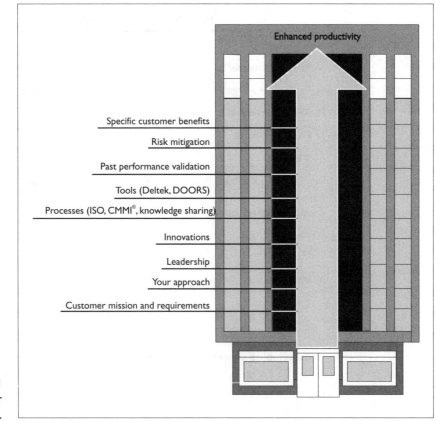

Enhanced productivity

Specific customer benefits

Risk mitigation

Past performance validation

Tools (Deltek, DOORS)

Processes (ISO, CMMI®, knowledge sharing)

Innovations

Leadership

Your approach

Customer mission and requirements

Figure 7.11
Building your
elevator speech.

 ## 7.5 **Kickoff meeting**

The formal kickoff meeting is your capture manager's and proposal manager's best opportunity to inspire and imbue their team with a winning attitude. The proposal manager can do this most effectively if he can demonstrate that he has planned carefully, organized a workable proposal response schedule, and made the best use of company resources. The attendance of your company's senior management during the kickoff meeting is highly desirable in order to ensure that the importance of the procurement to your firm's future is communicated in person.

7.6 **Writing**

The goal of each proposal author is to obtain the maximum score for each subsection. This is accomplished most effectively when writers follow the

Proprietary Bullet Draft Worksheet			
Proposal volume:	Subsection number:	Writer:	
Subsection author:		Page limit:	Due date:
RFP requirement:			
Discriminator:			
Themes:			
Features:		**Benefits:**	

Introductory paragraph or elevator speech (verbatim):

Major points (bullet format):

Concluding sentence (verbatim):

Approved by: _____   Date: _____

Figure 7.12
Sample bullet draft
worksheet.

storyboards or bullet drafts carefully. Proposal writing is not recording as much as one knows about a given technical or programmatic topic. Too often, technical staff who are assigned to write proposal sections merely provide a voluminous quantity of material that does not address the specific elements contained within the RFP, nor does it mesh effectively with the narrative and graphics prepared by the other authors. Quantity alone is totally inadequate to yield a winning response. Crisp, distilled writing that follows the RFP precisely complemented by appropriate graphics and photographs is a major step in the right direction.

One goal of preproposal intelligence gathering is to have assembled a hardcopy or CD-ROM technical library of documents relevant to the procurement at hand. Your proposal writers should consult this resource material in advance of the release of the RFP. In addition, the government

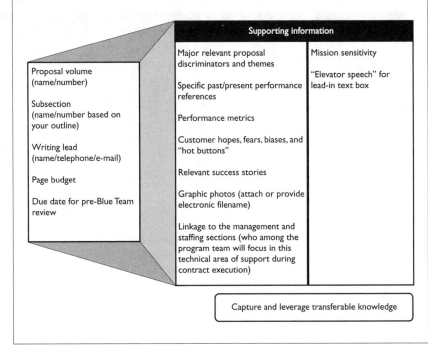

Figure 7.13
Storyboard
template.

often provides contractors with access to a technical library of project documents once the RFP is released. These resources can provide the basis for the detailed, client- and program-specific narrative that is a requirement for successful proposals. Papers and technical documents will help illuminate the types of technical issues of interest and concern to your client. You will want to demonstrate an understanding of these issues.

Writers should also keep in mind that in the case of multiyear projects, they must prepare a response that factors in evolutionary program growth and change, as well as *technology insertion*, the addition of advanced technological modules or elements into an existing infrastructure.

Metrics—explicitly *quantitative* measures of success—are an important means of validating the themes and the storylines of your proposals. As illustrated in Figure 7.14, metrics may encompass both *technical* (e.g., business system availability) and *programmatic* (e.g., schedule compression by *X* days) parameters. These metrics should be integrated into the text and graphics of your proposals, as appropriate.

Oftentimes, proposal writers need additional insights into the customer's mission, vision, leadership, strategic plan, and programmatic organization. This is where the Web is so valuable in today's proposal process. Each federal government organization maintains an extensive repository of such information at its agency Web site (see Figure 7.15), and

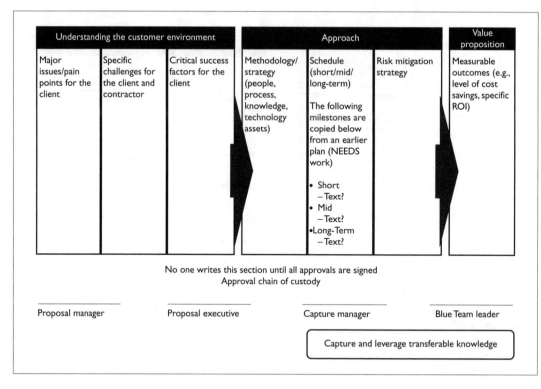

No one writes this section until all approvals are signed
Approval chain of custody

Proposal manager   Proposal executive   Capture manager   Blue Team leader

Capture and leverage transferable knowledge

Figure 7.13   (Continued.)

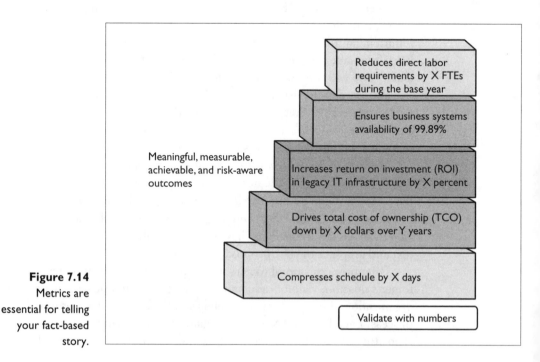

Figure 7.14
Metrics are
essential for telling
your fact-based
story.

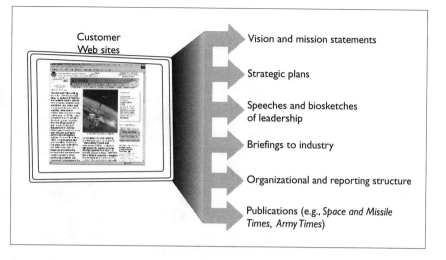

Customer
Web sites

Vision and mission statements

Strategic plans

Speeches and biosketches
of leadership

Briefings to industry

Organizational and reporting structure

Publications (e.g., *Space and Missile
Times, Army Times*)

Figure 7.15
Web-based
research is valuable
for proposal
success.

many agencies publish and make public important and insightful publications, such as the *Space and Missile Times* (http://www.vandenberg.af.mil/
space-missile-times/story_4/story4.htm) at Vandenberg Air Force Base in
California, the Navy's *Mayport Mirror* (Mayport, Florida; http://www.
mayportmirror.com), and the *Army Times* (http://www.armytimes.com/).

What can you learn and use from such Web-based information? In particular, you can begin to see patterns in your customer's *presentation of
itself* and its actions—patterns that include repeated keywords, images,
icons, color schemes, phrases, and priorities. Then, as appropriate, based
on specific marketing insight into this customer, these patterns can be
driven into your proposal across several key areas such as the Executive
Summary and oral presentation, as shown in Figure 7.16. *Your goal?* To
let your customers see themselves in your proposal. When an evaluator
reviews your submittal, she should be able to look at it and say, "Yes, I can
see my program in here. And this company really understands the issues
that are important to me and offers meaningful solutions in words and
graphics that I recognize."

To continue, the Web offers additional important resources for
proposal writers. Frequently, writers need some help getting started with
their sections that goes above and beyond the "elevator speech," annotated
outline, and populated "pain table." Figure 7.17 offers a number of
extremely beneficial Web sites that can be leveraged to provide key
jumping-off points for initial paragraphs as well as insight into the most
recent industry best practices. I find the Gartner Group site of particular
benefit. The white papers or position papers offered at that site for a
modest charge are invaluable as sources from which to glean sound ideas
for writing.

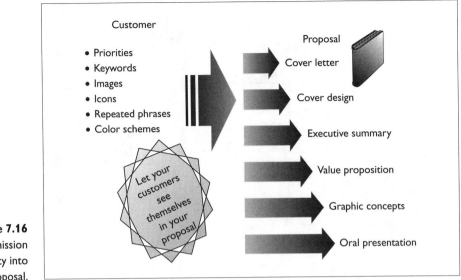

Figure 7.16
Driving mission sensitivity into your proposal.

7.7 Major contractor review cycles

All proposals your company produces should be subject to careful internal review before delivery to your client. Senior management endorsement of and participation in the review cycles is strongly encouraged. The purpose of the color team reviews is to assess the responsiveness of the proposal to the given RFP or RFS; to ensure that your company's themes are apparent throughout the entire proposal; to identify any statements that could be misrepresentative of your company or of the potential client; and to provide concrete suggestions for improvement in such areas as structure, strategy, technical approach, risk management, and pricing. Reviews are not exercises in fault-finding, but rather are the objective, solutions-oriented critique of vital company documents. Figure 7.18 shows a flow diagram of the internal review process. Note the close link between technical/management and cost volume reviews. Additional insight into the goals of internal proposal meetings and reviews is presented in Figure 7.19.

7.7.1 Blue or Pink Team

The Blue or Pink Team evaluations are generally early course correction reviews. The fundamental technical architecture and programmatic direction of your company's proposal should be evaluated at this stage, deficiencies identified and articulated, and corrective actions and specific

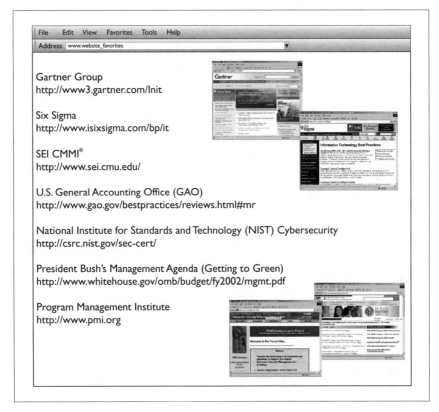

Figure 7.17
Web sites you
should use and
bookmark.

solutions offered. Avoid focusing on formatting, aesthetic, editorial, or page-count aspects of the proposal documents at this early stage.

If possible, electronic configuration control of the proposal documents should remain in the hands of the proposal technical writing team up through the response to and integration of Pink or Blue Team comments. An interactive meeting should take place between the reviewers and the proposal writing team once the review is complete in order to clearly communicate the strengths and weaknesses of the proposal effectively. The focus is on substantive corrective solutions. Such comments as "beef up this section," "weak," "needs work," or "rewrite this part" are not helpful. Comments beneficial to the proposal writers include "Enhance Section 4.6 by discussing specific automated cost and schedule control applications, particularly as they are compatible with …" or "Augment your discussion of technical and contractual interfaces with the Navy by including a graphic that clearly indicates points of contact. A rough sketch of this graphic is attached." Reviewers should attempt to provide specific guidance, direction, and solutions, as well as encouragement.

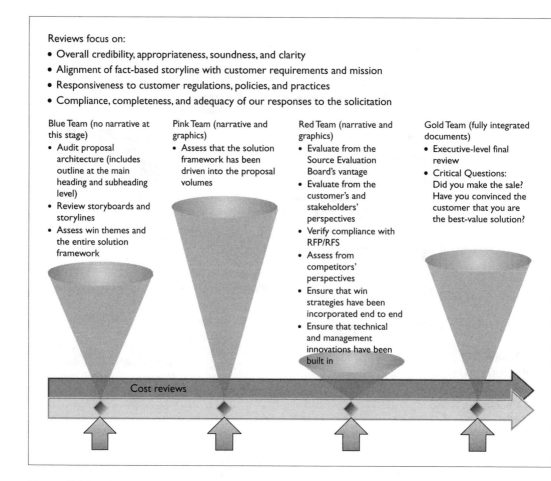

Reviews focus on:

- Overall credibility, appropriateness, soundness, and clarity
- Alignment of fact-based storyline with customer requirements and mission
- Responsiveness to customer regulations, policies, and practices
- Compliance, completeness, and adequacy of our responses to the solicitation

Blue Team (no narrative at this stage)
- Audit proposal architecture (includes outline at the main heading and subheading level)
- Review storyboards and storylines
- Assess win themes and the entire solution framework

Pink Team (narrative and graphics)
- Assess that the solution framework has been driven into the proposal volumes

Red Team (narrative and graphics)
- Evaluate from the Source Evaluation Board's vantage
- Evaluate from the customer's and stakeholders' perspectives
- Verify compliance with RFP/RFS
- Assess from competitors' perspectives
- Ensure that win strategies have been incorporated end to end
- Ensure that technical and management innovations have been built in

Gold Team (fully integrated documents)
- Executive-level final review
- Critical Questions: Did you make the sale? Have you convinced the customer that you are the best-value solution?

Cost reviews

Figure 7.18   Structured review process results in excellent end products.

7.7.2 **Red Team**

The Red Team functions as an internal SEB, critiquing the proposal for compliance with RFP instructions and evaluation criteria (as found in Sections L and M, respectively, of the RFP). The Red Team looks for consistency and continuity among sections and volumes (intervolume compatibility), inclusion of sales messages and win strategies, presence and substantiation of themes, clarity of text and artwork, and overall believability and persuasiveness. The proposal manager will determine how to respond to the Red Team comments, particularly when there is disagreement among reviewers. Figure 7.20 presents a sample proposal reviewer's comment sheet.

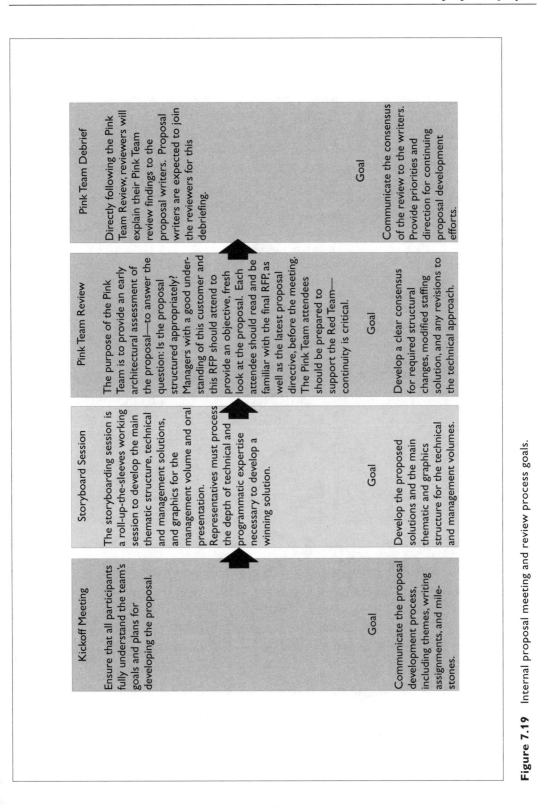

Figure 7.19 Internal proposal meeting and review process goals.

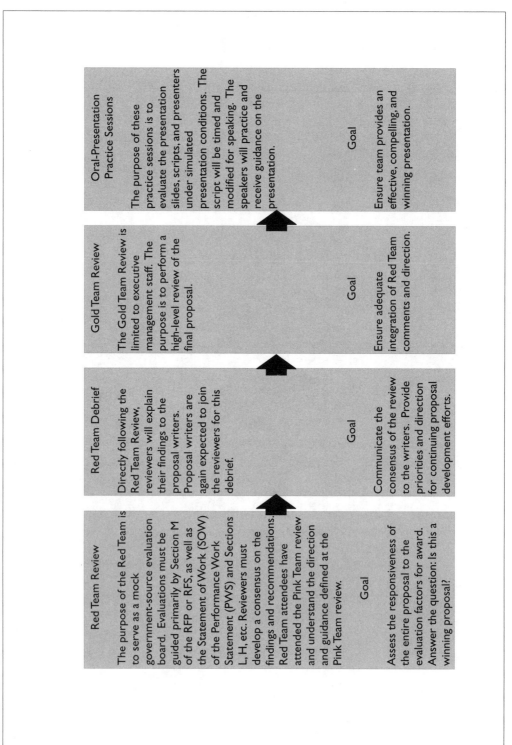

Red Team Review

The purpose of the Red Team is to serve as a mock government-source evaluation board. Evaluations must be guided primarily by Section M of the RFP or RFS, as well as the Statement of Work (SOW) of the Performance Work Statement (PWS) and Sections L, H, etc. Reviewers must develop a consensus on the findings and recommendations. Red Team attendees have attended the Pink Team review and understand the direction and guidance defined at the Pink Team review.

Goal

Assess the responsiveness of the entire proposal to the evaluation factors for award. Answer the question: Is this a winning proposal?

Red Team Debrief

Directly following the Red Team Review, reviewers will explain their findings to the proposal writers. Proposal writers are again expected to join the reviewers for this debrief.

Goal

Communicate the consensus of the review to the writers. Provide priorities and direction for continuing proposal development efforts.

Gold Team Review

The Gold Team Review is limited to executive management staff. The purpose is to perform a high-level review of the final proposal.

Goal

Ensure adequate integration of Red Team comments and direction.

Oral-Presentation Practice Sessions

The purpose of these practice sessions is to evaluate the presentation slides, scripts, and presenters under simulated presentation conditions. The script will be timed and modified for speaking. The speakers will practice and receive guidance on the presentation.

Goal

Ensure team provides an effective, compelling, and winning presentation.

Figure 7.19 (Continued.)

Red Team Evaluation Report

Proposal Title _____   Reviewer _____

Volume No./Title _____   Date _____

Section or Block No./Title _____

Criteria	Evaluation				
	Good	Average	Inadequate	Not Applicable	Cannot Judge
1. Overall impact (scoring potential)	❑	❑	❑	❑	❑
2. Readability and understandability	❑	❑	❑	❑	❑
3. Responsiveness to RFP requirements	❑	❑	❑	❑	❑
4. Responsiveness to evaluation criteria	❑	❑	❑	❑	❑
5. Theme inclusion	❑	❑	❑	❑	❑
6. Presentation format (including graphics)	❑	❑	❑	❑	❑
7. Technical content quality	❑	❑	❑	❑	❑
8. Believability and persuasiveness	❑	❑	❑	❑	❑

Specific suggestions for improvement:

Figure 7.20   Sample proposal reviewer's comment sheet.

The Red Team serves as a recommending group. Logistically, the Red Team reviewers should be fully dedicated to the effort, preferably colocated and isolated from the day-to-day demands of business during the course of their review. Red Team reviewers should include a number of people with appropriate technical competence and persons with high degrees of marketing and management competence. No reviewer should have participated in the proposal writing effort. Red Teaming is not merely an editorial exercise. Make available a copy of the full RFP to Red Team members, along with any amendments and a full kickoff meeting package (proposal directive) that includes your major and minor themes and completed storyboards or bullet drafts. The proposal manager should provide written instructions to the Red Team reviewers explaining what the expectations of the review process are and asking that comments be detailed and explicit so that the proposal team can respond effectively to those comments. The Red Team proposal draft should include a table of contents, compliance matrix, and list of acronyms.

The Red Team review is designed to assess the following:

- Overall impact and competitiveness of proposal;
- Readability and understandability;
- Compliance with RFP or RFS requirements—both overt and hidden;
- Inclusion of major and minor themes;
- Effectiveness of the presentation, including graphics and photographs;
- Quality of technical and programmatic content;
- Believability and persuasiveness of the story (are the specific benefits to the client very apparent?).

Because compliance with the RFP or RFS is *necessary but not sufficient* for proposal success, it is important that your Red Team review process take several additional steps beyond compliance verification. As portrayed in Figure 7.21, assign specific Red Team members to also assess whether your win strategies have been woven throughout your proposal, or do they appear on page 2 and not again until the end of the document? Another Red Team member could validate that your company's proposed innovations and forward-looking solutions indeed appear throughout the sections of the proposal. As an insightful DOE contracting officer once told me and my company's capture and proposal team at the postproposal debriefing, "Compliance + validated storyline + comprehensible presentation = high win probability." We had attended this particular debriefing following a $425 million contract award!

The recommendations and comments that emerge from the Red Team review should be presented to the proposal writing team in a briefing and weighed and integrated (as appropriate) into the proposal volumes under the direction of the proposal manager. Specific people should be assigned to completing all action items identified as essential.

Red Team reviews should be scheduled early enough in the proposal response cycle to allow sufficient time for the proposal manager and proposal writers to implement an adequate recovery. Typically, on a 30-day response cycle, this is 15 days prior to the final production and delivery of the proposal documents.

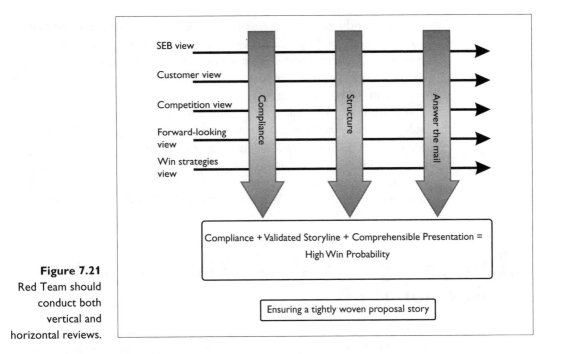

Figure 7.21
Red Team should conduct both vertical and horizontal reviews.

7.7.3 **Gold Team**

In some companies, the Gold Team is a final review of the proposal volumes by members of senior management. Other firms use the Gold Team as a final proofreading and quality assurance stage. Figure references, pagination, and table of contents should be verified at this point. Participants include your proposal manager and your senior executives.

7.7.4 **Black Team**

Black teaming is a low-profile, highly confidential activity designed to assess the strengths, weaknesses, and current business activities of the competition. A document known as a competitor assessment might be generated from black teaming. Black teams generally report to a member of senior management of the prime contractor. Black teams may include an outside legal firm or marketing intelligence consulting group who can investigate the competition without involving your company directly.

7.7.5 **Black hat review**

Black hat reviews of your proposal strategy and approach can be conducted by leveraging the knowledge and experience of recent hires within your company who came from your competitors. Check your résumé database for the names of these people. Then interview them regarding your competitors' (their former employers') business approaches and cultural biases. Document this information for future use. Reviewers attempt to identify gaps in your proposed solution set so that it can be strengthened with respect to the competition.

Black hat reviews can also be conducted successfully by outside consulting firms or individuals. These outside resources "pretend" that they are your competitors ("the bad guys" and hence the name "Black Hat") and look for ways to defeat your proposal. In effect, they predict how your competitors can win as a way to test and improve your win strategies. An appropriate time for the black hat review is after the Pink Team process but before the Red Team process.

7.8 Preparing for orals and Final Proposal Revision (FPR)

Proposal team activity does not stop following submittal of the volumes to the client. Once a company's proposal is determined to be within the competitive range based upon the evaluation factors, the client may elect to have the company deliver an oral presentation to the evaluation panel or to submit a Final Proposal Revision (FPR), or both. The FPR is, in effect, the opportunity to enhance your proposal. Generating an effective presentation and planning and rehearsing for orals are very important and require significant time and focus. Many times, FPRs involve providing the client organization with a revised cost volume that reflects the lowest possible cost for which your company can successfully perform the statement of work.

Increasingly important in the federal procurement process, oral presentations provide an excellent opportunity to reaffirm your company's distinctive capabilities, innovative technical solutions and their benefits, and risk-aware program management approach. Capabilities include qualified personnel (proposed project manager and key staff), phase-in experience, cost-effectiveness of your approach, and records of outstanding contractual experience, among others. The oral presentation cannot exceed the government-assigned time limit, which can range from 15 minutes to several hours. Multiple rehearsals or dry runs are crucial to help your company's staff gain confidence as presenters and to acquaint them with public speaking, tight time limitations for presentation, and smooth computer and audiovisual equipment operation. Some government agencies and private-sector clients are now requiring multimedia presentations, replete with color, sound, and highly professional imagery. Clients want to meet and interact with the proposed project team face-to-face. They want to be comfortable with the people they will work with during the contract life cycle.

7.9 Debriefings (refer to FAR 15.1003)

Contractors that have submitted proposals to the U.S. government for a particular procurement are legally entitled to debriefings. Debriefings may take the form of a face-to-face meeting, teleconference, or letter and are generally attended by representatives of the government's contracting office, project technical staff, and financial and legal offices. Request a debriefing in writing through the client's contracts offices. Lessons and information that your company can learn from a debriefing experience include the following:

- Breakdown of your technical score by proposal section (this is provided in very general terms);
- Process by which the scores were derived by the government evaluators;
- Strongest and weakest points of your proposal;

- How best to proceed with the next proposal to the same procuring agency; What specifically is important to that agency, for example, a reluctance to accept complex teaming arrangements, close and effective control of subcontractors and consultants that may be proposed, an emphasis upon a particular project management configuration, and incumbent capture strategy;
- Advisability of making further offers to the particular client.

Do not expect to have your proposal compared with your competitors' proposals during a formal debriefing process. Competitor proposal performance and cost strategies are strictly off limits. You may, of course, request a copy of the winning proposal volumes under the FOIA. Company-sensitive trade secrets and commercial or financial information contained in the competitor's proposal will not be released indiscriminately by the government. Instead, the proposal documents will undergo review both by the competitor (preliminary) and the government to ensure that only information in accordance with the FOIA is released.

We are now ready to take a closer look at the major parts of a proposal submitted in response to a formal RFP or RFS.

ENDNOTES

1. Holtz, Herman, and Terry Schmidt, *The Winning Proposal: How to Write It*, New York: McGraw-Hill, 1981.

2. The federal IFB, or sealed bid procedure, should result in a firm-fixed-price contract. This procedure is especially well-suited for government purchases of COTS items. Bids are opened publicly and the award is made immediately to the lowest priced, responsible, responsive bidder. No discussions are allowed, and no exceptions may be taken by the contractor.

   The federal government has been moving in the direction of procuring dual-use, COTS, and nondevelopmental items (NDI). An example of this procurement trajectory is in the purchase of intelligence and electronic warfare (IEW) equipment for the U.S. Army. (See "DOD's Airborne Recon Plan Pushes Sensors, Image Recognition," *Sensor Business News*, May 8, 1996.) In FY1997, DoD seeks a $250M *Dual-Use Applications Program* (DUAP). With COTS/dual-use embedded in DoD acquisition, MIL SPECS no longer provide giant defense firms with an artificial edge over agile new defense entrants into the marketplace. (See "DOD Acquisition Chief Pushes COTS/Dual-Use, Wants More Suppliers," *Technology Transfer Week*, Vol. 30, April 1996.)

   The concept of responsiveness was developed in sealed bidding, wherein a Contracting Officer is prohibited from considering any bid that deviates from the IFB. This prohibition does not apply to negotiated procurements. (See Cibinic, John, and Ralph C. Nash, *Formation of Government Contracts*, 2nd ed., Washington, D.C.: George Washington University, 1986, pp. 523–524.)

   An RFP, which results in competitive proposals for competitive acquisitions under P.L. 98-369 (Competition in Contracting Act of 1984) is a more flexible approach used in sole source or competitive situations. Any contract type from fixed-price to cost-reimbursement can be used. The government opens and reviews proposals privately. The number and identity of offerors are not revealed (immediately), and award may be based upon considerations other than price. Considerable discussion between the government and the offeror may or may not take place. The offeror may take exception to the solicitation terms or may submit alternate proposals. And modifications to proposals by the offerors are permitted. Upon acceptance, the proposal or offer becomes a contract in some government procurements wherein the technical and contractual requirements are tightly specified.

Chapter 8

Major proposal components

8.1 Overview

Your company's response to a competitive RFP or RFS will generally consist of a set of volumes: technical, management, and cost. These volumes may be called different names in different RFPs; technical approach, business management, and cost or price are several variations. In addition, there will probably be a transmittal, or cover, letter; an executive summary, which is often part of the technical volume but may be a stand-alone volume and should also appear in the cost volume; and various supporting volumes such as quality assurance plan, configuration management plan, and health and safety plan. Sometimes, the technical and management volumes are contained in one single document. A U.S. government RFP may require that SF 33s, "Solicitation Offer and Award," and SF 1411s, "Contract Pricing Cover Sheet," be submitted as separate volumes. Technical volume page counts for federal proposals can range from 10 to 15 sheets to hundreds or even thousands of pages. And certain major aerospace and defense proposals can contain more than 70 individual volumes, considering executive summary, technical volume, management volume, cost volume, make or buy policy, quality assurance plan, recruitment plan, health and safety plan, subcontractor plan, and so forth. The trend, however, even with billion-dollar procurements, is to require proposals on the order of 100 to 200. Focused, highly refined text and graphics are required to convey your messages, tell a fact-based, compelling story, and meet all of the RFP requirements in a concise manner.

8.2 Transmittal letter

The transmittal, or cover, letter should do more than simply transmit your proposal documents to the government or private-sector client. (Note, however, that there are occasions wherein a client will strictly preclude any letter other than a brief transmittal. The U.S. Coast Guard Operations Support Center is a case in point.) The letter should be viewed as one more opportunity to call your client's attention to the advantages to be gained from doing business with your company. This letter should be prepared on your company's letterhead with the appropriate office location address and be signed by your company's president or executive vice president. It should emphasize your president's personal commitment to the successful completion of the particular project or timely delivery of the specified product or service. It is advisable to attach a copy of the transmittal letter with each volume of the proposal because the volumes are often distributed to a committee of client evaluators.

Since the transmittal letter requires the signature of a member of senior management, the proposal manager should plan to have the letter completed for signature several days in advance of the proposal due date. The fluctuating schedule of business executives could result in their being out of the office at the very time their signature is required on the letter and cost volume forms.

Definitely avoid the following hackneyed wording in your transmittal letters:

We are pleased to submit this proposal in response to Air Force Solicitation No. FA7012-04-T-0085.

Be creative and inventive, yet to the point and in accord with any specific RFP instructions. Briefly convey your company's enthusiasm to contribute to your client's technical, programmatic, and budgetary success on this important project. Consider the following example:

When DOT partners with our company, you are harnessing the talents and focused energies of business people who are also information technology specialists. Four years of outstanding IT service provide the technical matrix in which fiscally prudent and market-driven business decisions are made every day. We very much appreciate that you and other DOT staff members have invested your time and effort in reviewing our submittal. From the vantage of our knowledge and experience base, we certainly see the challenges—and the genuine mutual successes—that are possible under this important program.

8.3 Technical volume

From a client evaluation perspective, the technical volume of your proposal is frequently weighted most heavily. This means that this volume is absolutely crucial to the success of any proposal effort. Sometimes, however, contractor technical staff and senior management place too much attention on responding to this volume, to the detriment of the remainder of the effort. Proposals are not only technical approaches and solutions, though to be sure, these items are critical. As noted previously, proposals are fundamentally sales documents—and every point counts.

The technical volume is the volume most often storyboarded, that is, broken down into logical subparts and "choreographed" to encourage written responses from the proposal team that are consistent thematically and compliant with the RFP requirements (see Figure 8.1). From a big picture vantage, the technical volume presents the following:

- What your company is selling;
- What the specific benefits to the client are;
- Why you are selling it;
- How your company will perform its function(s);
- How it is better than the competition.

8.3.1 Front cover

Your company's image,[1] or collective identity, should be conveyed in the cover design of your proposal documents, provided that the RFP does not stipulate otherwise. The cover should attempt to (1) attract positive

---

[1] Company image should be conveyed uniformly throughout all of your externally distributed materials, including business cards, letterhead, envelopes, marketing brochures, annual reports, marketing fliers, advertising, trade show booth, displays, and signage.

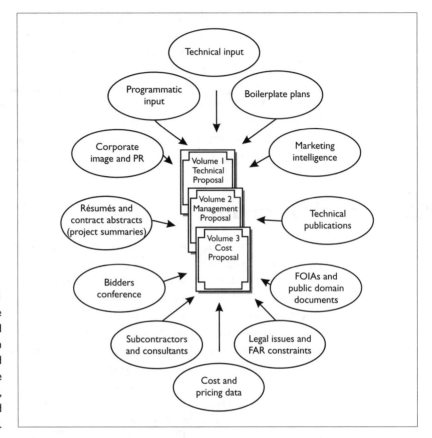

Figure 8.1
Proposals are choreographed exercises in information and knowledge management, assembly, and packaging.

attention with an appropriate high-impact presentation, (2) identify the RFP and the offeror (your company) in words and images, and (3) convey the most significant benefits or themes of your proposal. Almost every RFP states somewhere in the preparation instructions that attractive, colorful covers, diagrams, and such are uncalled for and, if used, show that the offeror is unresponsive and not very cost-conscious. Real-world surveys indicate, however, that attractive covers are noted positively by government evaluators. They are, after all, human beings who respond more favorably to attractive images than bland cover stock. In addition, some clients will assess your proposal as indicative of the caliber of deliverable your company would produce during the contract life cycle.

Your company's logo (and those of your teaming partners, if appropriate), photographs and artwork, and meaningful text can be combined to create a stunning cover design using CorelDRAW! or Adobe Illustrator or Photoshop, for example. Numerous companies offer stock photographs as electronic images (e.g., .jpeg, .gif, .tif, and .bmp formats) that are legally usable in your proposals from a copyright and intellectual property standpoint. To avoid unwanted and potentially costly litigation, never use photographs or artwork scanned from published books or magazines without the written consent of the photographer, artist, or publisher. Modifying a photograph is insufficient to reproduce it without permission. Useful repositories of public domain photographs include, http://www.firstgov. gov/Topics/Graphics.shtml (case sensitive), http://af_mil/photos, http:// nix.nasa.gov, and http://www.photolib.noaa.gov. Other Web-based sources of photographic images and clipart include Eyewire (http://www. eyewire.com/products/photo/photodisc/new.html); Ablestock, which offers high-resolution, royalty-free images (http://www.ablestock.com); and Jupiterimages, Inc.'s http://www.clipart.com, http://www.comstock. com/eb/default.asp and http://www.photos.com/en/index.You may require the support of an outside printing vendor to generate multiple copies of two- or four-color covers. If you elect to use three-ring notebooks to package your proposal volumes, be sure to have spines designed and produced, too. No back cover design is usually necessary, although some companies elect to prepare them to give their proposals a finished book-like appearance.

8.3.2 Nondisclosure statement on the title page

The nondisclosure statement should appear on the title page of each proposal volume. Designed to protect your company's proprietary information from illegal disclosure (see Figure 8.2), it serves notice that the content of your proposal documents is sensitive. Reference to the nondisclosure statement should be stamped or printed onto each page of each volume (e.g., Use or disclosure of data on this page is subject to the restriction on the title page of this document). The nondisclosure statement will not, however, prevent the government from releasing information from your

> The information contained in this proposal shall not be disclosed outside the government (or client organization, as appropriate) and shall not be duplicated, used or disclosed, in whole or in part, for any purpose other than to evaluate this proposal. If a contract is awarded to this offeror as a result of or in connection with the submission of the data herein, the government shall have the right to duplicate, use, or disclose the data to the extent provided by the contract. This restriction does not limit the government's right to use information contained in the data if it is obtained from another source without restriction. The data subject to this restriction comprise the whole of this proposal.

Figure 8.2
Sample nondisclosure statement for the title page.

submitted proposal in accordance with the FOIA. Sometimes an RFP or RFS will stipulate the exact verbiage for the nondisclosure statement.

Similarly, for commercial proposals, the nondisclosure statement identifies the document as containing proprietary information not to be disclosed outside of the client organization, unless authorized by the proposer. This obtains less protection than is desirable, but it at least puts the reader on notice.

8.3.3 Executive summary

The executive summary should be written so that even an uninformed reader will absorb and understand the principal ideas and major storylines expressed in each of your proposal volumes. The executive summary, which in many proposals is only one to two pages in length, should be structured to do the following:

- Introduce high-level themes that will reappear throughout the proposal volumes (technical, management, and cost).

- Enumerate your company's applicable credentials and applicable contractual track record.

- Demonstrate that your company understands the mission-level importance, as well as the technical requirements, of the project or product deliverable.

- Indicate that your company can perform the project successfully or provide the product or service on schedule and within budget.

- Provide an overview of your technical and management approaches/solutions and their benefits for this project in accordance with Section M of the RFP.

- Highlight your key project staff for service contracts in particular (by name if your client knows them).

- Focus on the primary reasons why the client should select your company as its partner on this important contract. How will the client's selection of your firm benefit them from a technical, programmatic, risk mitigation, and cost perspective? What specifically makes your company the best-value solution for this client?

- Demonstrate your local, established presence in the applicable geographic area that ensures your preparedness to meet client requirements in a timely manner and comprehend the local political and regulatory decision-making environment.

- Convey your firm's successful consulting work and your proven ability to manage multiple tasks simultaneously.

- Highlight your successful performance history with the client. Provide one or two success stories, as well as key lessons learned and best practices.

- Point to your solid knowledge base in applicable technical areas that will provide value-added technical support and direct benefit to the client.

- Outline your highly effective management strategies that ensure schedule adherence, cost control, and superior-quality deliverables. Introduce your project management leadership to the client, and demonstrate their pivotal programmatic, technical, and teamwork strengths.

- Bring to life your company's corporate commitment to partnership with the client to support its mission.

The benefits of your company must be sold right here.

Use photographs of related projects, equipment, and systems and installations to help authenticate your contractual experience and technical approach. A busy government decision-maker or private-sector executive should be able to scan your executive summary in about 5 minutes and understand at least the high points of what it is you are selling and why you are the company to be selected.

If at all possible and in accordance with RFP instructions, develop a full-page (8.5×11 inches) or foldout page (11×17 inches) graphic as part of the executive summary that conveys the primary benefits to the client of your company's approach. Tell your proposal story in one graphic image through the judicious use of photos, line art, clip art, and carefully selected words. If prepared well, this can be a powerful tool for capturing reviewer attention and interest in the remainder of your proposal.

8.3.4 Building a compliance (cross-reference) matrix

Compliance, cross-reference, or traceability matrices are all terms for tables that clearly map the RFP requirements (from Sections C, L, M, and H as well as the CDRL, DIDs, and RFP attachments as appropriate) to the specific sections within your proposal in which your company responded to a given RFP requirement. These matrices often appear in the front matter of your proposal volumes but may be placed at the very end of each proposal volume as an 11×17–inch foldout page that will allow the government evaluators to be able to refer constantly and easily to the matrix. Table 8.1(a–c) illustrates sample compliance matrices. Table 8.1(a) is a common, straightforward approach. Its format was dictated by the RFP. Table 8.1(b) includes reference to both text sections or paragraphs and figures contained in the proposal response. Table 8.1(c) presents a detailed mapping of elements from Sections C, L, and M of the RFP to proposal response sections and subsections. Note that in certain cases, the response to specific RFP elements is found in more than one proposal section or subsection. The number of columns in a compliance matrix varies depending upon whether you elect to delineate your responses by text sections and graphics. You might also consider adding a comments column in which important themes can be introduced, as shown in Table 8.1(c).

8.3.5 Narrative body of the technical volume

Translate your company's features into benefits to your client.

Table 8.1 Sample Compliance Matrices: (a) Common, Straightforward Approach; (b) With Reference to Text Sections, Paragraphs, and Figures Contained in the Proposal Response; and (c) Detailed Mapping of Elements from Section C, L, and M, of the RFP to Proposal Response Sections and Subsections

SOW Element	Proposal Section
1	2.1.1.a.
2	2.1.1.b.
3	2.1.1.c.
4	2.1.1.d.
5	2.1.1.e.
6	2.1.1.f.
7	2.1.1.g.
8	2.1.1.h.
9	2.1.1.i.
10	2.1.1.j.
11	2.1.1.k.
12	2.1.1.l.
13	2.1.1.m.
14	2.1.1.n.

(a)

RFP Number		Proposal	
		Part	**Part**
	Section/Paragraph	**Figure**	**Section/Paragraph**
L.2.A.5	Separation of technical and business proposals		Part I, Part II
L.2.B.	Technical proposal instructions	Part I	Part I
L.2.B.1.a(1)	Objectives		I, 2 all
L.2.B.1.a(2)	Approach		I, 3 all
L.2.B.1.a(3)	Methods		I, 5 all
L.2.B.1.b(1)	Experience	6-1, 6-3	I, 6.4; Appendix C
L.2.B.1.b(1)(a)	Pertinent contracts	6-1	I, 1 all; I, 6 all; and I, Foreword supporting materials (attached)
L.2.B.1.b(1)(b)	Principal investigator	6-4	I, Foreword; I, 3 all; I, 4 all
L.2.B.1.b(2)	Personnel	6-3	I, 6.4; I, Appendix C
L.2.B.4	Other considerations		I, Foreword; I, 3 all; I, 4 all
L.2.B.5	Special instructions		I, 5
L.2.C.1	Cost and pricing data	Part II	Part II
L.2.C.1.a(1)	Materials		II, SF 1411, Attachment
L.2.C.1.a(2)	Direct labor		II, SF 1411, Attachment
L.2.C.1.a(3)	Indirect cost		II, SF 1411, Attachment
L.2.C.1.a(4)	Special equipment		II, SF 1411, Attachment
L.2.C.1.a(5)	Travel		II, SF 1411, Attachment
L.2.C.1.a(6)	Other costs		II, SF 1411, Attachment
L.2.C.1.b(1)	Judgmental factors		II, SF 1411, Attachment
L.2.C.1.b(2)	Contingencies		None
L.2.C.2	Total plan compensation		II 4, all

(b)

Table 8.1 (Continued)

RFP Element			
Attachment A (SOW)	Proposal Section	Comments	
SOW 1.0	B.1.1	The team has the specific experience, capabilities, and knowledge to support all of the requirements of the SOW.	
SOW 2.0	B.1.2	Our staff, including our proposed senior scientist, have years of prior involvement with this project. We understand the mission objectives as well as the system requirements.	
SOW 3.0	B.1.3	Our knowledge coupled with our extensive experience supporting scientific projects provides us with an understanding of the critical project elements.	
SOW 4.0	B.1.4	The team has taken an integrated approach to all functions of the development effort.	

Table 8.1
(Continued)

(c)

If Section L of the RFP allows, each section (and major subsection) should be written to include an introduction, body, and summation. The body, or main narrative portion, might include a discussion of (1) your company's understanding of the client's requirements, including how they might evolve over the life cycle of the contract; (2) critical issues associated with the requirements or task performance; (3) your company's innovative approach to addressing the requirements, including lessons learned and success stories from similar contractual experience; and (4) success criteria for the task or effort. Strive to prepare a solution-oriented proposal—one that offers the least technical, schedule, cost, and political risks to your client. And weave your themes throughout the narrative sections of your proposal. Avoid phraseology like "Our major theme for this section is …" That is not effective proposal writing. I worked with one enthusiastic proposal manager who, upon learning about themes in my proposal training seminar, included major themes as a heading in his proposal. We had some restructuring to do!

Table 8.2 is a sample structure for a section of a technical volume in a task-order procurement. In responding to a particular RFP, you should structure your outlines and proposal volumes in accordance with the

Body	3.3	Technical task #3
	3.3.1	Introduction
	3.3.2	Understanding the EPA's requirements for Task #3 (include how requirements may evolve over the contract life cycle)
	3.3.3	Critical issues associated with task performance
	3.3.4	Your company's (or team's) innovative technical approach
	3.3.5	Success criteria for Task #3
	3.3.6	Summation

Table 8.2
Sample Technical
Volume Outline

numbering scheme and conceptual scenario provided in Section L. And use the exact wording from Section C as outline headings and subheadings, if possible. When clients ask for ABC, give them ABC, not BCA or XYZ. In other words, focus on and respond directly to the client's requirements. Translate your company's distinctive capabilities and extensive expertise into benefits to your client. This can be done in the narrative, in features and benefits boxes (as shown in Table 8.3), and in figure captions and table legends.

8.4 Management volume

The management volume of a proposal is many times the most overlooked and poorly constructed document in the entire proposal package. Too often, the management volume sections are viewed to be mundane boilerplate, in which each government agency is interchangeable. The management volume is a project- and agency-specific narrative that clearly demonstrates how your company will manage staff, maintain reasonable spans of authority, control schedules and costs, assess technical performance, address technical issues, meet performance thresholds, and access resources companywide as necessary—all to the benefit of your client. In general, your management volume must describe and illustrate the following clearly:

- How your company proposes to organize and marshal its resources to perform on the project or deliver the products or services;
- The roles and span of authority of your proposed project manager;
- How your company's project staff will communicate and interface with the client on an ongoing basis over the life of the contract;

Features	Benefits
Single-vendor solution with a lean organizational structure	Straightforward lines of authority and communication with the IRS, resulting in more effective and efficient service and problem resolution
Proven approach to transitioning incumbent personnel	Reduced risk through retention of institutional IRS knowledge and smooth contract start-up within 30 days of award
Project structure tailored specifically to the IRS's unique requirements	Responsiveness, flexibility, and value of service
Project-management team highly experienced and successful in the management of eight similar task order contracts	Proactive approach to meeting IRS needs due to complete familiarity with the contract process; the process functions as a mechanism to facilitate superior service
Three years of experience in facilities management	Effective maintenance of IRS baseline, with associated cost savings estimated to be $200,000 over the life of the contract
Four years of experience in prototyping and implementing advanced application technologies	Ability to plan and execute the IRS's migration to newer technologies within the contract life cycle

Table 8.3
Features and Benefits Box

- Why you will manage the effort more efficiently and effectively than your competition;
- Your company's identity (through staff résumés and related and past contractual experience);
- That you can accomplish the project successfully, within schedule constraints and under or within budgetary margins (including automated tools and industry-standard best practices);
- The rationale for forming your particular teaming arrangement, if there are subcontractor(s) involved;
- That you have successfully managed projects of similar size, scope, complexity, and contract type, along with relevant examples of such projects.

Here is a suggested outline for the management approach:

- Introduction
- Project Organization (include functional project organizational chart populated with staff from your firm, as well as with subcontractor staff, and indicate reporting relationships).
- Project Leadership
  - Project manager's role and span of authority
  - Accessing corporate resources (discuss your company's commitment of resources to project success, and the specific mechanisms that you employ to make this happen at the level of day-to-day business)
  - Project initiation and task order kickoff strategies
- Ongoing Communication with the Client Organization
  - Client and company technical and contractual interfaces and general topics of communication and interaction
  - Reporting schedules
- Project Execution
  - Staffing Management
    - Recruiting, retention, and turnover metrics
    - Recruiting policies and practices
    - Corporate benefits package
    - Capacity to meet workload fluctuations and accelerated schedules
    - Staff training, cross-training, and technical refresher practices and associated funding
    - Security clearances [e.g., secret, top secret, sensitive compartmentalized information (SCI), Q and L (DOE)]
  - Schedule Management—effective schedule control approaches and automated tools
    - Task prioritization methodologies
    - Demonstrated history of meeting interim and final project deadlines
    - Proposed schedule of project by task (Gantt charting)
  - Task Order Management
    - Metrics-driven work planning and monitoring
    - Response to evolving project requirements
    - Problem management and escalation processes

- Risk Management—cost, schedule, and performance risk mitigation approaches
- QA-QC Management
  - Continuous process improvement (CPI)
  - Application of industry-standard structured processes and methodologies [ISO, Capability Maturity Model Integration® (CMMI®), and IEEE]
  - QA metrics tracking
  - Interim and final deliverables
- Financial Management
  - Industry-standard cost control approaches
  - Automated cost tracking and control tools (e.g., Deltek modules, Primavera)
  - Efficient and compatible invoicing and reporting mechanisms
- Subcontractor Management
  - Rationale for teaming
  - Project manager's approval of subcontractor staffing
  - Ongoing communication with project team
  - Invoice verification
  - Integration of subcontractor staff into task activities
  - Work assignments
  - Issue escalation and resolution processes

You must demonstrate convincingly how your company will perform the work, manage and control the work, manage and control changes to it, and be responsive to the client's needs. You should demonstrate a staffing and facilities capability sufficient to handle a project of the size proposed. Depending upon the requirements set forth in Section L of the RFP, the management volume might include the following sections and subsections.

Introduction   An introduction should highlight the key aspects and associated benefits of your company's management approach. These might include a project organization that accommodates evolution and growth within the contract; assignment of seasoned and qualified personnel who understand the specific needs, goals, and metrics of their particular areas of support; clear definition of roles, responsibilities, and lines of communication; and the engaged support of your senior management for the contract.

Executive summary　An executive summary is similar to that used in the technical volume.

Project organization　Project organization addresses such items as project leadership, spans of authority and control of your proposed supervisory staff, and the clear lines of communication between your project manager and his or her technical and support staff. Emphasize your project manager's access to required company resources and this particular project's visibility and importance to your senior management. Demonstrate that your project organization can accommodate evolution and growth within the contract.

Corporate organization　The corporate organization section presents an organizational chart in graphic form, highlighting how the specific project under consideration is linked managerially and structurally to the company as a whole. Demonstrate a streamlined managerial structure with direct access to senior-level support and resources companywide.

Client-contractor interfaces　A section on client-contractor interfaces presents your company's proposed working relationship with the client at technical, programmatic, and contractual levels. Answer the question: If I am the client, whom do I contact within your company for project status in terms of technical accomplishment, schedule, cost, and deliverables, and so forth? Demonstrate clearly in narrative and graphic form the lines of interaction between your company's technical, programmatic, and contractual staff and the client's counterparts[2]. Discuss the clearly defined reporting pathways within your company.

Management plan　A management plan discusses your proposed project manager's authority, roles and responsibilities, and his or her relevant project management experience. Describe how your project manager will interface with your company's senior management and key government staff to manage the project effectively and efficiently. Demonstrate through reference to lessons learned and best practices from other similar contractual experience that your management approach can accommodate evolving, fluctuating, and unforeseen workloads; multiple task assignments; requirements creep;[3] and unanticipated employee turnover. Discuss work plan-

---

2　It is very effective to develop a graphic that illustrates your company's contracts group interacting with the client's contracts office, and your proposed project manager (and task leaders or delivery order managers) interacting directly with, for example, the client's contracting officer's technical representative (COTR) and assistant technical representative (ATR).

ning and monitoring, allocation of project resources, review of project work, and regular reporting to the client. The management plan may include contract phase-in and phase-out plans.

Cost and time controls In a section for cost and time controls, describe how your company plans to maintain control over the schedule and cost once the project begins. Illustrate specific and innovative control techniques proposed, emphasizing automated daily cost tracking and accounting, scheduling, and purchasing techniques (e.g., micropurchases), if appropriate. Explain what your company does to correct cost excesses in cost reimbursable contracts. For the U.S. government, clearly indicate that your cost accounting and purchasing systems are approved by the Defense Contract Audit Agency (DCAA) and that your company complies with cost accounting standards (CAS) and FAR cost principles. If a management information system (MIS) is used to forecast resources and track task progress and performance, articulate how it will be employed on this particular contract. State that your firm has the ability to control its workload through prudent and structured bid–no bid decisions.

Scheduling Demonstrate in graphic form the major schedule milestones and contract deliverables. Discuss automated scheduling tools that your company has employed successfully on other contracts of similar size, scope, complexity, and contract type.

Staffing plan Describe staffing levels required in each labor category over the life of the contract. Discuss your company's successful recruitment policies, ongoing staff training and technical refreshment programs, and internal programs to ensure employee motivation and retention (compensation plan).

Key and other résuméd personnel Present short biosketches that highlight experience relevant to the procurement.

Résumés Present résumés in the proper format as specified by the RFP, tailored to the specific requirements of the procurement. Map or link the individual's expertise and experience to the position description found in the RFP. Be certain that the government evaluators will understand precisely the benefits that each proposed staff member will bring to the particular project.

---

3   *Requirements creep* refers to the gradual expansion of the scope and complexity of the tasks or project.

Company capabilities and experience   Describe your company's quality, reliability, and timeliness record on projects of similar scope, schedule, budget, and contract type.

Clients often use past performance as the single most important, reliable predictor of future performance.

Related experience   Bring your company's relevant contractual experience to life. Focus on project successes such as cost savings (tell how many dollars), schedule adherence (tell how much time saved), repeat business from a client, value-added service, fulfillment of performance-based metrics, and the application of innovative technologies. Highlight lessons learned from previous and existing experience that can be applied quickly and transparently to the contract under consideration. Pay particular attention to preparing and tailoring the narrative of past and related experience because of the increasing importance of this area as an evaluation factor.

Subcontracting plan   For major subcontractors, include a letter of commitment from the president of the subcontracting firm(s). This is in addition to a formal teaming agreement. Provide evidence of sufficient subcontractor control. Discuss your project manager's oversight of the subcontractor(s)' staffing and invoices; issuance of subcontractor work assignments in accordance with performance metrics; planned methods for evaluating subcontractor performance; clear definition of subcontractor tasks and responsibilities, communicating and reporting practices and procedures between your company and the subs; meeting times and schedules, and problem identification, escalation, and resolution strategies. Ensure that your company's interactions with your subcontractor team are transparent to your client. Demonstrate that you will hold your subcontractors to the same rigorous quality standards and metrics for technical performance as you do your own company.

Company facilities and special resources   Describe the mechanism in your company to make resources and facilities available as necessary to support the project successfully. Indicate any government-furnished equipment (GFE) or contractor-furnished equipment (CFE) required. Describe how your facilities and equipment are applicable to the SOW or Statement of Objectives (SOO), and how you will use them in meeting critical project requirements. If you maintain a secure facility, mention that, as appropriate.

Risk management   Based upon marketing intelligence about your client's concerns and goals, demonstrate that your company will keep the project on schedule, control costs, retain key technical staff, maintain more efficient allocation of staff resources, respond to fluctuating workloads, conduct cross training, and so forth. You must convince your client that your company will make them look good in the eyes of their superiors, including possibly the U.S. Congress. Demonstrate that you are fully capable of risk identification, analysis, and mitigation. Given the growing emphasis on performance-based contracting (PBC; defined at FAR 37.101), this is a critical element in the management proposal.

Configuration management   Depending upon the technical scope of work, and particularly for software and hardware systems design, development, and maintenance, discuss how your company will address configuration control, design, development, and implementation.

Data and information management   Depending upon the SOW, discuss how you will apply automated tools to assist in data and information collection, analysis, archival, and retrieval. Also describe your techniques for ensuring data integrity and information assurance (IA) and for controlling data access. Indicate if your management information systems are compliant with government standards.

Work breakdown structure   The work breakdown structure (WBS), which is most often included in the management volume, is an organizational resource-allocation tool that can be used to assist with project management structuring, cost estimates, and cost substantiation. (See Figure 8.3.) Sometimes referred to as a work element structure, the WBS helps to "interrelate the technical, management, and cost aspects of the work" [1]. Specifically, the WBS is developed to ensure that all required work elements are included in your company's technical approach and management plan without duplication. Lower-level tasks and subtasks that can be more easily scheduled and estimated in terms of staffing are identified and arranged in hierarchical fashion to assist in the overall accumulation (or roll up) of resource requirements. During actual project performance, the WBS serves as a tool for cost control, along with a monthly estimate-to-complete (ETC) analysis.

Technical performance measurement system   Highlight any performance metrics that your company may employ to estimate, monitor, assess, and report project performance in an effort to identify potential problems before they devolve into major issues. Discuss performance metrics that this client will apply on the contract. Metrics might include system availability, time to fill technical position openings, and response time to user help requests. NASA, the Air Force, and DOE have moved proactively toward PBC.

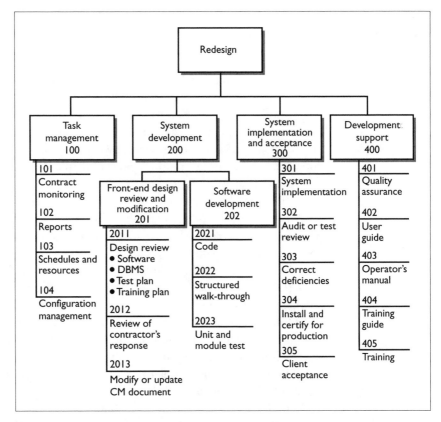

Figure 8.3
Sample WBS.

Deliverable production and quality assurance   Demonstrate through reference to other successful project execution that delivering superior-quality deliverables which have undergone internal quality review prior to submittal to the client is a top priority to your company. Discuss formalized QA-QC procedures and proactive process planning and review mechanisms that have been established within your company. For example, perhaps you have implemented a program that includes document design meetings and senior technical review (STR) for all contract deliverables. Describe this program and how it will benefit your client on the particular contract at hand. You might also discuss your deficiency tracking program and include a sample deficiency tracking form.

Summation   Reinforce your primary management theme statements in a set of strong closing paragraphs.

8.5 Cost volume

Cost is always pivotal in competitive procurements. It is, therefore, important for your company to develop an early technical design or approach upon which cost strategies and preliminary bottom-line prices (bogeys) can be built. The costing effort will also be facilitated by generating a staffing and management plan as soon as possible after RFP release. Develop your cost volume in close association with the technical and management volumes. The technical solution and staffing plan are integrally linked with the pricing strategy.

The cost volume must contain resource estimates, costing methodology, and a rationale sufficient to allow the client to establish the fact that the product or service to be provided will be cost-effective and that the cost estimates are realistic to perform the contract work successfully [1]. Resource estimates include direct costs (primarily labor[4]), other direct costs (ODCs) (for example, material, travel, computer costs, and consultants), and indirect costs [overhead and general and administrative (G&A)]. Credibility is essential in deriving the optimal balance of performance and cost. Cost volumes also include terms and conditions and responses to representations and certifications (Section K of the RFP). Provide sufficient detail to satisfy the client. Ensure that your marketing efforts have provided this kind of detailed guidance from the client.

There are automated tools available on the market to assist your company in the preparation of various costing estimates. MicroFusion Millenium v5.0 for Windows, which is produced by Integrated Management Concepts [(805) 778-1629; http://www.intgconcepts.com] in Thousand Oaks, California, is one such application. With MicroFusion, you can develop multiple rate structures, complex burden structures, and escalate program costs.

Cost proposals are analyzed by the government to assess realism and the probable cost to the government. Cost realism is the process of independently reviewing and evaluating specific elements of each offeror's cost estimate to determine if the estimated proposed cost elements are realistic for the work to be performed, reflect a clear understanding of the requirements, and are consistent with the methods of performance described in the offeror's technical proposal.

If total or cost element ceilings are included, the evaluation will also assess the maximum probable cost based upon such ceilings. The client is interested in proposed methods to control costs that do not have a negative

---

4   Because of holidays, vacation, and sick leave, the total number of labor hours in one staff year equals approximately 1,860.

impact on contract performance. Cost and pricing data must be current, accurate, and complete.

To expand further regarding cost realism, this analysis refers to the government's evaluation of the system of logic, assumptions about the future, and reasonableness of the historical basis of your cost estimate. That is, cost realism focuses on the elements that make up the foundation of the estimate.

Cost realism analysis answers questions such as the following:

- Are the assumptions used in your estimating process reasonable?
- Has the historical database used been normalized to account for environmental parameters such as inflation?
- Is the cost estimate logical? Does it make sense in the context of the hardware or software product or service being estimated?
- Does the estimate display a bias toward being too low or too high? If so, how is this bias displayed in the estimate?
- Is the cost estimating organization motivated to produce an inordinately high or low estimate in order to serve the contractor's own purposes?
- If the product is fixed-price sole source, has the historical basis data been hand-selected and therefore skewed to ensure the cost estimate obtained is unreasonably high or unreasonably low?
- If the program is competitive, has the contractor or government program office created program expectations that are far too optimistic?

A forward pricing rate agreement (FPRA), as defined in FAR 15.801, is a written agreement negotiated between a contractor and the government to ensure that certain rates and factors are available during a specified period for use in pricing contracts or contract modifications. Such rates and factors represent reasonable projections of specific costs that are not easily estimated for, identified with, or generated by a specific contract, contract end item, or task.

The trend in cost volumes is to make that document more sales-oriented—with more narrative, themes, graphics, and attractive formatting. They too can contain a copy of the executive summary.

The cost volume has a higher level of confidentiality than the other proposal volumes. Fewer copies should be made, and they should be distributed for internal company review on a need-to-know basis; however, build in time for thorough review and quality checking. Clients often require that the cost volume be submitted separately from technical and management. Many companies produce their cost volumes within the contracts or procurement departments rather than through a publications group.

8.6 Government contract requirements

The federal government imposes standards, procedures, guarantees, and documentation requirements on companies with which it does business; and it exacts significant penalties for noncompliance [2]. The Truth in Negotiations Act (TINA) was enacted in 1962 to protect the government from unscrupulous DoD contractors who supported their cost proposals with erroneous information. Congress extended the Act in 1984 to include all government contracts. On all negotiated contracts and subcontracts in excess of $500,000, your company will be required to furnish certified cost or pricing data. The government will include a "Price Reduction for Defective Cost or Pricing Data" in any resulting contract.

Small businesses must also be aware and knowledgeable about federal CAS. Created by Congress in 1970, the CAS was designed to "promote uniformity and consistency in the way defense contractors measure, assign, and allocate contract costs" [3]. The applicability of the CAS was extended in 1988 to include all negotiated government contracts in excess of $500,000. The "contract clauses involving CAS also restrict changes in company accounting practices.... Cases exist where a company has made an accounting change only to learn later that it owed the government money" [3]. As an entrepreneur, you will be well served to implement and energize a well-organized, highly trained, and watchful contract administration group.

The proposal effort is not over when the technical, management, and cost sections are written. There still remains editing, proofreading, configuration management of multiple proposal iterations, integration of iterations and review comments, word processing and publishing, graphics generation, outside printing coordination, photoreproduction, collation, assembly, letters of intent to obtain, volume coordination and consistency cross-checking, and delivery in hardcopy or appropriate electronic format.

ENDNOTES

1. Stewart, Rodney D., and Ann L. Stewart, *Proposal Preparation*, New York: John Wiley & Sons, 1984, p. 126.

2. Wall, Richard J., and Carolyn M. Jones, "Navigating the Rugged Terrain of Government Contracts," *Internal Auditor*, April 1995, p. 32. This section draws upon Mr. Wall and Ms. Jones's article.

3. Wall, Richard J., and Carolyn M. Jones, "Navigating the Rugged Terrain of Government Contracts," *Internal Auditor*, April 1995, p. 33.

Chapter 9

Acquisition/capture and proposal team activities

W HEN A COMPANY is tracking a particular marketing opportunity, advanced planning and strategizing are crucial to its ultimate success. One mechanism that companies use in this up-front marketing planning is the concept of an acquisition/capture team.

9.1 Formation and function of acquisition/capture teams

The acquisition/capture team is formed in advance of the release of the formal RFP or RFS by the client organization. Under the direction and guidance of an acquisition manager, capture manager, or campaign manager, team members (usually three to five people) participate in preproposal intelligence gathering (including FOIA requests), strategizing, briefing the potential client in a structured manner through your call plan process, interviewing incumbent staff and facilitating open houses to attract qualified staff, developing strategic proposal themes, and generating bullet drafts or storyboards to use as storyline planning and compliance-verification tools during the proposal process. Fundamentally, the capture or campaign manager "owns" the successful pursuit and award of the particular program or project. He or she must focus every day on doing what it takes to win. The essence of capture management is "the winning solution" and resource management. Specific capture or campaign manager responsibilities include the following:

- Shouldering overall responsibility for the successful pursuit of selected programs or proposal efforts for the business development group.

- Maintaining the highest level of program accountability of all capture team members. The capture manager is the holder of the vision and the winning solution for the program.

- Working closely with business development in structuring the overall approach and win strategies and managing the capture plan, which is an action-oriented, structured plan that defines how the company will win a given marketing opportunity.

- Coordinating Freedom of Information Act (FOIA) requests for past contracts, monthly technical progress reports, and so forth. Ensuring the dissemination of that information to appropriate proposal team members.

- Orchestrating teaming arrangements and nondisclosure agreements (NDAs) with subcontractors, in conjunction with the business development vice president.

- Participating in incumbent capture initiatives (open houses, job fairs, face-to-face meetings with incumbent staff).

- Developing and monitoring B&P budgets, in conjunction with the proposal manager.

- Attending the preproposal conference.

- Marshalling corporate and division-level resources to support the proposal development effort.

- Communicating with executive management regarding proposal status and resource requirements.

- Facilitating communication across the technical, programmatic, and cost elements of the proposal team, both within the company and across subcontractors.

- Leading the pricing strategy process and participating in developing the cost model.
- Participating in the development of the staffing plan and organizational structure for the proposal.
- Attending government debriefing.

The acquisition/capture team is joined by and augmented by the proposal team when the RFP is released.

Crucial to proposal success is the transfer of relevant marketing knowledge and intelligence to appropriate members of the proposal team. And from there, that knowledge must be translated and driven into the actual proposal documents—into narrative, theme statements, graphics, captions, and the compliance matrix. If your marketing intelligence does not find its way into the proposal, then it cannot be evaluated and point scored. Therefore, it is valueless.

9.2 Prekickoff activities

Time spent in effective planning and strategizing will be critical to the success of the overall proposal effort.

Once an RFP or RFS to which your company will be responding is released, the proposal manager and key members of the Proposal Team should dedicate at least one if not more days to detailed analysis of that RFP along with in-depth proposal planning. Don't panic or become overly anxious, because the time spent in effective planning and strategizing will be critical to the success of the overall proposal effort. Too many times, proposal drafts emerge from review cycles with the recommendation for fundamental restructuring of the overall outline, proposed staffing solution, related contractual experience, primary strategic themes, and so forth. Building concurrence, based on the validated marketing intelligence you have collected, early in the proposal planning process will save innumerable hours later in the proposal life cycle.

The goal of prekickoff planning is to ensure that the formal kickoff meeting is an efficient and effective forum of information exchange and assignment making. During the prekickoff planning phase of the proposal response life cycle, you should do the following.

Perform a detailed RFP analysis.

Deduce probable weighting of evaluation criteria from Section M of the RFP. The goals of RFP analysis are to (1) become familiar with the proposal structure and format (Section L), (2) understand each technical and programmatic requirement in detail, (3) understand the source evaluation criteria and client success criteria (Section M), (4) evaluate your company's pre-RFP assumptions about the procurement and the client, and (5) develop written questions for submission to the client.

Prepare a draft executive summary.

The draft executive summary will be invaluable in assisting writers to fully understand the scope of the procurement and your company's primary sales messages. The executive summary should be one to two pages in length. It should be built directly from the win strategy document.

Develop and hone your strategic win themes.

Review the proposal debriefings files, related proposals, project descriptions, and résumés in your company's proposal data center or library (knowledge base).

Learn what you did right on previous proposals, and where your company missed the mark.

Plan to get the right people to write proposal sections at the right time; obtain senior management's support to make them available.

Prepare the Proposal Directive or kickoff package.

Appendix A presents a modified example of an actual completed package. The outline for the proposal does not follow the standard A, B, C format; rather it was mapped directly to the numbering scheme in the client's RFP. In this particular case, technical and financial information was included in one volume.

When the Proposal Directive has been prepared, it is then appropriate to convene the formal kickoff meeting.

Another example of a technical volume outline is presented in Table 9.1. Again, this outline was built using the numbering scheme presented in the client's RFP. This type of outlining configuration effectively serves multiple purposes and is highly recommended. First, it assists in the generation of a cross-reference or compliance matrix for the proposal volume. Second, it links themes and critical issues with specific proposal sections and subsections so that your company's proposal writers know what to infuse into their narrative. Clearly defined points of contact for

Technical Volume Outline Procurement Name						
Company Outline Number	**RFP Section**	**Technical Item**	**Person Responsible**	**Page Allocation**	**Themes/ Critical Issues**	**Status**
—	—	Executive summary	John Smith	2		Closed
1.	L.5.1	Mission suitability factors	Jane Jones	0.5 (Intro)		Open
A.	L.5.1.A	Key personnel	James Goodfellow	0.5 (Intro)		Open
A.1		Project factors	Sally Bates	1.5		Open
A.2		Deputy project manager	John Smith	1.5		Open
B.1	L.5.1.B.1	Overall understanding	James Goodfellow	0.5		Open
B.1.1	SOW 1	Data center operations	Sally Bates	3		Open

Table 9.1 Effective Outline Configuration Tracks Status and Ensures Compliance

each section and subsection are noted, and the status of each can be tracked. This overview provides an excellent starting point for the proposal kickoff meeting discussion.

9.3 Proposal kickoff meeting

The kickoff meeting for a proposal activity is cochaired by your company's capture manager and proposal manager. This is the opportunity to build upon the successful activities of the capture team.

Invite all members of the proposal team, including major subcontractors[1] if appropriate.

- Principal writer and contributors;
- Representative of senior management to demonstrate personally corporate commitment to the proposal and the project;
- Representative from contracts department;
- Publications group manager;
- Human resources manager;
- Key reviewers (Red Team participants).

Distribute key materials (Proposal Directive or kickoff package).

- Full RFPs, not only Section C, the SOW (now is not the time to save money on photocopying costs);
- Complete listing of the proposal team membership and their roles;
- Evaluation factors for award;
- Hopes, fears, biases; success criteria; and critical issues (see Figure 9.1);
- SWOTs (see Figure 9.2);
- Competitive analysis (see Figures 9.3 and 9.4);
- Proposal summary sheet (see Figure 9.5);
- Milestone schedules in calendar format for each proposal volume;
- Proposal volume outlines and writing assignments (see Table 9.1), including page and word allocations;
- Validated win themes and discriminations;
- Thematic and writing guidelines;
- Draft executive summary;
- Program organizational chart;
- Résumé format requirements;
- Documentation and production requirements;
- Action items sheet;
- Telephone and e-mail contact list.

---

1 You may want to hold a separate meeting with the subcontractors since they may not have exclusive teaming agreements with your company. You do not want anyone with divided loyalties gaining access to competition-sensitive proposal information.

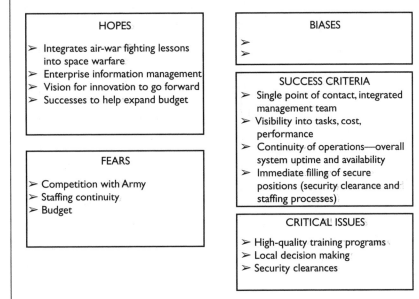

Figure 9.1
Hope, fears, biases; success criteria and critical issues.

Figure 9.2
SWOT analysis results.

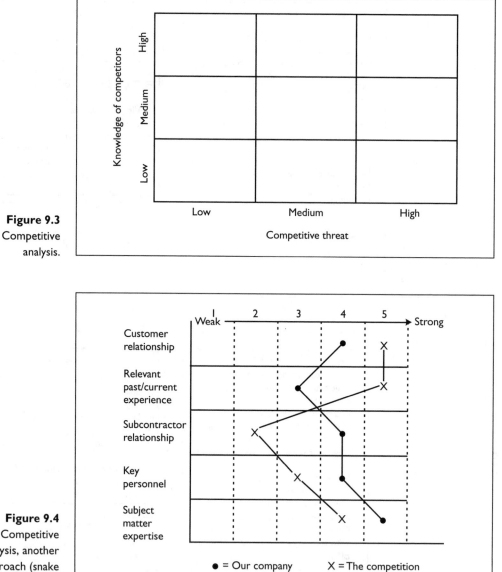

Figure 9.3
Competitive
analysis.

Figure 9.4
Competitive
analysis, another
approach (snake
diagram).

Emphasize senior management commitment to this important project. This is the reason for the member of senior management to be in attendance.

Build motivation. Emphasize your company's proposal win bonus policy.

Proposal Summary Sheet	
Proposal title: Internal B&P charge code: Proposal manager:	

1. Solicitation or RFP #:
 2. Proposal due date and time:
 3. Client name (point of contact):
   Telephone:
   E-mail:	

4. Dollar value 5. Level of effort (FTEs) 6. Type of contract: 7. Period of performance: 8. Place of performance: 9. Contract start date:	

10. Proposal copies required
   Technical:
   Management:
   Cost/price:
 11. Page limitations:	

12. Overview of scope of work: 13. Offer acceptance period (proposal validity period): 14. Packing label delivery address:	
15. NAICS code:	
16. Subcontractors/teammates: Specialty vendors: 17. Key personnel: 18. Other résuméd personnel:	
19. Project summaries your company will use:	
20. Defense Priorities and Allocation (DPAS) rating:	
21. Security clearances required (staff and faculty):	
22. Oral presentations:	
23. Source evaluation process and staff:	

Figure 9.5
Proposal summary sheet is an integral part of the proposal directive package.

Highlight the nature of the procurement and what marketing intelligence your company has collected regarding the opportunity.

Review your competitors' as well as your own company's strengths and weaknesses (SWOT analysis).

Discuss the concrete reasons for bidding on this procurement (i.e., sales hooks).

Highlight the strategies for winning (technical, management, cost). Why your company and not your competition?

Discuss any major or unexpected changes in the RFP from what had been expected during your intelligence-gathering efforts (capture plan activities).

Discuss the respective roles of your teaming partners and subcontractors.

Review the proposal outlines.

Review the proposal milestone schedules. Emphasize the importance of adhering to the schedule from start to finish.

Make well-defined writing assignments; assign specific action items for tracking to appropriate staff.

Discuss writing guidelines (such as style, page limitations, and words per page).

Review your major validated sales messages (themes) and storyline for this proposal.

Discuss the résumés and related contractual experience that will be included with your proposal.

Obtain approval of the concept for the cover design of the proposal volumes.

Review the staffing plan and WBS.

Identify special needs (such as binders, tabs, photography, and colored paper, CD-ROMs, and specialty software applications).

Determine who will deliver the proposal to the client and by what means (for example, driving and air travel).

9.4 **Postkickoff activities**

Prepare storyboards

- They are vital for subcontractors.
- They help technical authors overcome their initial writer's block.
- They ensure that all RFP requirements are addressed in your outline.
- They foster the smooth integration of writing from various authors.
- They link text and graphics or photographs.
- They are primarily associated with technical proposal volume.

In the several days following the storyboarding sessions, the proposal manager should spend sufficient time with members of the proposal writing team in order to do the following:

- Make certain they understand precisely what is expected from them during the proposal effort;
- Help the writers understand how their input relates to and fits with the entire proposal;
- Reinforce the need to incorporate appropriate proposal themes throughout their writing;
- Impress upon them the importance of adherence to the milestone schedule.

Keep in mind that all of these activities are designed and intended to assist your company and your staff to remain on track over the entire course of the proposal response life cycle.

We will now turn our attention to the role of the proposal manager, the individual charged and empowered with the responsibility and authority to oversee and orchestrate the entire proposal and postproposal response life cycle.

Chapter 10

The role of the proposal manager

The job of leaders is to paint an actionable vision …
—Roger N. Nagel, *Industry Week,* July 7, 1997

10.1 Overview

The proposal manager appointed to plan, organize, schedule, manage, arbitrate, and execute a given procurement response for a small business is the pivotal individual in the entire proposal process. He or she must skillfully balance the business opportunity against severely limited B&P monies and scarce human and material resources. And if those elements were not enough, the entire proposal response cycle unfolds in a matter of a few weeks—indeed, most smaller procurements have only 15- or 45-day response fuses. Certain state and local RFPs have response times measured in several days, as do task order proposals under certain GWAC vehicles, such as COMMITS. Amazingly, some responses are due in 24 hours from time of RFP receipt by industry!

Your company needs several people (in very small firms, two people may be sufficient) to be well trained in the art and science of proposal management and development. Candidate staff need to possess (and ultimately enhance) the ability to plan and organize their own activities as well as those of a wide variety of other staff. Proposal managers should be effective leaders and positive motivators; exercise tact and diplomacy; display drive and enthusiasm; and practice thoroughness, organization, and attention to detail. Most importantly, they must take full ownership of the proposal development process.

Specific knowledge required for proposal manager candidates includes (1) a broad understanding of your company, including its history, core competencies, products, services, personnel and human talent, facilities, technologies, and costing policies; (2) familiarity with the particular client's organization, technical requirements, key staff, acquisition policies, buying habits, and special methods of evaluating proposals; (3) broad exposure to the operations of the particular marketplace, such as the federal regulatory and statutory structure; (4) awareness of the potential competitors for a given procurement; (5) knowledge of public relations and marketing techniques; and (6) appreciation of documentation and computer-based publication methods so that the proposal is a well-presented sales document and knowledge product.

The proposal manager is charged with complete ownership of, responsibility for, and authority over the technical, management, and cost volumes of a proposal.

In many small companies, the proposal manager for a given procurement is drawn from the technical ranks because of the individual's knowledge of the project. Often, the proposal manager will also be proposed formally as the project or program manager. Because of the lack of depth of human resources in small firms, the proposal manager will likely have to provide the client on an existing contract with 40 hours per week of technical and programmatic support in addition to his proposal management responsibilities and obligations. Midsized and large companies have the comparative luxury of a full-time proposal management infrastructure, which can be drawn upon for proposal support as required. The proposal manager will most likely interact directly with the proposal coordinator or specialist within the business development group. The proposal coordinator or specialist can augment and complement the skills, knowledge, and time resources of the proposal manager. Small companies should avoid the very natural tendency of allowing the proposal coordinator or specialist, because that individual is probably a full-time overhead person, to assume too many of the proposal manager's functions and obligations. Both individuals should work together closely, but tasks should not merely be transferred from one person to the other. Proposal coordinators often participate in more than one proposal effort at a time. In fact, there are instances in small firms of one person coordinating 10 or more proposals simultaneously.

10.2 Generalized job description

The job description of the proposal manager is extremely open-ended, but in general he or she must provide technical, programmatic, cost, and organizational direction for the entire proposal effort, extended to include the following:

- Organizing, directing, and taking ownership of the internal company proposal team and the overall proposal development processes, which should be structured and repeatable;
- Attending preproposal conference;
- Monitoring B&P budgets, in conjunction with the capture manager;
- Developing proposal directive and orchestrating the proposal kick-off meeting;
- Facilitating interactive storyboarding sessions. Being proactive in developing strategic win themes and discriminators and the overall proposal storyline;
- Developing and controlling proposal development schedule;
- Developing targeted questions and interviewing technical and programmatic staff to extract from them critical information to include in the proposal;
- Remaining current on all amendments and modifications to the solicitation;
- Overseeing incorporation of feedback from formal color review teams (Blue, Pink, Red, and Gold);
- Coordinating all aspects of proposal documentation and production;
- Coordinating all Evaluation Notice (EN) responses and Final Proposal Revision (FPR) logistics;
- Orchestrating oral presentation slide development and oral practice sessions;
- Attending government debriefings;

- Conducting postaward debriefings internally, regardless of outcome, and revising business development and proposal development strategies as appropriate;
- Developing proposal boilerplate (reuse) material to incorporate in future RFP opportunities through an automated, text-searchable knowledge base;
- Training additional staff on proposal management skills;
- Developing and continuously improving the company's proposal development processes to streamline and improve proposal development.

The proposal manager must also translate purpose into specific actions, as illustrated in Figure 10.1.

A detailed position description for a proposal manager is provided in Figure 10.2. Figure 10.3 shows the sample text for a classified advertisement designed to attract top-flight candidates for a proposal manager position.

Recruit, organize, and direct the proposal team.

Arrange for necessary human, material, and computer resources. Requisition or obtain dedicated, secure floor and wall space to prepare the proposal.

Getting enough of the right resources for a proposal at the right time can be challenging in both small and large businesses. This is one of several points at which senior management's direct involvement is critical.

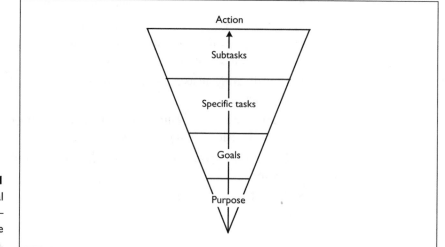

Figure 10.1
Proposal management—translating purpose into action.

Position Description

Title: Proposal Manager
Organization: Proposal Development Group

Nature and Scope of Position
The Proposal Manager organizes, directs, and takes ownership of internal proposal teams and the overall structured and repeatable proposal development processes. The manager develops and updates Proposal Directives [including points of contact (POCs), outlines, schedules, and so forth], orchestrates kickoff meetings, and facilitates storyboarding sessions. The manager takes a proactive role in the development of strategic solutions, win themes, and discriminators. The manager is responsible for coordinating all aspects of proposal documentation and production, which may include submission of responses to requests for information, submission of questions to contracting officers, and preparation of material for formal reviews. The manager is expected to contribute to the continuous improvement of proposal development processes and take a hands-on approach to proposal tasks.

Reporting Relationship
The Proposal Manager reports to the Proposal Director.

Responsibilities
• Take end-to-end ownership and command of the proposal process for assigned proposals, including oral presentations, ENs, and FPRs.
• Create outline and schedule for proposals; remain current on all proposal amendments and modifications.
• Provide detailed guidance to proposal writers and contributors both within the company and among subcontractors; develop targeted questions and interview technical and programmatic staff to extract critical proposal-related information.
• Write significant portions of proposal/ marketing narrative (for technical, past performance, and management sections of proposals) in real time.
• Edit, organize, and refine entire proposal for final delivery.
• Provide requested information for proposal development to companies hiring the company as a subcontractor.
• Contribute to knowledge base and stores of "boilerplate" material for effective reuse in proposals.
• Lead storyboarding, kickoff, and review meetings; assist proactively in development of strategic win themes and discriminator.
• Communicate proposal-related requirements and status to senior management as appropriate.

Critical Attributes
• Exceptional communication skills;
• Leadership and management skills;
• Ability to work independently or as part of a team;
• Flexibility;
• Ability to multitask on concurrent projects;
• Superior organizational skills;
• Motivator;
• Positive, proactive attitude.

Qualifications
• Bachelor's degree in English or Communications;
• Four to seven years of progressive experience in a fast-paced, high-volume proposal environment;
• One to two years of management experience;
• Hands-on knowledge of electronic file management;
• Good familiarity with standard MS Office software, e-mail applications, and Web browsers.

Physical Requirements (per ADA [1992])
• Significant level of repetitive motions, such as fingering/keyboarding;
• Ability to differentiate colors on paper and on a computer monitor;
• Tasks may involve lifting packages less than 20 pounds;
• Majority of tasks performed while sitting in nonadverse environmental conditions.

Figure 10.2 Detailed description of the proposal manager position.

Figure 10.3
Sample classified ad
for proposal
manager
candidates.

We are currently seeking a full-time Proposal Manager to become part of our Corporate Proposal Development Team. Candidate needs to be a solutions-oriented, high-energy individual to serve as a hands-on leader for multiple proposal efforts of varying size. Responsibilities will include development of strategic solutions, win themes, outlines, storyboards, and schedules. Critical attributes include exceptional communications skills, leadership, teamwork, flexibility, organization, motivational skills, and a "do-whatever-it-takes" attitude. Must have proven ability to write proposal narrative, conceptualize graphics, and provide guidance to proposal writers and contributors. Knowledge management and document production experience a plus.

Give every proposal that your company designs and prepare the resources necessary to win, or do not bid.

Attend preproposal (bidders) conference (see FAR 15.409).

- Identify competitors and potential teaming partners;
- Obtain the client's official definition of evaluation criteria;
- Note any last-minute changes to the RFP or RFS;
- Listen for hidden messages from the government (e.g., We are happy with the incumbent);
- Obtain details about the contract job site.

Ensure adequate planning of the proposal effort.

Time spent on planning and strategizing early in the proposal response cycle before the formal kickoff meeting is crucial to the overall success of the proposal team. Participate in RFP analysis, schedule and outline development, page allocation limits, writing assignments, and storyboard preparation and review. Talk often with key technical writers, reviewers, and your publication staff to ensure that people understand your expectations and your role and responsibilities, as well as theirs.

Prepare and then negotiate B&P budget estimates with senior management; monitor and control B&P and ODC expenditures along with the capture manager.

Prepare B&P projections to ensure select key technical and programmatic staff are available to prepare the best possible proposal.

It is highly advisable that your company develop and implement a method for projecting and tracking B&P-related costs. Direct labor for marketing a potential business target as well as for the technical writing, cost development, proposal development, and production of the proposal, oral presentation, and FPR should be estimated in detail for the entire proposal life cycle. In addition, ODCs, including travel, communications initiatives (billboards, radio spots, print ads, and Web sites, for example), classified advertising for new hires, subcontractor costs, temporary staff, and proposal consultants should be estimated as well. These estimates should be completed by the assigned proposal manager in advance of a proposal tracking number (for time sheets) ever being assigned and approved by a member of senior management. The approved forms (hardcopy or electronic) are then used as an extremely effective tool to manage the cost of the entire proposal process in a manner analogous to managing a project. B&P projections can then be approved or disapproved by senior management based upon what the company had allocated for B&P expenditures for a given fiscal year.

The benefits of this structured, planned approach are that actual costs for proposal development, including marketing and postproposal activities, are more accurately projected as real data are collected and that new estimates are calibrated against these real data. Proposal managers should be held accountable for any significant proposal-related cost overruns. Another benefit is that there is an up-front commitment to make select key technical and programmatic staff available to prepare the best possible proposal.

Due to staffing constraints, small companies may need to bring temporary publication staff on board for a portion of the proposal response cycle. In addition, outside professional proposal consultants with technical or proposal development specialties and expertise may be required to augment your company's internal resources. There are companies throughout the country that offer end-to-end proposal development services.[1] They will write, review, and produce your proposal for a negotiated fee.

Define and present proposal themes and strategies.

Ensure adherence to proposal themes within the proposal documents. The marketing intelligence that your company's acquisition/capture team

---

1   Full-service firms are located coast to coast, including Organizational Communications, Inc. (OCI) in Reston, Virginia (http://www.orgcom.com); DSDJ in Martinsburg, West Virginia (http://www.dsdj.com); Steven Myers and Associates (SM&A) in Newport Beach, California, and Herndon, Virginia, a wholly owned subsidiary of Emergent Information Technologies (http://www.emergent-it.com); The Proposal Group in Towson, Maryland (http://www.lenduffy.com); Hyman Silver and Associates (HSA) in Los Angeles, California (http://www.hsilver.com); and Shipley Limited in the United Kingdom (http://www.shipleylimited.com).

has collected for a particular procurement must be translated into the thematic messages that are embedded throughout the proposal volumes. As proposal manager, you will need to set the marketing tone for the entire proposal response. Ensure that the documents are designed to be, and ultimately become, sales documents—authenticated by technical, programmatic, contractual, staffing, institutional, and technological strengths, lessons learned, metrics, and innovations.

Organize and conduct the preproposal activities and proposal kickoff meetings.

The proposal manager should communicate with the acquisition/capture team that has done the direct marketing with the client to ensure that all of the intelligence gathered has been communicated and is understood thoroughly.

Provide needed information to your company's contracts and pricing staff.

Include information regarding direct labor, travel, equipment purchases, property rentals and purchases, government-furnished equipment (GFE) and contractor-furnished equipment (CFE), subcontractors and consultants to be used, documentation requirements for contract deliverables, proposed cost center, recommended fee and profit margin, and pricing strategies. Develop the technical and management volumes in close coordination with the cost volume. Communicate regularly with costing and contracts staff.

Review and critique proposal input from the authors.

Whether received in hardcopy form, on diskette, or via electronic mail, a proposal manager must review the sections and subsections of the proposal for a myriad of factors including RFP compliance, thematic continuity, storyline integrity, technical accuracy, programmatic appropriateness, risk minimization, and client awareness. Provide writers with interactive, ongoing feedback so that they stay on target from both a technical and schedule standpoint. Help them to understand the importance of their contributions.

Plan the project (organization, schedule, and budgets).

One of the most important aspects of proposal planning is selecting the staff whom you will propose to the government as your company's project or program team. Early staffing selection and management approval of this selection will be extremely beneficial to the entire proposal team. Definitive staffing decisions will make many processes easier, from résumé development and organizational chart generation to cost structuring and management plan preparation. Too many times, staffing decisions are delayed so that technical solutions can be developed and technical writing can be accomplished. Gathering detailed information for each résumé can be extremely time-consuming. Every time a staffing change occurs, a new résumé must be developed. Name changes also present editorial challenges in both text and graphics in the sense of finding every instance in which a person's name had appeared and putting in the new person's name.

The proposal manager and capture manager must ensure that an appropriate project schedule is developed that accounts for all client-required deliverables. This project schedule must be tightly linked with the details of the management plan.

- *Maintain ongoing communication and liaisons with senior management to select subcontractors, identify key project personnel, review the proposal progress, and obtain assistance as needed.* Use every opportunity to brief managers with the proposal status and issues that they can help to resolve.

- *Maintain client contact in conjunction with the business development or advanced planning groups.* Although your company's marketing staff cannot discuss the particular procurement for which a proposal is being prepared with the client, ongoing relationship building, customer relationship management (CRM), and discussions must continue.

- *Coordinate subcontractor liaison with the contracts department.* Ensure that technical information, résumés, project summaries, and cost data are submitted in a timely manner. Many times small disadvantaged businesses (SDBs), disadvantaged business enterprises (DBEs), and small, woman-owned businesses (WOBs) do not have adequate administrative staffs to prepare proposal input for your company (the prime) in electronic format or compatible electronic format. That means additional word processing tasks for your company's publication group.

- *Arbitrate Red Team and other review cycle comments.* Provide written instructions to the Red Team reviewers explaining what the expectations of the review process are and asking that comments be detailed and explicit so that the proposal team can respond effectively. During the recovery period following a review cycle, the proposal manager may have to assign "tiger teams" to ramrod the effort to fix any major deficits identified during the review process.

Comprised of a small number of very senior staff and subject matter experts, these teams will need to work quickly and effectively to address these major weaknesses. Senior management support for adequate resources is vital at this stage.

- *Keep the proposal team members motivated.* Attend to such needs as meals and special transportation requirements. Do not forget your support staff. They are the ones called upon to produce proposal after proposal on a regular basis. Demonstrate clear, concrete appreciation for a job well done.

- *Ensure adherence to the proposal response milestone schedule.* As proposal manager, you must be relentless in adhering to the schedule. Time spent in prekickoff strategizing, planning, communicating, and decision making will serve you well downstream in the response cycle. Ensure that even the early proposal response milestones are met. There is no need, indeed it is counterproductive, to take every proposal schedule down to the final hour. Manage the process effectively from beginning to end. It will be exhausting, but everyone will benefit from a well-executed process, and your proposal will have a much better probability of being a winner.

- *Review and respond to all amendments to the RFP or RFS.* The government will often release one or more amendments to the original RFP in response to questions from offerors or to make a correction. Each of these amendments must be reviewed carefully because they may, but certainly do not have to, extend the due date of the proposal. In some cases, amendments can have significant impact on staffing, technical approach, costing, page count,[2] and font size, for example. Amendments must be acknowledged in the proposal response, sometimes in your company's transmittal letter, as well as in the completed Representations and Certifications (Reps & Certs), Section K.

- *Ensure timely delivery of the requisite number of proposal volumes.* Avoid packaging all of the proposal documents in one box when delivering. Divide the copies among several boxes to minimize the loss if a box is dropped or misplaced.

- *Ensure that the transmittal letter and cost volume forms have all the necessary signatures.*

- *Evaluate the performance of personnel assigned to work on the proposal.* Once the proposal has been submitted to the government,

---

2 For example, an amendment to a USACE-Sacramento RFP changed the original technical volume page count from 80 to 65. This change had a significant effect on the structure of the final technical volume.

take the time to evaluate the performance of both technical and support staff during the proposal response period. More than likely, you will be serving with these people again. Identifying who requires proposal development mentoring will be beneficial to them and to you. Express your appreciation. Hold a party, recommend people for bonuses, and arrange for proposal team T-shirts or other team-building items to enhance morale.

- *Prepare for orals and FPR.* Ensure that your entire proposal—text, graphics, and cost data and justifications—are archived in a secure manner and easily retrievable. You will have to refer to your submittal frequently as you prepare for an oral presentation in front of the client or a written FPR.

Ensure that your presentation team is reinforcing your company's answer to the question why us and not our competition.

Depending on what Evaluation Notices (ENs) have been identified in your original proposal, FPR responses can be involved and time-consuming. Also, there may be more than one FPR, as was the case with an $800 million Air Force Space Command proposal in 2004.

In many cases, you as proposal manager will have to gain management support in reassembling the initial proposal team to prepare a solid response. Oral presentations will also require dedicated time to strategize your response and then conduct dry-run practice sessions to hone the public speaking skills and cooperative interaction of the project team. Individuals who critique these practice sessions should ensure that the presentation team is reinforcing your company's answer to the question, "Why us and not our competition?" Staff should practice a sufficient number of times to present their material effectively while standing calmly looking at their audience. Make sure that the presentation is timed during the dry runs so that all of your material can be presented well in the time allotted by the government. Verify with the government exactly what audiovisual and computer equipment will be provided for you and also inquire as to the size of the room. Perhaps you could visit the presentation room beforehand and photograph it. If your company is taking its own computer equipment, slide projector, or overhead projector, make sure to take extension cords, spare bulbs, and backup CD-ROMs or Zip disks. Test the actual equipment you will be taking before arriving for the presentation.

Many government agencies are now requiring presentations as part of the proposal response. NOAA, DOE, the U.S. Coast Guard, Air Force, and Army, and the Federal Deposit Insurance Corporation (FDIC) are several such agencies in which specific organizations often require presentations. Today, with multimedia presentation software, such as Microsoft PowerPoint 2002, HyperStudio 4, CyberLink's Media Show, Harvard Graphics Advanced Presentations version 3.0, and Harvard Graphics Pro Presentations II, presentations can be more professional and alive than ever [1, 2]. Multimedia software, speakers, sound cards, and cabling will be a capital investment for your company; or multimedia-equipped computer systems may be rented for occasional use as well.

Request and attend the client debriefing.

A debriefings file within your knowledge base is an excellent training tool for novice proposal managers within your company. It is important to understand exactly why your company was not selected for award of a given procurement. Ask questions of the government contracting and technical staffs in attendance at the debriefing. The most important activity is to document and communicate the lessons learned from the loss to other proposal managers and senior management within your company. Maintain a debriefings file and refer to that file before beginning the next proposal.

Guard overall proposal security.

To help ensure confidentiality, proposals should not be discussed outside the office. Proposal-related paper, including graphics, cost tables, and narrative, must be shredded and disposed of when it is no longer needed through a reputable waste management firm. Companies have been known to purchase competitors' trash! And network or PC-level password protection and encryption are essential to guard the electronic proposal files.

If at all possible, the proposal manager should have no writing assignments. In small firms, however, this is very difficult to avoid. Yet if the proposal manager is to be effective in that capacity, he needs to attend to a myriad of technical details and human and organizational issues and not be burdened with writing proposal sections.

10.3 Changing focus of proposal management

Traditionally, proposal managers—both full-time staff members within a federal contractor organization as well as freelance and agency-based consultants—bring solid experience and a knowledge base that are focused toward compliance, process, and documentation. Both today and in the future, however, compliance- and process-driven proposal management approaches alone cannot and will not produce winning proposal documents and stellar oral presentations on a sustainable basis. Instead, proposal managers must combine these important baseline approaches with solution development and facilitation; idea generation and genuine creativity; and proactive, engaged leadership. In this way, they become more valuable to their organization and also become more sought-after knowledge workers in the marketplace at large. The need for an expanded and renewed focus on solution development competencies within the ranks of proposal management professionals has arisen in large part from the migration toward performance-based contracting (PBC) and performance-based acquisition (PBA) within the federal government marketplace.

Solution development refers to the critical activity of building and articulating fact-based *storylines* or *story arcs* that convey a federal contractor's specific approaches to providing meaningful, measurable, achievable, and risk-aware outcomes for its federal government customer. Robust solution sets must encompass such proposal areas as technical, program management, staffing, phase-in, and past performance. The storylines must link together such major elements as "Understanding the Customer Environment," "Approach," and "Measurable Outcomes" in a manner that ensures continuity of major sales messages or themes.

To amplify further, solution sets and their associated storylines are fact-based, quantitatively validated articulations of the framework and spirit in which a federal contractor will approach providing services and/or products to the federal government. These services and products must be designed *to directly benefit the customer* given the contractor's understanding of the government environment, requirements, and mission. The "customer" will most likely include the specific government agency and its users, customers, and stakeholders.

In today's extraordinarily competitive and solution-driven federal marketplace, the *delta,* or difference, between winning and coming in second borders on the microscopic. Proposal managers can provide very tangible value to their organizations as measured by an enhanced proposal win rate by fortifying their skill sets to span all of the key elements presented in Figure 10.4. These important proposal management competencies include detailed compliance with the solicitation document (#1); process flow coordination (#1); documentation and production, editing, and quality control (#2); and configuration control. In addition, and most importantly, proposal managers must develop and hone their skills in solution development and leadership (#3 and #4) to attain proposal management success.

Let's take a closer look at performance-based acquisition, or PBA. Migration toward PBA is a key and growing element within the federal marketplace. The Request for Solution (RFS) is becoming as common, if not more prevalent, than the traditional Request for Proposal (RFP). And even many RFPs have performance-based components, including quantitative metrics, performance standards and thresholds, options to propose most efficient organization (MEO) staffing options, and performance-based

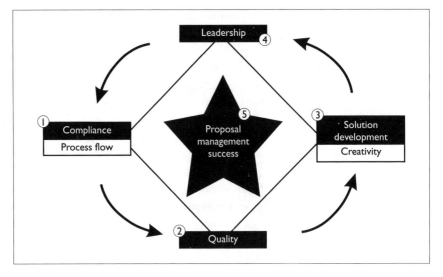

Figure 10.4
Solution development, creativity, and leadership are critical competencies for proposal management professionals now and in the future.

work statements (PBWSs). Now, the focus is on *innovative, risk-aware solutions.* These must be presented in proposal documents and oral presentations in clear and concise words and illustrations that build on such industry best practices and disciplined approaches as the Software Engineering Institute's Capability Maturity Model Integration® (CMMI®) structured software engineering methodologies, ISO 9001:2000 and 9002:2000 standards, and Institute of Electrical and Electronics Engineers (IEEE) guidelines. Importantly, performance-based service contracting is a major component in President Bush's Management Agenda, which emphasizes market-based, results-oriented, citizen-centered government.

Performance-based contracting is a procurement strategy defined and described at FAR Subpart 37.6 and in the Office of Federal Procurement Policy's (OFPP's) Policy Letter 91-2 (dated April 9, 1991). That strategy structures all aspects of an acquisition around the *purpose* and *results* of the work to be performed, as opposed to either the manner in which the contractor or offeror must perform the work or the processes that must be used. The *WHAT* is emphasized over and above the *HOW.* (In general, if the level of effort, staffing levels, skill mix of staff, or educational levels of staff are specified in the RFP, then the contract is *not* performance based.) The key elements of a PBA performance work statement for *services* are (1) a statement of the required services in terms of output, and support for the agency's mission; (2) a measurable performance standard for the output; and (3) an acceptable quality level (AQL) or allowable error rate. The government may employ a variety of measurement methods, including a project surveillance plan or an award-fee evaluation plan. The primary elements of a PWS for *hardware* or *end-item deliverables* are (1) a specification of what the end product must do in terms of performance, along with any critical constraints such as weight or footprint, and (2) measures of quality that are related directly to the end products' capacity to perform its intended use.

This innovative federal PBA strategy leverages the ingenuity of industry while providing the government with access to the best commercially available products, services, processes, knowledge, and technologies. From the government's perspective, use of performance-based contracting strategies reduces acquisition cycle times and costs, because contractors are not compelled to perform to detailed design-type specifications that can inhibit creativity and efficiencies. Most importantly, performance-based contracts help to ensure contractor *accountability for mission-focused results.*

Fundamentally, all federal contractors, including small businesses, must be able to prepare proposals and oral presentations that demonstrate *innovative* performance approaches for managing *risk* proactively while ensuring superior *quality*. In return for performance excellence, contractors earn positive monetary incentives for meeting technical, schedule, and cost standards and thresholds.

10.4 Effective solution development

So how can proposal managers become integrally involved in solution and storyline development? Proposal managers can provide significant benefit to their organizations by proactively leveraging diverse resources such as customer strategic plans, technical library documents, Web sites, and call plan results. *Why?* To conceptualize, develop, and produce results-focused storyboards, solution sets, storylines, proposal narrative, and graphics concepts in concert with other capture and proposal staff. In the course of a proposal manager's self-learning process, graphics concepts, background information on government leaders, and other useful information emerge. For example, in researching IT standards that are applicable to a given customer, one might locate a compelling graphic concept that is already known to that customer because it is an integral part of their Web site. That graphic may be used as is, or modified, to convey a key concept in a federal contractor's proposal narrative or oral presentation. Or a listing of critical-path activities that a proposal manager locates through the federal customer's technical library can be built into a table in the proposal that demonstrates understanding of that customer's operational environment.

Web-based research on recent awards by a given customer to competitors can also shed considerable light on that customer's buying habits and preferred technical solutions. Speeches and biosketches or *curricula vitae* of key federal leaders as well as published papers of those individuals can prove to be invaluable in proposing meaningful, on-target solutions to those same customer leaders. Given the level of executive review and approval, customer news releases are excellent sources of knowledge. Specific words and phraseology that are used can be incorporated into the proposal or oral presentation. Customer telephone directories as well as Web sites can help proposal managers understand how the customer's organization is structured and also provide insight into key graphic or word concepts that the customer chooses to use frequently. Documented trip reports that follow from executive call plan execution can be rich reservoirs of direct insight into a customer's hopes, fears, biases, and critical issues and success factors. Federal government publications such as Air Force Base (AFB) papers again can shed light on the latest points of interest to the customer community. Specific examples include the *Space & Missile Times* published on behalf of the 30th Space Wing at Vandenberg AFB in California, the *Space Observer* at the 21st Space Wing at Peterson AFB in Colorado, the *Astro News* at the Space & Missile Center at Los Angeles AFB, and the *Army Times*. Another important resource for proposal managers is customer briefings to industry.

In the process of researching and thinking and learning about the customer, the program, and the mission, the proposal manager should focus on how the knowledge and information being collected contributes to overall understanding of the customer, the strategy of his or her federal contracting company, the approach and solution that follows from that strategy, and the tangible and intangible benefits or "value proposition" that particular solution set brings to the customer and its program as well as its overall mission (see Figure 10.5). The storylines for the proposal and the oral presentation emerge from the critical thinking and synthesis of ideas that are leveraged to populate such diagrams as that shown in Figure 10.5.

Let's examine a sample storyline:

Our corporation will work together in close partnership with the Department of the Army to enable greater knowledge sharing among Army communities in support of the overall Army Transformation. Collaborative Web-based knowledge portals, e-learning technologies, systems integration, intelligent data mining, intelligent agents, and expert systems technologies will be assessed and deployed with an enterprise-wide focus targeted to maximize the Army's investments in its people and knowledge assets. Such knowledge sharing will facilitate improved, fact-based decision dominance by commanders and business stewards in the battlespace, organizations, and Army's mission processes. This decision dominance will contribute directly to

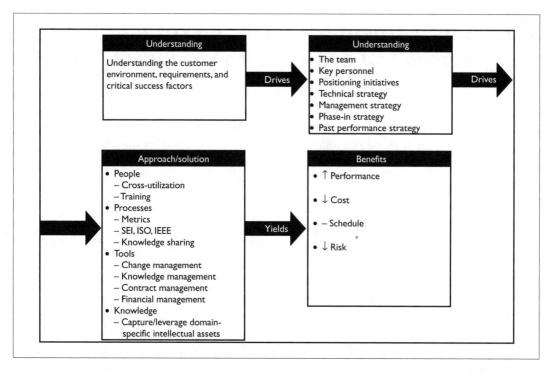

Figure 10.5   Example of customer-focused, benefits-based solution pathway.

the effectiveness and day-to-day safety, security, and quality of life of America's warfighters worldwide.

This sample storyline conveys the tools, technologies, and approaches that the federal contracting firm will employ in delivering significant benefits to the Army mission of transformation and to the men and women warfighters around the globe. Such a storyline would be used at an overarching level (1.0 or 1.1) in a management or technical approach, with much greater qualitative and quantitative detail provided in the subsections that would follow. Of note is that in a recent government debriefing on a major winning proposal, my company learned that *Compliance + Validated Storyline + Comprehensible Presentation* translates into a very high probability of winning.

Complementary roles and responsibilities of proposal and capture managers

Capture managers and technical leads are integrally involved in solution set and storyline development, but it would behoove proposal managers to become deeply immersed in this process as well. Of particular note is that important dovetailing occurs between the capture manager and proposal manager functions. Proposal managers and capture managers are most effective together when (1) they communicate multiple times each day, (2) brainstorm solution sets and storylines together, (3) use each other as sounding boards for key ideas and concepts, and (4) share what they know about the customer, program, key staff, and applicable technologies openly and in an atmosphere of mutual trust. This author worked from February to October 2002 on a half-billion-dollar proposal for the U.S. Air Force Space Command as my company's proposal manager. Every morning, the capture manager and I met to plan the actions and follow-up issues for the day, identify problem areas in the proposal process, share ideas about pivotal, solution-conveying graphics, and ensure that the major storylines were being articulated throughout each one of the seven proposal volumes.

Contemporary proposal development is about articulating and illustrating a fact-based, validated storyline that provides tangible value for the federal government customer as well as its users, customers, and stakeholders. That proposal storyline must be told convincingly, while simultaneously being in full compliance with Sections L and M of the RFS from an infrastructure and requirements standpoint. Current and future proposal management professionals must prepare themselves to be an integral part of the proposal solution set and storyline development process. Compliance is necessary, but certainly not sufficient. Innovative, cost-effective, risk-aware solution sets and storylines that span people, processes, technologies, and knowledge assets that federal contracting firms can provide are what is needed to win proposals now and in the coming years.

10.6 The growing importance of oral presentations

During the past several years, the federal government has used oral presentations to augment and even largely replace written proposals submitted by the contractor community for competitively negotiated procurements.

Since 1994, various agencies of the federal government, including the Department of Energy, Nuclear Regulatory Commission, NASA, Centers for Disease Control, Federal Aviation Administration, Internal Revenue Service, and Bureau of Engraving and Printing, have experimented with selecting contractors on the basis of oral presentations instead of hardcopy or electronic technical and management volumes. And in 1997, the Federal Acquisitions Regulations included—for the first time—coverage of oral presentations.

A major goal for the government is to reduce the level of written materials associated with the procurements process. In addition, oral presentations permit government evaluators to receive relevant information directly from the key members of the contractor's proposed team who will perform the work on the contract. And expect evaluators to address past performance during oral presentations. In this manner, communication and the exchange of information between the government and the contractor can be enhanced. The government is, in effect, conducting job interviews of key staff. Proposal-related costs could be reduced for the government as well as the contractor with the implementation of oral presentations. In addition, procurement lead times can be reduced significantly.

From both a legal or contractual and practical perspective, debate continues over exactly what portions of a contractor's proposal will be integrated into the resultant contract. Various federal agencies have videotaped the oral presentations for later review, although there is no regulatory requirement that prescribes that a record of the oral presentation be maintained. Both video and audio tape may become available through the Freedom of Information Act (FOIA). Time limits on oral presentations, number of participants from each contractor, number of presentation materials, and type of presentation materials all vary from agency to agency.

Although no standards for oral presentations have emerged, there is a distinct movement in the direction of such presentations throughout the federal government. A proven provider of outstanding oral presentation conceptualization and development is The 24 Hour Company in northern Virginia. You can contact them at (703) 533-7209.

10.6.1 Outsourcing oral presentation support

As a small business, you might consider using the services of an outside oral presentation consultant to assist your own internal presentation team. One proven resource about whom this author has direct working knowledge is Mr. Gregory W. Pease. He delivers professional oral presentation coaching and proposal consulting services to customers in all industries. Mr. Pease combines 20 years of technical, management, and proposal experience and graduate education in management to help teams deliver powerful presentations. You can reach Mr. Pease at http://www.mycoach-greg.com or in Maryland at (301) 317-5038 or (301) 237-3727.

10.6.2 Oral presentation development process

Your company will benefit by following a structured, repeatable process for developing your oral presentations. As illustrated in Figure 10.6, begin building your oral presentation slides with an outline mapped directly to

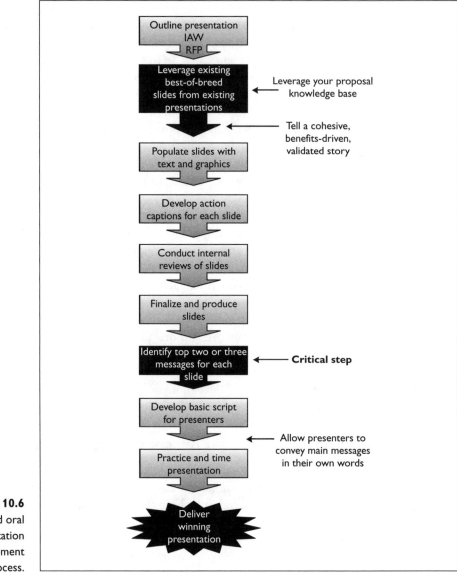

Figure 10.6 Structured oral presentation development process.

the RFP or RFS. It is advisable to include specific references to the RFP on each slide of the presentation (e.g., L.2.1), as well as embed frequent road-map slides that let the evaluators know exactly where you are with respect to their RFP during your oral delivery. Be sure to build on and leverage the best-of-breed slides from previously prepared oral presentations. These should be an integral part of your proposal knowledge base.

When populating slides for the new presentation with text and graphics, add an action caption (sometimes called a tag line or bumper sticker) to each one that completes and closes the thought and focuses on direct benefits to the client. Plan to review the slides once they are populated with text, graphics, mapping to the RFP, and action captions several times prior to finalizing and producing them in hardcopy and electronic format on CD-ROM or Zip disk. Be sure to review the slides using the actual projection device for the formal presentation. Sometimes color palettes or photos that appear crisp on a computer screen become pixilated and jagged when projected onto a large screen.

When preparing for the oral presentation practice sessions, identify the top two or three major messages found on each slide. This is a vital step because it facilitates the development of scripting and also ensures that each member of your presentation team tells your company's proposal story clearly and consistently. Because most people are not public speakers, it is often helpful to develop basic scripts for them, such as the following:

Sample script for X slide   During the fulfillment of my project management function in support of DOE in Washington, D.C., I have gained an in-depth knowledge of your current work and resource requirements. My daily, face-to-face interactions with DOE technical and contractual staff have contributed directly to this understanding. We have designed our site organization to fully support all current DOE requirements. Our proposed organization is flexible to accommodate growth and evolutionary changes in requirements.

Sample script for Y slide   I previously mentioned our proven automated tool. Now let me tell you how it will help me—and you—every day of this contract. It will be the primary operational management tool. Interactive, secure, and Web-based, this tool has proven itself on our mission-critical contracts with USACE in Baltimore and at the U.S. Coast Guard in West Virginia. We funded the entire development of this automated tool, so it will be available to you at no direct cost.

Sample script for Z slide   To reinforce a major point, we will also retain complete responsibility for performance in each and every task area of this contract. In my role as group manager, I will ensure that prime contract terms, conditions, and performance measures flow down and are followed by both of our subcontractors. The members of our integrated team will all work to the same standards. We will communicate regularly with you through programmatic reviews and with each of our teammates to review their performance. You will benefit directly from our approach with our subcontractors, who are fully accountable to us for their performance. We, in turn, will be fully accountable to you for the contract's overall success.

However, as long as team members communicate the major message for each of their slides, allow them to speak in their own words rather than memorizing or reading their text. Add real-world examples and anecdotes that validate and bring your major points to life. Conduct several practice sessions, including timed dry runs, to ensure that the physical hand-off from one presenter to another is smooth, that the presentation can be delivered within the allowed time frame, and that each presenter establishes eye contact with his or her audience. In addition, practice manipulating the computer or acetate vugraphs (yes, these can still be required; NASA Ames Research Center is a case in point) to ensure that the right cadence is maintained during the presentation. In most cases, the dress for presentations is formal business attire. Ensure that the team takes several small bottles of water with them into the actual presentation.

Increasingly, oral presentations include an ad hoc response to a sample task or problem, or series of government questions. Your presentation team members may be tasked with developing a follow-up presentation in real time after just having given their formal briefing. The government will assess how the team members interact with each other to solve the issue and develop a credible presentation with the materials at hand. You should know in advance of the oral presentation date whether this step will occur. Prepare accordingly by taking blank acetate slides, colored writing markers, and flip charts, as appropriate.

10.6.3 Specific oral presentation guidelines for success

Presentation software involves four distinct communication modes: (1) speech, (2) text, (3) graphics, and (4) electronic. For your audience—the government evaluation team—meaning is image-driven and received via their eyes and ears in a group setting, which is subject to group reaction. That meaning is textured by the presenters' body language, gestures, eye contact, tone of voice, voice inflection, pauses in speaking, and choice of words. Note that presenters will speak at about 120 to 180 words per minute. Conversely, people read at rates from 250 to 1,000 words per minute.

Points of emphasis

- Tell a cohesive and fact-based story.
- Think graphically—limit the amount of text on any given slide and make the graphics jump off the page.
- Punctuate the oral presentation with real-world examples—bring it to life!
- Include a benefits-oriented action caption at the bottom of each slide (e.g., Collaboration Through Clearly Defined Communication Pathways, or Directly Applicable Experience Will Minimize Program Risks).
- Ensure legibility of all text and graphics callouts.
- Personalize the presentation—avoid use of the words "the government," but rather talk in terms of yourself or other individuals' names.
- Convey enthusiasm and positive passion during the presentation.
- Go for the close—ask for the sale.

10.7 Attending to the details

Your company may want to consider establishing a petty cash fund to handle unforeseen proposal requirements such as photocopying, color laser copying, photography, graphic enlargements for foldout pages (11×17 inches), and small office supplies. The proposal manager should have access to this fund. He or she should also be given keys and access codes to the building and appropriate offices, file cabinets, and personal computers. Evening and weekend proposal work often requires ready access to hardcopy materials, diskettes, hard disk drives, and networks.

At the formal kickoff meeting, it is advisable to generate and distribute a list of key proposal participants, including subcontractors and consultants. The list should provide names and telephone numbers (office, home, cellular, and beeper).

The proposal manager must ensure timely delivery of the proposal. Often, the mailing address is different from the hand-carried delivery address. For air carriers, such as Federal Express, military bases will need barracks or building numbers and not simply mail stops. Allow sufficient time for contingencies such as inclement weather, heightened security levels at government installations, automobile breakdowns, construction delays, accidents on the road, and delays in public transportation. Know exactly who will deliver the proposal to the government site. Deliver the required number of copies of the proposal document to the correct address. Avoid the use of couriers for proposal delivery because they do not have a vested interest in making certain your proposal documents are delivered undamaged and on time. Finally, obtain and retain a receipt from the government that is stamped with the time and date of your delivery.

10.8 Control of the schedule

Proposals are challenging to complete, because for most of us they are an effort above and beyond our normal contract-related responsibilities. In your small company, you probably do not have a comfortable depth of staff in either the publications department or in the technical ranks. But adhering to the agreed-upon schedule will help to make the process smoother and help to ensure that the finished product is one that your company can be proud to submit. Winning competitive proposals requires a substantial amount of planning and guidance, well before the release of the RFP, during the actual proposal-preparation process, and in the postproposal phase.

Seasoned proposal managers come to realize that, in general, everything about the proposal process will take longer than planned. A novice proposal manager will quickly ascertain that many participants in the process—from the technical writers to the subcontractors to the publications group—feel that they are being pressured unnecessarily. And supporting departments, such as human resources or engineering, may be indifferent to even a well-planned proposal milestone schedule. Subcontractor technical, staffing, and cost information is often late or delivered in an electronic format that requires conversion and cleanup.

When generating the schedule for your proposal, be sure to take into account the effects of weekends and holidays, important company social functions, and competing priorities within the publications group. Generating a calendar-type schedule as shown in Figure 10.7 that notes the highlights of the proposal schedule is one effective tool in keeping the overall proposal effort in perspective. A Gantt or bar chart is another such tool that indicates the many simultaneous proposal activities. Effective implementation of a proposal milestone schedule is challenging and often thankless work. People tend to procrastinate, or want to keep polishing their writing, or have competing priorities. Frequent but brief proposal status meetings coupled with the full support of senior management will definitely assist a proposal manager to keep his or her proposal effort on schedule. A word about management support. Senior management involvement is critical to ensure adequate resources are committed at the right times

Sunday	Monday	Tuesday	Wednesday	Thursday	Friday	Saturday
	1 Release of RFP. Copy and distribute RFP to proposal team.	2 Analyze RFP. Generate kickoff package.	3 Analyze RFP. Generate kickoff package.	4 Kickoff meeting. Writing assignments made.	5 Proposal writing. Ongoing communication between proposal manager and writers.	6 Proposal writing.
7 Proposal writing.	8 Preliminary assessment of writing by proposal manager.	9 Proposal manager's comments back to writers.	10 Proposal writing.	11 All proposal sections delivered to proposal manager by writers in electronic form. Proposal manager reviews all sections and electronic files before sending to Publications Department. Production of Red Team draft.	12 Production of Red Team draft. Distribution of Red Team draft by close of business.	13 Red Team review.
14 Red Team review.	15 Interactive meeting between Red Team reviewers and entire proposal writing team.	16 Red Team recovery period (proposal manager prioritizes sections that require reworking).	17 Red Team recovery period.	18 Red Team recovery period.	19 Proposal manager reviews all sections that have been reworked and provides comments back to writers.	20 Additional writing or modification.

Figure 10.7 Calendar provides the highlights of the proposal schedule for easy review.

Sunday	Monday	Tuesday	Wednesday	Thursday	Friday	Saturday
21 Additional writing or modification.	22 Proposal manager reviews all modified sections again.	23 All materials provided to Publications Department for generation of final Gold Team draft.	24 Production and editing of Gold Team draft.	25 Gold Team review.	26 Final production.	27
28	29 Photore-production and final quality-control. packaging.	30 Deliver proposal volumes to client.				

Figure 10.7 (Continued.)

throughout the proposal response life cycle. Management buy-in to your technical solution, staffing plan, management approach, and costing strategy is also pivotal early in the response life cycle. However, senior management must also be willing to entrust and empower the appointed proposal manager to orchestrate the proposal response activities. They must allow the proposal manager to take ownership of and responsibility for the entire task. Too many times there is significant micromanagement of the process or senior management gets involved late in the response process and significantly revises the staffing plan, costing strategy, or technical solution.

10.9 Training additional staff in proposal-management skills

Particularly within small companies, the number of experienced proposal managers is very low, sometimes less than three. One way out of the quandary of always using the same people to manage proposals is to have novice proposal managers directly assist their more experienced colleagues. After two or three assists, perhaps the novice could be appointed as the primary proposal manager with full responsibilities and obligations, but with access to the support of the more senior proposal staff. Outside consultants can also be used for just-in-time proposal training and mentoring.

In-house proposal management and preparation seminars are another option to be considered. Curricula can be developed in-house, and seminars can be held in the evening or on a weekend. The point is to expand the base of proposal management experience within your company. This helps to prevent personnel burnout and promotes better proposal efforts, which in turn translates to winning more often.

10.10 Finish the job at hand

A proposal manager's duties are not over once the proposal documents have been delivered to the government. The proposal manager should either personally review or assign someone to review the entire proposal to identify, for example, any omissions, critical errors, and areas that require improvement. If the procurement cycle includes discussions or an FPR, the proposal manager and his team need to prepare a meaningful response to written questions as well as rehearse oral responses for the face-to-face meeting with the client.

Finally, the proposal manager should prepare a written lessons learned of the entire proposal effort for senior management. Examples of outstanding commitment by members of the proposal team should definitely be noted along with observations, suggestions, and recommendations to improve your company's proposal process. Senior management should then review and act appropriately on the recommendations. "[L]ess successful performers treat every opportunity as a discrete and unique event. These companies spend little time examining and improving their approach…. The proposal process is left unchanged. Each proposal team must relearn how to put together a proposal and experiences the same problems over and over again" [3]. Take the time as a company to learn from each proposal experience and apply those lessons to improving the process on a continuous basis. If ever there was an application for continuous process improvement, the proposal and oral presentation response life cycle is it!

The successful proposal manager has provided leadership for his or her proposal team throughout the entire proposal response life cycle.

10.11 Successful proposal managers

The most successful proposal managers with whom I have worked and been associated possess a marked ability to motivate and communicate with people of all levels—from company presidents to technical proposal writers to publications staff to financial and administrative personnel. They command and instill confidence in others by their technical knowledge, business development acumen, self-confidence, and encouragement. Often, they hold advanced degrees in a relevant technical discipline and have an intimate working knowledge of the particular client organization and its business processes and sell points. Successful proposal managers know when to push their proposal team, when to empathize, when to encourage, when to be flexible, and when to relax. They have a well-developed sense of humor and can laugh even during the most intense proposal development and production periods. These individuals talk with senior management and are forthright in that communication. Senior management, in turn, better understands the technical, costing, and logistical issues at hand that must be addressed and solved. Finally, successful proposal managers seek work-arounds. They demonstrate a certain proposal agility. If one way does not produce the desired result, they are very willing to try another and another until they achieve a resolution. Fundamentally, they are leaders, facilitators, communicators, and listeners—and owners of the proposal development process to the smallest detail.

ENDNOTES

1. Wodaski, Ron, "Planning and Buying for Multimedia: Effective Multimedia for the Small Office," *Technique: How-To Guide to Business Communication*, October 1995, pp. 16–25.

2. Joss, Molly W., "Authoring Alchemy: Ingredients for Brewing Up a Multimedia Masterpiece," *Desktop Publishers*, January 1996, pp. 56–65.

3. O'Guin, Michael, "Competitive Intelligence and Superior Business Performance: A Strategic Benchmarking Study," *Competitive Intelligence Review*, Vol. 5, 1994, pp. 8–9.

Chapter 11

Pursuing international business and structuring international proposals

The time and space buffers that used to limit a company's exposure to changes are gone. Business now is vulnerable to anything that happens anywhere instantly.
—Marc J. Wallace, Jr., Center for Workforce Effectiveness,
   quoted in *Industry Week*

11.1 Overview

The benefits of doing business in the international market can be substantial; however, "[c]ompanies that want to operate globally must have a global mind-set" [1]. In the new millennium, your firm's business opportunities will probably be linked very closely to unfamiliar cultures and customs. Advances in information and telecommunication technology have intensified competition in product and financial markets during the past 20 years, as noted by Marina Whitman and Rosabeth Moss Kanter [2]. And new and different trade, economic, and business models are likely to become necessary to understand and implement within your own company. Professor Vijay Govindarajan of Dartmouth College's Amos Tuck School of Business Administration suggests that "companies must open themselves to cultural diversity and be ready to adopt best practices and good ideas regardless of their country of origin" [1].

Many large U.S. high-tech firms have doubled their size in the global marketplace. And the Peoples' Republic of China, the Republic of Korea (South Korea), and Mexico are assuming leading economic roles, according to Franklin Root, author of *Entry Strategies for International Markets* [3]. Root sees the transformation of the international economy as "a geographical extension of the industrial revolution that started in Great Britain more than two centuries ago" [3]. He asserts that "[t]o survive and prosper in the 1990s, companies will need to develop new strategies. For the truth is that today all business firms—small or large, domestic or international—must strive for profits and growth in an international economy that features enormous flows of products, technology, intellectual and financial capital, and enterprise among countries. In this economy, no market is forever safe from foreign competition. And so even when companies stay at home, sooner or later they learn from hard experience that there are no domestic markets, only global markets" [4].

The losses associated with international business, however, can be devastating. Contract awards can actually result in severe financial losses and liabilities for the selling firm, depending upon the type of the contract and terms and conditions of that contract. Time delays in RFP or tender release dates are frequent, so booking and sales projections should be conservative [5]. A significant challenge associated with conducting international business in many countries (particularly developing nations) is supplying the technical and programmatic expertise, as well as the funding—all as a package to the host country. Two examples of funding sources are the World Bank, headquartered in Washington, D.C., and the Asian Development Bank, centered in Manila in the Philippines. The World Bank receives its resources from industrialized nations and from tapping the world's financial capital markets. Projects funded by the World Bank are not limited geographically, whereas the Asian Development Bank is a lending institution serving only the Pacific Rim [6].

11.2 Where in the world to begin?

As a small U.S. business, the international marketing arena can certainly appear overwhelming. There are, however, several important starting points that can maximize your time and energy when preparing to market your company's goods, services, and knowledge overseas:

- *Export.gov,* the U.S. government's export portal (http://www.export.gov). The Export.gov community is comprised of 19 federal government agencies involved in opening foreign markets and promoting the export of U.S. products and services. These 19 federal agencies are known as the Trade Promotion Coordinating Committee (TPCC). The Export.gov Web site is the TPCC's gateway for the U.S. public to access a broad range of export assistance, export finance, and trade advocacy services.

  By registering for the Export.gov community, you will enable the TPCC agencies to access basic data, enabling them to contact you with targeted information on potential trade opportunities, market research, and government services. Registration also allows U.S. companies and researchers to access a vast library of country and industry market reports and trade leads.

- *BuyUSA.gov* (http://www.buyusa.gov/home/). The U.S. Commercial Service helps U.S. companies find new international business partners in worldwide markets.

- *The Americas* (http://www.buyusa.gov/americas/). The Americas Web site brings together the resources of U.S. Commercial Service offices in 21 markets throughout the region, providing your company with a single point of access to regional trade events, their extensive services, and research covering markets throughout the region. The Americas site also provides information on the existing and proposed free-trade agreements throughout the region, market research, best prospects in the region, trade event lists, industry-specific information, business service providers, useful links, and key contacts.

- *The Economist Intelligence Unit (EIU)* (http://www.eiu.com). The Economist Intelligence Unit is a leading provider of country, industry, and management analysis. Its mission is to provide timely and reliable analyses for making successful global decisions.

  General export information and development assistance is located on the SBA's Web site (http://www.sba.gov/oit/info/index.html). A good place to begin is the SBA7 publication *Breaking into the Trade Game: A Small Business Guide* (http://www.sba.gov/oit/info/Guide-To-Exporting/trad5.html).

- *A Basic Guide to Exporting* (http://www.unzco.com/basicguide/index.html). This 1988 publication of the U.S. Department of Commerce was created in cooperation with Unz & Co., Inc.

11.3 The importance of the World Bank Group

The World Bank is not a "bank" in the common sense, but rather is one of the United Nations' specialized agencies. The World Bank centers its efforts on reaching the Millennium Development Goals, which were agreed to by UN members in 2000 and are aimed at sustainable poverty reduction.

With $20 billion in new loans each year, the World Bank is the largest global provider and coordinator of developmental and infrastructure assistance, which takes a financial, as well as technical, form. The World Bank is owned by its 184 member countries, with the largest industrialized nations—the Group of Seven or G7[1]—controlling about 45% of the Bank's shares. The board of governors and the Washington-based, 24 full-time-member board of directors of the World Bank, however, represent all member countries. Of note is that a central tenant of the evolving role of the World Bank is to build it into a world-class knowledge bank. In FY1998, prototype knowledge management systems (country-specific best practices, lessons learned, success stories) in education and health were established under the direction of the bank's president, James D. Wolfensohn [7]. The bank's current and proactive indigenous knowledge (IK) initiate is a vital case in point. The World Bank finances more than 200 new projects each year, which involve nearly 30,000 individual contracts. Projects financed span AIDS awareness in Guinea, education of girls in Bangladesh, health care delivery in Mexico, and helping East Timor rebuild following independence. Specific projects include the African Virtual University; the Millennium Science Initiative in Chile, designed to boost Chile's science sector; and the SEECALINE nutritional program in Madagascar. Together, the International Bank for Reconstruction and Development (IBRD), the International Development Association (IDA), and the Multilateral Investment Guarantee Agency (MIGA) comprise the World Bank Group, which also works closely with the International Finance Corporation (IFC) which is headquartered in Washington, D.C.

The publication called *UN Development Business* (UNDB), which is released by the United Nations (UN) Department of Public Information

---

1   The G7 members are Canada, France, Germany, Italy, Japan, the United Kingdom, and the United States. The addition of Russia has led to the emergence of the G8. These countries meet annually to address major economic and political issues facing their domestic societies as well as the international community as a whole.

and sells for $590 per year (print) and $550 (online), contains comprehensive World Bank and UN procurement information. (See http://www.devbusiness.com.) The paper version is printed and mailed twice monthly. The online version is updated several times a week. This publication includes invitations to bid on World Bank and UN projects as well as information about projects financed by the African, Asian, Caribbean, and Inter-American Development Banks; the European Bank for Reconstruction and Development; and the North American Development Bank. [Established in 1959, the Inter-American Development Bank (IDB) is an international financial institution that finances economic and social development in Latin America and the Caribbean. Its loans exceeded $6.8 billion in 2003.]

UN Development Business provides a monthly operational summary report listing all of the projects being considered for financing by the World Bank; general and specific procurement notices to supply the products and services required to carry out World Bank–financed projects; and contract awards for major projects. General Procurement Notices (GPNs) alert potential bidders to the overall types of goods and services needed for a project and invite firms to express interest in being placed on lists to receive further announcements. Specific Procurement Notices (SPNs) identify the specific goods and services needed for a project including how to obtain bidding documents, their cost, and the bid deadline date. Your company can subscribe to this important publication at:

http://www.devbusiness.com
Tel. (212) 963-1516
Fax (212) 963-1381
E-mail: dbsubscribe@un.org

The UN Business Directory (http://www.devbusiness.com/directory.asp) lists Web sites of companies and consulting firms involved in international business. This list can be used for advertising products and services and for locating partners already doing business overseas.

The International Bank for Reconstruction and Development and the International Development Association provide average annual lending commitments for investment projects of $10 billion to $15 billion a year. These funds are used by recipient countries to purchase goods and equipment, construct civil works, and obtain the consulting services needed to implement these projects. Each project may involve many separate contracts and business opportunities for suppliers, contractors, and consultants worldwide. Companies, academic institutions, nongovernmental organizations (NGOs), and individuals from member countries of the World Bank are eligible to compete for these business opportunities. Within any given project, there can be literally hundreds of business opportunities varying in size from as little as a few thousand dollars to as large as tens of millions of dollars.

When pursuing business opportunities in projects financed by the World Bank, it is essential to understand that *the implementing agency in the recipient country is responsible for procurement.* All contracts are between the borrower (usually the government department that is its implementing agency) and the supplier, contractor, or consultant. The World Bank's role is to make sure that the borrower's work is done properly; that the agreed procurement procedures are observed; and that the entire process is conducted with efficiency, fairness, transparency, and impartiality.

More information on this process can be found in the *Guidelines: Procurement under IBRD Loans and IDA Credits,* dated May 2004 (http://web.worldbank.org/WBSITE/EXTERNAL/PROJECTS/ PROCUREMENT/0,,contentMDK:20060840~menuPK:84282~pagePK :84269~piPK:60001558~theSitePK:84266,00.html]. For comprehensive guidance, you should refer to the *Resource Guide to Consulting, Supply and Contracting Opportunities in Projects Financed by the World Bank* (http://web.worldbank.org/WBSITE/EXTERNAL/PROJECTS/ PROCUREMENT/0,,contentMDK:20109705~menuPK: 84282~pagePK:84269~piPK:60001558~theSitePK:84266,00.html).

Occasionally, the World Bank hires consulting firms to support its development work. These opportunities are posted at *Requests for Expressions of Interest for Consulting Work* (http://web.worldbank.org/WBSITE/ EXTERNAL/PROJECTS/PROCUREMENT/0,,contentMDK:20127689 ~menuPK:84282~pagePK:84269~piPK:60001558~theSitePK:84266,00. html).

World Bank borrowers are required to submit timely notification of bidding opportunities and to advertise these opportunities and expressions of interest. Invitations to bid or express interest for contracts under World Bank-financed projects can be found at the Development Gateway Market (http://www.dgmarket.com/). Currently, dgMarket publishes tender notices for projects funded by the African Development Bank, the Asian Development Bank, Europe Aid, European Bank for Reconstruction and Development, European Investment Bank, European Union (EU) members states, Phare/Tacis, and the World Bank. For more information, contact info@dgmarket.com. Current tenders are categorized into topic areas such as transport, energy, urban development, and agriculture and food.

The Phare Programme is an EU initiative that supports the development of a larger democratic family of nations within a prosperous and stable Europe. Phare does this by providing grant finance to support its partner countries in central and eastern Europe through the process of economic transformation and strengthening of democracy to the stage where they are ready to assume the obligations of membership of the European Union. The Tacis Programme is an EU initiative for the New Independent States and Mongolia that fosters the development of harmonious and prosperous economic and political links between the European Union and these partner countries. Tacis does this by providing grant financing for know-how to support the process of transformation to market economies and democratic societies.

Finally, the World Bank's Web site also provides country-specific procurement forecasts (http://www-ds.worldbank.org/servlet/WDS_IBank_Servlet?ptype=advSrch&psz=20&pcont=results&dt=540617).

To assess the qualifications of firms and to assist in-country borrowers who are trying to establish a short list of potential contractors, the World Bank and Inter-American Development Bank maintain an automated database of consulting firms interested in doing business on World Bank–financed projects. This database is called the Data on Consulting Firms (DACON) system and is open only to consulting companies with five or more staff as opposed to manufacturers or suppliers. You can contact the DACON Center at http://www.adb.org/consulting/dacon.asp.

Of note is that a private company cannot approach the World Bank directly to request financial assistance for a project in which it and the host country government are interested. Only the government of an eligible World Bank member can request such financing.

The Data on Individual Consultants (DICON) system contains the names of persons who are citizens of the Asian Development Bank's (ADB) member countries, are at the forefront of their respective fields, and may be (1) practicing consultants, (2) employees whose organizations are prepared to release them for short periods to serve ADB, or (3) retired persons with special experience. The names, expertise codes, and other details of the qualifications and experience of several thousand consultants are found in the DICON system. Only a limited number of these consultants are hired by ADB each year.

When it requires the services of a particular type of consultant, ADB searches DICON to determine whether appropriate candidates are available, by matching expertise codes to the desired skills and experience. It lists all registered individuals with the required expertise and reviews their files, and then progressively shortens this list by removing less qualified candidates until only the three best qualified remain. ADB contacts these three shortlisted candidates in the order of ranking to determine their availability and interest in the project.

11.4 Your company's participation in United Nations procurements

The United Nations is comprised of a variety of organizational entities such as agencies, organizations, commissions, programs, and funds. Each entity has a distinct mandate that addresses political, economic, social, scientific, or technical fields. Although each organization is responsible for its own procurement requirements, there are many commonalities in the procurement processes across the UN system. As part of Secretary-General Kofi Annan's reform of the United Nations, the procurement activities of the UN system have undergone considerable change. *Why?* To make the process more efficient and cost-effective, as well as to maximize equity, openness, and integrity. To this end, a single portal, http://unbiz.un.int, has been established to link all of the Web sites of all participating UN organizations.

The United Nations, including its many affiliated organizations, represents a global market of approximately $4.6 billion annually for all types of goods and services. The *UN Global Marketplace* (http://www.uncsd.org/AboutUncsd.aspx) acts as a single portal through which the business community may register with the UN system, providing an excellent springboard to introduce your goods and services to many UN organizations, countries, and regions. The database facilitates the interchange of supplier information within the UN system as information is made available to all UN organizations, and it acts as a procurement tool to shortlist suppliers for competitive bidding.

Also, the Inter-Agency Procurement Service Office (IAPSO) has developed booklets with practical tips on how to do business with the United Nations. *The Business Guide* (http://www.iapso.org/pdf/gbg_master.pdf) for potential suppliers of goods and services is a 148-page booklet that describes the procurement needs and procedures of all UN organizations (with addresses and phone/fax numbers) and indicates the dollar value of goods and services purchased during the year. *Doing Business with the UN System* can be downloaded at http://www.iapso.org/pdf/tips.pdf.

In addition, the UN Procurement Division posts its acquisition plan at http://www.un.org/Depts/ptd/2004plan.htm. This plan offers a projection of requirements from offices and departments at UN Headquarters and offices away from Headquarters.

IAPSO was established by the United Nations Development Program (UNDP) in 1978 to assist its partner organizations within the UN system in the most economical acquisition of essential equipment and supplies.

Originally based in New York, and designated the Interagency Procurement Services Unit (IAPSU), it moved to Geneva in 1982. In July 1989, the office was relocated to Copenhagen and renamed the Interagency Procurement Services Office (IAPSO).

11.5 European Bank for Reconstruction and Development (EBRD)

Established in 1991, the European Bank for Reconstruction and Development (EBRD; http://www.ebrd.com) assists 27 countries in central and eastern Europe and the former Soviet Union [now Commonwealth of Independent States (CIS)] with privatization and entrepreneurship efforts and promotes structural and sectoral economic reforms. The EBRD encourages cofinancing and foreign direct investment from both the private and public sectors. The mandate of the EBRD stipulates that it must only work in countries that are committed to democratic principles. The EBRD's *Procurement Policies and Rules*—derived from Standard Bidding Documents (SBDs) developed and in use by the World Bank, the Asian Development Bank, and the Inter-American Development Bank—articulates the procurement guidelines to be followed in EBRD-financed operations and establishes tendering procedures for use by clients when procuring goods and services following open tendering in operations financed by the EBRD. In addition, the EBRD *Procurement Policies and Rules* outlines how tenders are received, opened, and evaluated under open-tendering procedures for contracts financed by the European Bank for Reconstruction and Development. Invitations to tender, expressions of interest, contract award information, and other essential information regarding EBRD-funded contracts is published monthly by the Bank in *Procurement Opportunities,* which is also available on the Bank's Web site at http://www.ebrd.org/oppr/procure/index.htm.

The term *EU* denotes the European Union (http://www.europa.eu.int), which now consists of 25 member nations that include Austria, Belgium, Denmark, Finland, France, Germany, Greece, Ireland, Italy, Luxembourg, the Netherlands, Portugal, Spain, Sweden, and the United Kingdom. This group of European countries has decided to collaborate on a variety of significant areas that encompass a single economic market as well as a common foreign policy and legislative and judicial policies. The European Union came into existence as a result of the ratification of the Maastricht Treaty in November 1993; however, its roots can be traced back to the Treaty of Paris in 1951 and the Treaty of Rome in 1957. Formally, the EU consists of three pillars: the European Parliament, the

Council of the European Union, and the European Commission. The European Investment Bank (EIB) serves as the European Union's financing institution.

Official Journal of the European Communities

The *Official Journal of the European Communities* (OJ or OJEC; http://www.ojec.com) provides a gateway to specialized resources for purchasers who wish to publish tender notices in the S series of the OJEC and suppliers who wish to search for business opportunities. The *Official Journal of the European Communities* is the only periodical published every working day in all 11 official languages of the European Union. It consists of two related series (the L series for legislation and C series for information and notices) and a supplement (the S series for public tenders).

The Tenders Direct Web site (http://www.tendersdirect.co.uk/default.asp) provides access to more than 30,000 current government and utility company contracts in the United Kingdom and Europe. The Office for Official Publications of the European Communities (EUR-OP) provides between 400 and 900 new tender notices every day. Tenders Direct is provided on a subscription basis and you must register for one of the services before you can search for tenders. You can try out the service free of charge for 30 days before you decide to subscribe.

11.6 Asian Development Bank (ADB)

The Asian Development Bank (ADB), a multilateral development finance institution whose capital stock is owned by 63 member countries including the United States, is engaged in promoting the economic and social progress of its developing member countries in the Asia-Pacific region. Since the ADB began operations in December 1966, it has been a catalyst in promoting the development of one of the most populous and fastest growing regions in the world. ADB makes loans and equity investments, and provides technical assistance grants for the preparation and execution of development projects and programs, and also for advisory purposes. It promotes investment of public and private capital for development purposes. You can visit ADB on the Web at http://www.adb.org. Headquartered in Manila, Philippines, the ADB is led by a board of governors and board of directors. At the time of this writing, Tadao Chino was serving as the bank's seventh president, a seat he has held since 1999.

Like the World Bank, ADB's mandate to assist developing member countries in the Asia-Pacific region has resulted in the need to maintain an inventory of suitably qualified firms or organizations from its member countries that could act as consultants to provide services required for various projects assisted by the bank. ADB's information on the capabilities and experience of firms or organizations is maintained in the ADB DACON system, the acronym for Data on Consultants. It is a computer inventory used by ADB to record the eligibility of such firms for possible engagement as well as their experience and qualifications for easy retrieval by ADB when needed. To be eligible for registration in the ADB DACON system, a company must have a minimum of five permanent full-time professional staff, must have been incorporated for at least 2 years, and must have completed a minimum of three major projects.

11.7 International market planning

In developing international marketing entry and marketing plans, keep in mind that your company is entering both a new country and a new market [8]. The entry mode (export, contractual, or investment) determines the amount of a company's control over the marketing program in the target country. For example, indirect exporting and pure licensing allow little or no control over the marketing program. Your company's international marketing plan should include objectives for sales volumes, market share, profits, and return on investment as well as resource allocations and a time schedule [9]. Market potential and growth, as well as risk associated with market entry, can be evaluated through such published sources as the *Price Waterhouse Country Guides* and Dun and Bradstreet's *Exporters' Encyclopedia*.

If you are importing or exporting from various countries, then D&B *Exporters' Encyclopedia* can help you gain insight into the trading environment of more than 180 countries so you can decide to do business safely and profitably. The *Exporters' Encyclopedia* provides you with the following details:

- *Trade regulations:* Licensing, tariffs, value for duty, inspection requirements, import taxes;
- *Documentation:* Shipping requirements, invoices, bill of lading, certificates of origin;
- *Key contacts:* Foreign consulates, chambers of commerce, government departments;
- *Transportation:* Post entry, warehousing, foreign trade zones, shipping restrictions, inland transport;

- *Legislation affecting export commerce:* Consumer and environmental protection, safety regulations.

The *Exporters' Encyclopedia* also provides tips on foreign business travel, business etiquette climate, and holidays.

The U.S. Commercial Service provides market research that is available free of charge to U.S. citizens and U.S. companies. Market research publications are available online at http://www.export.gov/marketresearch.html. The first time you go to the site you will need to register with the site. Registration is free of charge. U.S. Commercial Service publications include the following:

- *Country Commercial Guide:* A comprehensive annual report that includes an overview of each country's commercial environment, economic analysis, best prospects for products and services, and valuable contact information.

- *Industry Sector Analysis:* In-depth reports (10 to 15 pages) that cover key sectors in a country and give companies a more complete overview of some of the best market prospects in the country.

- *International Market Insight:* A short and timely report on market developments and opportunities. Includes reports on a variety of topics, including regulatory updates, major projects, trade leads, hot opportunities, industry profiles, market developments, and trade show information.

The Economist Intelligence Unit (EIU) out of London, England, publishes special reports on specific international marketing topics [10]. The EIU, http://www.eiu.com, is an information provider for companies establishing and managing operations across national borders anywhere in the world. Established 50 years ago in London, the organization now has a worldwide network of offices in London, New York, Hong Kong, Vienna, Singapore, and Tokyo. The EIU produces analyses and forecasts of the business and political environments in more than 200 countries. The EIU's new Industry Forecasts are 5-year economic forecasts for eight key industries. The service covers the world's 60 major economies, which account for more than 95% of global output and trade. The EIU is part of the Economist Group, which also publishes *The Economist* magazine.

Many small companies simply do not have the staff to dedicate to the time-consuming task of international marketing. There are local and regional trade programs designed to assist small businesses with identifying international markets. One such program is the State University of New York (SUNY) at Albany's International Marketing Assistance Service

(IMAS). This service pairs SUNY-Albany graduate students in marketing with local New York companies to identify and seek out international markets. The IMAS has also conducted several trade missions, which have included trips to Mexico, Chile, Argentina, Venezuela, and Brazil. In New York State, the Department of Economic Development has participated in several trade shows coordinated by the U.S. Department of Commerce that link small companies with representatives and distributors in international countries [11, 12].

Because of the critical shortages of resources in small companies, international business must be a carefully planned and orchestrated process. Small companies successful in this field often have foreign shareholders who know the local environment and its pitfalls and processes. International market entry must become part of your strategic plan and marketing plan if it is to be pursued with the diligence that successful international business requires.

11.8 In-country partnerships

Creating ongoing business partnerships with the host country's key decision makers is essential to long-term marketing, proposal, and contractual success. Partnerships can be strengthened through scheduled in-country visits [be sure to work around major holidays and holy days such as Ramadan in Islamic countries, Setsubun in Japan, Carnival in Brazil, and Spring Festival (Chinese New Year)] that include social and community activities, reciprocal invitations to the United States, and establishment of offices in country. Well-managed representatives, agents, or distributors who understand the language, culture, and procurement laws and processes and are sensitive to the in-country political directions can be invaluable to your company's success. Cross-cultural considerations must be taken into account, so it is suggested that these representatives be employees or consultants of the buying country's nationality. Also, do not try to cover the whole world—develop a profitable beachhead in a country that your company knows already.

In-country partnerships can result in your company defining the technical requirements that will appear ultimately in the RFP or tender.

Your company might also consider establishing formal business partnerships with several select consulting firms in strategically important countries throughout the world. These firms can then function as in-country partners, providing important and personalized local support as well as facilitating worldwide coverage for your multinational clients.

11.9 Host country procurement environments

As in domestic U.S. business development efforts, long-term marketing and relationship building on a global scale are critical to your company's contract award success and long-term profitability. Understanding your international client includes his technical expectations, procurement and source selection cycle, business decision-making processes and framework, import-export regulations, cultural environment, language requirements, and political agenda. Detailed understanding extends to potential tax exposure, countertrade requirements, work visas required, host country law, U.S. treaties with the host country, quality assurance parameters, delivery and acceptance restrictions, payment parameters, and currency restrictions [13]. In certain countries (e.g., Brazil) and on specific types of procurements, your competitors will have the opportunity to openly and thoroughly review your proposal documents for the purpose of attempting to disqualify your submittal on technical grounds or procedural or administrative technicalities.

11.10 Import-export considerations and technology transfer

Time must be allocated in delivery schedules for any necessary U.S. Department of State approval or Department of Commerce licensing, such as in the case of advanced electronic equipment. "The international customer community wants to obtain as much technology from U.S. corporations as possible" [14].

The United States has export controls for two primary reasons: (1) national security (terrorism and weapons control) and (2) trade protection (supply shortages and crime control). The U.S. Directorate of Defense Trade Controls (DDTC), Bureau of Political-Military Affairs, in accordance with 22 U.S.C. 2778-2780 of the Arms Export Control Act (AECA) and the International Traffic in Arms Regulations (ITAR) (22 C.F.R. Parts 120-130), is charged with controlling the export and temporary import of defense articles and defense services covered by the United States Munitions List (USML). It has among its primary missions (1) taking final action on license applications for defense trade exports and (2) handling matters related to defense trade compliance, enforcement, and reporting. ITAR regulations prohibit the export of military articles and information as well as defense services to countries such as Iran, Iraq, Libya, North Korea, Syria, and Belarus.

11.11 Risk assessment

Your company may want to consider establishing a separate legal entity for the purpose of minimizing corporate exposure in the international arena. Carefully consider the following: taxes and duties, import-export quotas, host country legislation and regulations, language, host country culture, quality and audit standards, countertrade requirements, packaging or labeling standards, political issues and stability, currency exchange rates, foreign direct investment, customs paperwork and protocol, weather, religion, inspection and acceptance guidelines, in-country banking restrictions, and host country law.

The U.S. Small Business Administration has teamed with the U.S. Chamber of Commerce and American International Group, Inc. (AIG) to launch a Web site to help small business owners determine what insurable exposures they may encounter in doing business overseas. The URL is http://www.AssessYourInternationalRisk.org.

11.12 Terms and conditions

Developing restrictive, risk-aware terms and conditions (Ts & Cs) is absolutely vital for international proposals. There are many challenges associated with resolving legal problems at an international level. Questions arise as to where to try a particular case, which law should apply, and how to apply foreign laws. To help mitigate the problems inherent in such issues, include clauses in your Ts & Cs that address in detail which specific laws will obtain in any contractual disputes [15].

There are certainly risks associated with international payments, two of which are nonpayment and variations in the foreign exchange rate [16]. Such risk can be mitigated by stipulating cash in advance or an irrevocable confirmed letter of credit. This letter of credit is issued by the importer's bank and then confirmed by a bank in the exporter's (your company) country. Thus, your company does not have to depend on the importer for payment [17]. Be aware, however, that your competitors may be offering easier, more flexible, and therefore more attractive payment terms. U.S. exporters can obtain export credit insurance that protects against the risks of nonpayment for both commercial and political reasons from the Foreign Credit Insurance Association (FCIA). This is an independent association of insurance companies. Having this insurance can allow your company to extend credit on more favorable terms to overseas buyers and thus be more competitive.

Ts & Cs will vary depending on the product or service being provided to the host country or industry. A listing of frequently used line items found in Ts & Cs includes the following:

- Warranty terms, such as commencement of warranty and warranty for replacement parts;
- Warranty exclusions (for example, improper unpacking, installation, or treatment of equipment or unauthorized attempts to repair, reconfigure, or modify equipment);
- Extended warranty and maintenance;
- Limitations on liability;
- *Force majeure* (protects supplier from casualty or cause beyond reasonable control, such as strikes, floods, riots, acts of governments, and war);
- Pricing for tasks beyond scope of work;
- U.S. government approvals;
- Taxes and duties;
- Dispute resolution (for example, arbitration under the Rules of Procedure of the International Chamber of Commerce, London, England);
- Excusable delays;
- Shipping and insurance;
- Contract termination;
- Acceptance criteria;
- Payment terms, schedule, and currency;
- Training;
- Documentation (it is critical to specify the language and dialect in which training manuals will be provided).

11.13 Ex-Im Bank of the United States assists small businesses

The Export-Import Bank of the United States (Ex-Im Bank; http://www.exim.gov) is the official export credit agency of the United States. With nearly 70 years of experience, the Ex-Im Bank has supported more than $400 billion of U.S. exports, primarily to developing markets worldwide. The bank's mission is to assist in financing the export of U.S. goods and services to international markets. Led by Philip Merrill, president and chairman, the Ex-Im Bank enables U.S. companies—large and small—to turn export opportunities into real sales that help to maintain and create U.S. jobs and contribute to a stronger national economy. In fiscal year 2001, the Ex-Im Bank authorized financing to support $12.5 billion of U.S. exports worldwide. More than $1.6 billion of this financing supported U.S. small business exports.

The Ex-Im Bank and the U.S. Small Business Administration (SBA) have signed a memorandum of cooperation to increase coordination to promote exports by the U.S. small business community. The initiative is also intended to increase awareness and use of the financing products offered by Ex-Im Bank and the SBA that facilitate export-related working capital loans to small businesses.

The Ex-Im Bank does not compete with private-sector lenders but provides export financing products that fill gaps in trade financing. It assumes credit and country risks that the private sector is unable or unwilling to accept. The bank also helps to level the playing field for U.S. exporters by matching the financing that other governments provide to their exporters.

Ex-Im Bank provides working capital guarantees (pre-export financing), export credit insurance (postexport financing), and loan guarantees and direct loans (buyer financing). No transaction is too large or too small. On average, 85% of the bank's transactions directly benefit U.S. small businesses.

In 2002, the Ex-Im Bank and the SBA agreed that their representatives would meet regularly to strengthen institutional ties and identify ways to improve delivery of trade assistance programs. The two agencies will work together to develop a joint marketing campaign for their respective working capital programs, including publications, a program guide, and lender and small business training.

Importantly, there are U.S. Export Assistance Centers located in major metropolitan areas throughout the United States. These centers serve as one-stop shops that are ready to provide your small- or medium-sized business with local export assistance. You can receive personalized assistance by professionals from the SBA, the U.S. Department of Commerce, the Ex-Im Bank, and other public and private organizations. For a listing of Export Assistance Centers near you, visit http://www.sba.gov/oit/export/useac.html.

General export information and development assistance are located on the SBA's Web site (http://www.sba.gov/oit/info/index.html). A good place to begin is the SBA publication *Breaking into the Trade Game: A Small Business Guide* (http://www.sba.gov/oit/info/Guide-To-Exporting/trad5.html).

11.14 Helpful Web-based resources and in-country support infrastructures for small businesses

The Web provides small businesses with a host of business development sites to support international marketing initiatives.

MERX: Canada's national electronic tendering service

http://www.merx.com. MERX is the most complete source of public tenders available in Canada. This Internet-based electronic tendering system was established and launched in 1997 by Cebra, the Bank of Montreal's electronic commerce company. MERX is now a subsidiary of Mediagrif Interactive Technologies, Inc. The service is designed to improve access, increase competition, and provide a level playing field for small businesses competing for bidding opportunities within the public sector in Canada. At the present time, MERX provides the opportunity to bid on most Canadian federal government contracts. The system also posts opportunities for eight provincial governments: Ontario, Quebec, New Brunswick, Nova Scotia, Manitoba, Prince Edward Island, Alberta, and Saskatchewan. MERX replaced Canada's existing Open Bidding System in October 1997.

Government Logistics Department, Government of the Hong Kong Special Administrative Region, Hong Kong, China

https://www.ets.com.hk/English/GeneralInfo/info.asp. The Electronic Tendering System (ETS) Web site is hosted by the Government Logistics Department (GLD) of the Government of the Hong Kong Special Administrative Region through Computer and Technologies International Limited (C&T). This Web site is mainly designed for displaying tender notices, contract award notices, and general terms and conditions for GLD tenders. By subscribing to the ETS service operated by C&T, you can log onto the system and enjoy additional functions such as applying online as a GLD supplier, updating your company information, downloading tender documents and clarifications, submitting queries on tenders, and submitting tender offers.

GLD is the central purchasing, storage, and supplies organization that services more than 80 government departments, subvented organizations, and certain nongovernment public bodies.

European Tenders Direct Procurement Gateway

The subscription-based Tenders Direct service (http://www.tendersdirect. co.uk) has been supplying information on tenders and contracts to companies since 1993. The company is officially endorsed by Eur-Op, the Office for Official Publications of the European Communities, who license this service to publish public sector tender notices from Supplement S of OJEC.

This Web site provides access to more than 30,000 current government and utility company contracts in the United Kingdom and Europe. The database is updated with between 400 and 900 new tender notices every day. Predefined categories help users quickly find relevant tenders. Keyword searching is also available.

The National Euro Info Centre Network (EIC)

Euro Info Centres (EICs; http://www.euro-info.org.uk/index_new.htm) provide local access to a range of specialist information and advisory services to help companies develop their business in Europe. The Euro Info Centre network can inform you about all tendering opportunities that appear in your area of interest.

Public procurement represents a major part of the European Union (EU) economy. Under European legislation, contracts issued by the public sector, local authorities, government bodies, and utility companies, among others, must be advertised and awarded according to strictly regulated criteria. In 2002, the total EU procurement market was worth €1.5 trillion or 16% of EU gross domestic product (GDP) and in Ireland public procurement represents 13.3% of our GDP. Competitive public procurement is essential for efficient public spending and for this reason the procurement process is strictly regulated by EU and national legislation. It is estimated that the existing Public Procurement Directives have reduced by 30% the prices paid by public authorities for goods and services. The EU is making an ongoing effort to further improve the procurement procedures with such guidance as Directive 2004/18/EC and Directive 2004/17/EC, which are aimed at promoting the use of electronic procurement.

Under Irish and European law, all public service contracts above certain minimum thresholds have to be advertised in the Supplement to the *Official Journal of the European Union,* where they can be found on the European tenders Web site TED (Tenders Electronic Daily) at http://ted. eur-op.eu.int.

Department of Trade and Industry (DTI) (United Kingdom)

The Department of Trade and Industry (DTI; http://www.dti.gov.uk) focuses on "prosperity for all" by working to create the best environment for business success in the United Kingdom. DTI, formerly the Board of Trade, helps people and companies become more productive by promoting enterprise, innovation, and creativity. DTI champions U.K. business at home and abroad. It invests heavily in world-class science and technology. In addition, DTI stands for fair and open markets in the United Kingdom, Europe, and the world.

According to a statistical report (http://www.sbs.gov.uk/content/analytical/statistics/news162.pdf) issued in August 2004 by the Small Business Service (SBS), an executive agency of DTI, 99.2% of British business enterprises were small (0 to 49 employees; businesses with 0 employees were sole proprietorships or self-employed owner-managers). Together, small and medium-sized enterprises accounted for more than half of the employment (58.2%) in the United Kingdom.

Middle East Association

The Middle East Association (http://www.the-mea.co.uk/) is an independent nonprofit organization established in 1961 to promote trade and investment in the Arab world, Iran, Turkey, and Afghanistan, on behalf of its members. The association maintains close relations with British government departments, particularly the Foreign and Commonwealth office and Trade Partners UK. In addition, the association has connections with overseas government ministries and departments and other official bodies as well as links with businesses and business organizations in the region.

The new definition of small firms within the European Union

Micro, small, and medium-sized enterprises (SMEs) are very socially and economically important, because they represent 99% of all enterprises in the EU. These firms also provide approximately 65 million jobs and contribute to entrepreneurship and innovation. However, they face particular difficulties that the EU and national legislation try to redress by granting various advantages to SMEs. A legally secure and user-friendly definition is necessary in order to avoid distortions in the single market. A single market is a customs union with common policies on product regulation and freedom of movement of all the factors of production (goods, services, capital, and labor). Examples of single markets include the European Community (EC) and the European Economic Area (EEA).

In May 2003, the European Commission adopted a new definition of micro as well as SMEs. This definition was aimed at promoting entrepreneurship, investment, and growth; facilitating access to venture capital; cutting administrative burdens; and increasing legal certainty. This new definition was shaped by two rounds of extensive public consultation in 2001 and 2002. It maintains the different staff thresholds that define the categories of micro and SMEs. However, it provides for a substantial increase of the financial ceilings (turnover or balance sheet total), in particular as a result of inflation and productivity increases since 1996, the date of the first European Community SME definition.

Various provisions mean that the benefit of access to national SME support mechanisms and European programs supporting SMEs is reserved exclusively for those enterprises that have the characteristics of real SMEs (without the economic strength of larger groupings). To allow a smooth transition at EU and national level, the new definition will be used as of January 1, 2005. This modernization of the SME definition will have an impact on promoting growth, entrepreneurship, investments, and innovation. It will favor cooperation and clustering of independent enterprises.

At the EU level, almost every policy has an SME dimension. SMEs' special needs and concerns are incorporated into most EU policies and programs. For 2002–2006, nearly €2.2 billion has been allocated to support SMEs' research and innovation efforts.

The revision ensures that enterprises that are part of a larger grouping—and could therefore benefit from a stronger economic backing than genuine SMEs—do not benefit from SME support schemes. The increase of the financial ceilings is designed to take into account subsequent price and productivity increases since 1996, however the head count ceilings remain fixed (see Table 11.1).

The European Charter for Small Enterprises was approved by EU leaders at the Feira European Council in June 2000. The charter calls on member states and the commission to take action to support and encourage small enterprises in 10 key areas:

1. Education and training for entrepreneurship;

2. Cheaper and faster start-up;

3. Better legislation and regulation;

4. Availability of skills;

5. Improving online access;

Enterprise category	Headcount	Turnover or Balance sheet total	
Medium-sized	< 250	≤ € 50 million	≤ € 43 million
Small	< 50	≤ € 10 million	≤ € 10 million
Micro	< 10	≤ € 2 million	≤ € 2 million

Table 11.1 New Categorization of Small and Medium-Sized Businesses Within The European Union

6. Getting more out of the single market;

7. Taxation and financial matters;

8. Strengthening the technological capacity of small enterprises;

9. Making use of successful e-business models and developing top-class small business support;

10. Developing stronger, more effective representation of small enterprises' interests at European Union and national levels.

The Multiannual Programme for Enterprise and Entrepreneurship (Council Decision 2000/819/EC) of December 2000 is a multiannual program for enterprise and entrepreneurship, and in particular for SMEs (2001–2005). This plan of activities aims at:

- Enhancing the growth and competitiveness of business in a knowledge-based internationalized economy;
- Promoting entrepreneurship;
- Simplifying and improving the administrative and regulatory framework for business so that research, innovation, and business creation in particular can flourish;
- Improving the financial environment for business, especially SMEs;
- Giving business easier access to European Community support services, programs, and networks and improving the coordination of these facilities.

UK Trade & Investment

UK Trade & Investment (http://www.tradepartners.gov.uk) is the British government organization that supports both companies in the United Kingdom trading internationally and overseas enterprises seeking to locate in the United Kingdom. This organization's trade services offer support to companies based in the United Kingdom to achieve their export potential. For those firms exporting for the first time or businesses experienced in international trade expanding into new markets, this organization can help develop export capabilities and provide expert advice, reliable data, and professional research. Its international trade teams are active in more than 200 posts worldwide and are available in more than 45 U.K. offices. UK Trade & Investment works closely with the English regional development agencies and the national development agencies in Scotland, Wales, and Northern Ireland.

European Procurement Information Network (EPIN) (Ireland)

The European Procurement Information Network (EPIN; http://www.epin.ie/epinie.htm) is a business development tool that provides companies with access to a database notice of contracts. The EPIN database contains all contracts published in the S-Supplement of the *Official Journal of the European Communities* (OJEC), that is, all the notices in Tenders Electronic Daily (TED). EPIN provides access to the following: (1) £1 billion worth of new contracts a day, (2) market research facilities, (3) competitor analysis, and (4) links to your company's Web site.

Irish Government Public Sector Procurement Opportunities Portal

This site at http://www.e-tenders.gov.ie has been developed as part of the Irish Government's Action Plan on Implementing the Information Society in Ireland. The site is designed to be a central facility for all public-sector contracting authorities to advertise procurement opportunities and award notices. The site is managed by the National Public Procurement Policy Unit (NPPPU) of the Department of Finance.

On a daily basis, the site displays all public-sector procurement opportunities currently being advertised in the *Official Journal of the European Communities* (OJEC), and the national and local press and other opportunities directly uploaded to the site from awarding authorities. At any given time, it will contain all open opportunities in the form of tender notices, prior indicative notices (PINs), and contract award notices (CANs). It also provides associated tender documents (where available) that can be down-loaded from the site. Awarding authorities are encouraged to publish information on opportunities not currently being advertised in other media. In addition, the site has the functionality to allow awarding authorities to publish notices on the site that will then be sent to the OJEU automatically. The http://www.etenders.gov.ie site also provides background information on procurement rules and guidelines. These include European Directives and National Guidelines on the Public Procurement Process. The site offers the opportunity to widen the net of potential suppliers to the Irish public sector. There is no charge to contracting authorities for this service; the site is also freely available for use by the public.

The public-sector tender market in Ireland is worth about £11 billion annually. All of the procurement opportunities advertised in this market by Central and Local Government in Ireland (North and South) are published on the Tenders Ireland Web site (http://www.tendersireland.com).

As EU and national directives continue to change, more and more of these contracts are being advertised. They offer great market opportunities for small and medium-sized companies as well as for the larger organizations that have traditionally supplied this market.

Office of Public Works (Ireland)

For more than 160 years, the Office of Public Works (OPW; http://www.opw.ie) has provided the government in Dublin and the public sector with services in the areas of procurement, property, and construction. The OPW is a government office responsible for the procurement of supplies and services in common use in government departments, such as office supplies, print, publications, uniforms, vehicles, furniture, and so forth.

The Government Supplies Agency, which is part of the OPW, has responsibility for the central management of government procurement and publications. The agency is responsible for ensuring the most cost-effective means of meeting departmental and office needs for goods and services. It prescribes those goods that should be supplied by the agency itself and those that departments may procure directly from the private sector. The agency centralizes the purchasing process by means of its technical expertise.

Système d'Information pour les Marchés Publics (European Commission)

The Système d'Information pour les Marchés Publics (SIMAP; http://simap.eu.int/) project was designed to provide the information system infrastructure needed to support the delivery of an effective public procurement policy in Europe by providing contracting entities and suppliers with the information they need to manage the procurement process effectively. The European Commission launched the project in order to encourage best practice in the use of modern information technology for public procurement. Initially, the project is intended to improve the quality of information about European Union procurement opportunities and to ensure that information is made known to all potentially interested suppliers.

SIMAP aims to support an effective single market by encouraging suppliers and contracting entities to adopt best practices and use electronic commerce and information technology to provide all of the information needed to deliver value for money in public procurement.

THEMiS: the System for Regulated Procurement

Public and utility procurement in the European Union is subject to regulation that articulates detailed procedures for awarding contracts. Achilles Information Ltd. is a specialist in this area of procurement, and provides both consultancy and software to help purchasers meet their commercial needs while complying with the legislation. Because of the complexity of the requirements of the EU procurement legislation, and its U.K. implementation, Achilles has developed THEMiS (http://www.achilles.no/services/themis/index.html), a specialist decision support package that provides the tools to comply with the legislation easily and with minimum cost. THEMiS is updated both with new documents and court cases at least every 4 months. The system also gives access to the latest news and information in the United Kingdom, Europe, and beyond, where legislation on public procurement is becoming increasingly important.

Procurement Information Online (Germany)

The Procurement Information Online (PIO) Web site (http://www.procurement.de) is maintained by the law firm Arnold Boesen in Bonn and provides a list of links that potential tenderers in the German procurement market may find useful. The following are online databases established for enterprises that want to tender for public contracts. In general, access is for paying subscribers.

- *Ausschreibungs-ABC:* This is a specialized database for public procurement in Germany. It can only be used by registered paying subscribers. Registration is possible online.

- *Bundesausschreibungsblatt Online*: This is the official and specialized organ for the procurement of public institutions in Germany.

Procurement opportunities are published here according to the German *Verdingungsordnungen*. Moreover, there are other institutions such as NATO and institutions of the European Union that publish procurement opportunities here. The database can be used only by subscribers of the *Bundesausschreibungsblatt*. A subscription can be ordered online directly.

- *bi online:* This is a specialized database for public procurement concerning building and construction in Germany. It can only be used by registered paying subscribers. Registration is possible online.
- *TIL Tenders Information Library:* This database is specialized for public procurement in Germany.

In Germany special advice centers (*Auftragsberatungsstellen*) impart practical experience and information to enterprises that plan to tender for public contracts. Usually, these centers are organized as registered societies or assigned to the chambers of commerce of the several *länder* (states). Each *land* (state) keeps a center of advice to guide enterprises that have a presence in a given *land*. Some centers already have a site on the Internet, while others can be contacted by mail or telephone or telefax only.

The *Vergabe-News* is a monthly German-language information service updating tenderers and contracting authorities on the latest developments in German and European procurement law and practice. It is edited by the specialized law firm Arnold Boesen.

Confederation of German trade fair and exhibition industries

The Association of the German Trade Fair Industry (AUMA; http://www.auma.de) represents the interests of the trade fair industry on a national and international level. AUMA, in close cooperation with the German Federal Ministry of Economics and Labour and also the Ministry of Consumer Protection, Food and Agriculture, prepares the official German trade fair program abroad. Within the framework of this program, the German federal government provides financial assistance for German companies' joint participation at foreign events, but also for independent presentations by German industry abroad. AUMA's members include the central associations of German business (industry, trade, skilled trades), trade associations of the exhibiting and visiting industries, and the German trade fair and exhibition organizers, as well as the companies organizing foreign fair participation.

Germany is a leading site for international trade fairs. Approximately 140 major international trade fairs and exhibitions attract 9 to 10 million visitors each year. For German companies, trade fairs constitute a vital vehicle for business-to-business communication.

French Committee for External Economic Events (CFME)

The Agency for the International Promotion of French Technology and Trade (http://www.Ubifrance.com; CFME ACTIM), a nonprofit association based in Paris, develops technical and commercial collaboration between France and her foreign partners. The agency brings together French business people and their foreign counterparts through such forums as trade shows; distributes information on French products and technology to the international trade press; and develops partnerships between French companies and companies abroad through joint ventures, subcontracting, and technology transfer.

Japan External Trade Organization

Japan External Trade Organization (JETRO; http://www.jetro.go.jp) is a nonprofit, Japanese government–related organization dedicated to promoting mutually beneficial trade and economic relations between Japan and other nations. Established in 1958 by the Japanese government as a special public institution to implement trade policy, JETRO spent its first two decades essentially promoting exports to help build up the Japanese economy. Later, as Japan's trade surplus grew to unprecedented levels, JETRO concentrated on helping foreign companies enter the Japanese market. In 1998, the organization substantially upgraded its research ability by merging with the Institute of Developing Economies (IDE).

On October 1, 2003, JETRO carried out a major reorganization to become a new type of publicly funded organization. JETRO's reorganization reflects new economic realities. In response to ongoing trends toward economic globalization and regional integration, Japan must achieve increased trade and investment, develop closer economic relationships with other countries, and expand its involvement with international economic cooperation. To help Japan achieve these objectives, JETRO has adopted a new approach that is both user oriented and achievement oriented. As a result, services are closely tailored to the specific needs of targeted groups, including foreign direct investors in Japan and Japanese small and medium-sized companies.

Backed by a worldwide network comprising JETRO Headquarters Tokyo, JETRO Osaka, IDE, 36 branches throughout Japan, and 77 overseas, JETRO is helping Japanese enterprises, particularly small and medium-sized enterprises, build stronger business ties in the Association of Southeast Asian Nations (ASEAN) countries and the rest of East Asia, thereby stimulating the Japanese economy.

JETRO focuses on promoting foreign direct investment in Japan—including greater inflows of capital, technologies, know-how, and intellectual assets—to stimulate the Japanese economy, increase employment, and enable Japan to continue demonstrating leadership in the global economy. This organization also assists Japanese small and medium-sized enterprises in the export of high-quality products, technologies, and designs to the global market.

In addition, JETRO continues to carry out economic research overseas and collects extensive information for dissemination to business circles. JETRO also provides a variety of other valuable services and support, including the organization of exhibitions and trade fairs and the dispatch of international trade and investment missions. At the JETRO Web site, users can locate the Trade Tie-up Promotion Program (TTPP), an online service that helps companies find each other for international business tie-ups.

Asia-Pacific Economic Cooperation

The Asia-Pacific Economic Cooperation (APEC; http://www.apecsec.org.sg) was established in 1989 in response to the growing interdependence among Asia-Pacific economies. Begun as an informal dialogue group, APEC has since become the primary regional vehicle for promoting open trade and practical economic cooperation in the Pacific Rim. Member nations include the Philippines, Australia, Peru, Japan, Canada, the United States, Mexico, People's Republic of China, and Vietnam. The goals of this cooperative, multilateral economic and trade forum are to advance trade and investment liberalization, business facilitation, and economic and technical cooperation. Today, APEC's 21 member economies bring a combined gross domestic product of more than $47 trillion in 2003 and account for 44% of global trade. Ambassador Mario Artaza from Chile is the 2004 executive director of the APEC Secretariat.

Women in Small Business Information Site

Launched in September 2000, the Women in Small Business Information Site (WISBIS; http://www.wisbis.qut.edu.au) portal provides access to information for women small-business owners and managers in Australia. This portal is a gateway to a wide variety of information on business tools, courses, exporting, grants, professional organizations, management, finance, marketing, and networking.

Tender Information Network Fully Online (I.N.F.O.) (Electronic Tendering and Procurement Network)

Since its conception in 1989, Information Network Fully Online (I.N.F.O.; http://www.tenders.net.au/), an Australian organization, has been a world leader in electronic tendering and procurement systems. Tenders.Net offers a complete tendering and procurement system for firms seeking to purchase goods and services. In addition, the site offers a thorough information and submission service for firms seeking to supply goods and services.

Australian Trade Commission (Austrade)

The Australian Trade Commission (Austrade; http://www.austrade. gov.au) is the Australian government agency that helps Australian companies win overseas business for their products and services by reducing the time, cost, and risk involved in selecting, entering, and developing international markets. Operating an international network of offices in 109 locations in 57 countries, Austrade is able to identify potential buyers, accurately match Australian suppliers with interested overseas contacts, and arrange introductions. Austrade maintains an export hotline, provides export consulting services, coordinates Australian stands at more than 100 international trade fairs annually, and also offers financial assistance to Australian exporters.

Arabnet: reaching the Arab World in the Middle East and North Africa

Found at http://www.arab.net, Arabnet offers country-specific business, political, and cultural information about 22 nations in the Arab world in the Middle East and North Africa.

STAT-USA GLOBUS Information System

STAT-USA (http://home.stat-usa.gov), an agency in the Economics and Statistics Administration, U.S. Department of Commerce, provides vital economic, business, and international trade information produced by the U.S. government to you so you can make an important decision that may affect your business. STAT-USA collates information for you that is produced by hundreds of separate government offices and divisions.

11.15 British-American Business Council

An association of British-American chambers of commerce and business associations, the British-American Business Council [BABC; http://www.babc.org/index5.php; 52 Vanderbilt Avenue, 20th Floor, New York, NY 10017; tel. (212) 661-5660; fax (212) 661-1886] is based in major cities throughout the United States and the United Kingdom. It is the largest transatlantic business organization, representing more than 4,000 companies. The BABC's objectives are to support the unique business partnership between the United States and the United Kingdom and to ensure its continuing vitality; to help its member associations provide business development and business intelligence services to their member companies; and to provide a broader, transatlantic business network for these member companies. Founded in 1993, the BABC now includes two associate organizations in Canada and Mexico. The commercial interactions between the United States and United Kingdom represent the most substantial business relationship between any two countries in the world, including more than $460 billion in two-way investment and some $73 billion annually in two-way trade. The United Kingdom is the largest overseas investor in the United States. U.S. companies provide employment for about a million people in the United Kingdom, and U.K. companies employ about a million people in the United States.

The BABC has been playing a more active role in supporting a positive environment for the further growth in transatlantic business. It has established an ongoing dialogue and partnership with the U.S. and U.K. governments and participated actively in a number of business initiatives.

The BABC is governed by a board of directors, on which all its member associations are represented, and an executive committee. The BABC's secretariat is based in New York and led by its executive director.

11.16 U.S. Trade and Development Agency

The U.S. Trade and Development Agency [USTDA; http://www.tda.gov; 1000 Wilson Boulevard, Suite 1600, Arlington, VA 22209; tel. (703) 875-4357; fax (703) 875-4009] assists in the creation of jobs for Americans by helping U.S. companies pursue overseas business opportunities. Through the funding of feasibility studies, orientation visits, specialized training grants, business workshops, and various forms of technical assistance, TDA enables American businesses to compete for infrastructure and industrial projects in middle-income and developing countries.

Procurement notices for USTDA direct contracts, such as definitional missions, can be accessed through the Federal Business Opportunities Web site (http://www.fedbizopps.gov). Small and minority-owned U.S. firms that wish to be included in USTDA's consultant database and considered for future Desk Study solicitations should register online at https://www.tda.gov/consultantdb/index.html.

11.17 U.S. Agency for International Development

Established in 1961 by President John F. Kennedy, the United States Agency for International Development (USAID; http://www.usaid.gov) is the independent government agency that provides economic development and humanitarian assistance to advance U.S. economic and political interests overseas. The Office of Small and Disadvantaged Business Utilization and Minority Resource Center (OSDBU/MRC) is the initial point of contact at USAID for U.S. small businesses, small disadvantaged businesses, women-owned small businesses, HUBZone small businesses, and service-disabled veteran-owned small businesses. OSDBU/MRC is a small business advocacy and advisory office with the responsibility for ensuring that these enterprises receive access to USAID programs. The office serves as an information clearinghouse for U.S. small businesses, counsels small businesses on how to do business with USAID, and examines USAID buying needs for possible set-asides for 8(a) and other small businesses prior to publication on the FedBizOpps Web site. In addition, OSDBU/MRC reviews all prime contracts to identify subcontracting opportunities for small businesses and maintains the Small Business Resource Database (http://www.usaid.gov/business/small_business/vendordb.html). This is an electronic database for U.S. small businesses and organizations that are interested in participating in agency sustainable development programs. OSDBU/MRC also sponsors outreach conferences on "How to Do Business with USAID," and participates in national, regional, and local conferences sponsored by both private and public organizations.

An Office of Small and Disadvantaged Business Utilization exists in all U.S. federal agencies in accordance with Public Laws 95-507 and 100-656 to ensure the participation of U.S. small businesses, small disadvantaged businesses, women-owned small businesses, HUBZone small businesses, and service-disabled veteran-owned small businesses in federal procurement opportunities.

The USAID business site (http://www.usaid.gov/business/) is your one-stop place for finding important information to get you started.

USAID awards approximately $4 billion each fiscal year in federal contracts and grants. Contracts are awarded primarily for technical assistance but also for commodities and/or equipment, transportation services, and occasionally construction. Grants are awarded for a variety of programs—some recurring (e.g., Food for Peace grants and other grant programs exclusively for Private Voluntary Organizations)—and some for unique nonrecurring programs. All contracts and grants issued ultimately support objectives of that part of the U.S. foreign assistance program managed by USAID.

Established in 1994 as a U.S. Agency for International Development program, the Global Trade and Technology Network (GTN; http://www.usgtn.net/) has worked to facilitate sustainable economic growth in developing countries and emerging markets through business links and technology transfer. In 2002, GTN launched an Internet-based trade facilitation platform to further spur trade between companies located in the United States and developing nations. Since that time, the GTN trade platform has become a leading worldwide provider of international trade lead services. GTN's trade services include trade and investment business matching services, technology transfer, trade lead follow-up services, trade financing referrals, and market information. Since 2002, GTN has achieved successes, including these:

- Facilitated more than $345 million in potential transactions;
- Generated more than 4,000 trade and investment leads;
- Assisted with more than 350,000 matches.

International business planning resources for small companies in the United States include the following:

- *Rutgers University Resouces for International Business.* Rutgers University Libraries (http://www.libraries.rutgers.edu/rul/rr_gateway/research_guides/busi/business.shtml) provides a comprehensive listing of international business resources. These resources span current periodicals and news, demographics and statistics, and regional as well as country-specific listings.
- The *National Trade Data Bank* (NTDB; http://www.stat-usa.gov/tradtest.nsf). The NTDB site offers Trade Opportunity Program leads, international marketing insight reports, country-specific reports, Asia Development Bank business opportunities, foreign exchange rates, World Bank international business opportunities, and much more.

- *Virtual International Business & Economic Sources* (VIBES; http://library.uncc.edu/display/?dept=reference&format=open&page=68). VIBES provides more than 2,800 links to Web resources of international business and economic information. Maintained by the University of North Carolina at Charlotte, VIBES' links are in English and are free of charge. Links include full-text files of recent articles and research reports, portals, and statistical tables and graphs.

United States International Trade Commission

The United States International Trade Commission (USITC; http://www.usitc.gov) is an independent, nonpartisan, quasijudicial federal agency that provides objective trade expertise to both the legislative and executive branches of government, determines the impact of imports on U.S. industries, and directs actions against certain unfair trade practices, such as patent, trademark, and copyright infringement. USITC analysts and economists investigate and publish reports on U.S. industries and the global trends that affect them.

International Trade Administration

The International Trade Administration (ITA; http://www.ita.doc.gov), which is an organization within the U.S. Department of Commerce, helps U.S. businesses participate fully in the growing global marketplace. ITA provides practical information to help you select your markets and products. It also ensures that you have access to international markets as required by U.S. trade agreements.

Central & Eastern Europe Business Information Center

The Central & Eastern Europe Business Information Center (CEEBIC; http://www.mac.doc.gov/ceebic/) is a business facilitation program for U.S. firms interested in expanding into the Central and East European markets. Established in 1990 by congressional legislation under the Support for East European Democracy (SEED) Act, CEEBIC is a one-stop shop and the U.S. government's clearinghouse for the most recent economic, commercial, and financial information on the 15 countries of Central and Eastern Europe. CEEBIC offers a wide array of services, business counseling, and information products designed to help primarily small and medium-sized U.S. companies. CEEBIC's Washington-based trade specialists and dedicated overseas staff in 15 countries of the region work together to implement this unique program for U.S. firms.

CEEBIC's experienced trade specialists provide individualized counseling over the phone or by appointment. CEEBIC helps U.S. companies to enter the markets of Central and Eastern Europe by advising them of export and investment issues, including those related to market access. CEEBIC staff continues to offer advice and guidance to companies once they have entered these markets.

CEEBIC's overseas staff in the region gathers information and business opportunities, which are then distributed through CEEBICNet and CEEBIC publications. CEEBIC's trade specialists in Washington work together with the overseas employees to provide detailed answers to U.S. firms' inquiries about business opportunities.

We now turn our attention in the next chapter to innovative and successful proposal production and publication techniques.

ENDNOTES

1. From "Test Your Global Mindset," *Industry Week*, November 2, 1998, p. 12.

2. Whitman, Marina, and Rosabeth Moss Kanter, "A Third Way? Globalization and the Social Role of the American Firm," *Washington Quarterly*, Spring 1999.

3. Root, Franklin R., *Entry Strategies for International Markets*, revised and expanded, New York: Lexington Books, 1994, p. 1.

4. Root, Franklin R., *Entry Strategies for International Markets*, revised and expanded, New York: Lexington Books, 1994, p. 19.

5. Flagler, Carolynn, "The ANY Aspects of International Proposals," *Contract Management*, March 1995, p. 13.
    Franklin Root suggests that "[f]or most companies the entry strategy time horizon is from three to five years, because it will take that long to achieve enduring market performance." Root, Franklin R., *Entry Strategies for International Markets*, revised and expanded, New York: Lexington Books, 1994, p. 22.

6. Conceived in 1944 at the United Nations Monetary and Financial Conference held in the United States, the World Bank is a multilateral development institution whose purpose is to assist its developing member countries further their economic and social progress. The term *World Bank* refers to two legally and financially distinct entities:

the International Bank for Reconstruction and Development (IBRD) and the International Development Association (IDA). The IBRD and IDA have three related functions: to lend funds, to provide economic advice and technical assistance, and to serve as a catalyst to investment by others. The IBRD finances its lending operations primarily from borrowing in the world capital markets. IDA extends assistance to the poorest countries on easier terms, largely from resources provided by its wealthier members. (See *The World Bank Annual Report 1994*, Washington, D.C.: The World Bank.)

7. *The World Bank Annual Report 1998*, Washington, D.C.: The World Bank; http://www.worldbank.org/html/extpb/annrep98/overview.htm.

8. Root, Franklin R., *Entry Strategies for International Markets*, revised and expanded, New York: Lexington Books, 1994, pp. 25–26.

9. Root, Franklin R., *Entry Strategies for International Markets*, revised and expanded, New York: Lexington Books, 1994, pp. 41–42, 44.
   An entry mode is "an institutional arrangement that makes possible the entry of a company's products, technology, human skills, management or other resources into a foreign country." Root, Franklin R., *Foreign Market Entry Strategies*, New York: AMACOM, 1987, p. 5. See also Erramilli, M. Krishna, and C. P. Rao, "Choice of Foreign Market Entry Modes by Service Firms: Role of Market Knowledge," *Management International Review*, Vol. 30, 1990, pp. 135–150.

10. Douglas, Susan P., C. Samuel Craig, and Warren J. Keegan, "Approaches to Assessing International Marketing Opportunities for Small- and Medium-Sized Companies," *Columbia Journal of World Business*, 1982, p. 27.

11. Farrell, Michael, "Many Small Firms Want to Do Business Internationally but Lack the Wherewithal," *Capital District Business Review*, Vol. 22, September 1995, p. 15.

12. DerGurahian, Jean, "Albany MBA Candidates Earn Real-World Success," *The Business Review*, March 31, 2000.

13. Flagler, Carloynn, "The ANY Aspects of International Proposals," *Contract Management*, March 1995, pp. 14, 16, 19.
   A common area of tax liability exposure relates to sending technical staff into the buying country to provide technical support and training. This type of technical support is often subject to taxes.

Countertrade refers to a mandated purchase requirement imposed on the foreign seller by the buying community or its government.

14. Flagler, Carolynn, "The ANY Aspects of International Proposals," *Contract Management*, March 1995, p. 17.

15. See McCubbins, Tipton F., "Three Legal Traps for Small Businesses Engaged in International Commerce," *Journal of Small Business Management*, Vol. 32, July 1994, pp. 95–103.

16. Root, Franklin R., *Entry Strategies for International Markets*, revised and expanded, New York: Lexington Books, 1994, p. 97.

17. Root, Franklin R., *Entry Strategies for International Markets*, revised and expanded, New York: Lexington Books, 1994, p. 98.

Chapter 12

Proposal production and publication

PUBLICATION OF A SET of superior-quality proposal volumes is a professional-level, time-intensive, dedicated effort. It is not a lower level clerical function that can be accomplished adequately and consistently by marshaling the secretarial and administrative support available within your company. Even small businesses benefit from a dedicated core publications staff of at least one desktop publisher or word processing operator and one graphic artist.

Senior management is well served to recognize and support the publication professionals they have on staff, even if they number only two or three people. Support is most meaningful in the form of senior management's judicious, targeted involvement in assisting the proposal manager to maintain the proposal response schedule at every milestone. Otherwise, publication staff are faced with compensating for schedule slippages along the way. And when schedules are missed on a regular basis and significant levels of night and weekend publication time become routine, publication staff morale can be at risk. Turnover among publication staff can be highly detrimental to your company's ability to prepare outstanding proposal documents every time.

Proposal efforts are not complete when technical and management volumes are written and reviewed, and the costing is completed after careful integration with the technical and management approaches and staffing plan. There remain the following critical production-related activities:

- Editing of both text and graphics;
- Creation or generation of tables of contents, lists of figures and tables, and compliance matrices;
- Proofreading of text and graphics (pay extra attention to proposal covers; they make a first impression and you want it to be positive);
- Electronic and hardcopy configuration management and version control;
- Word processing and desktop publishing;
- Generation of graphics and preparation of photographic prints for photocopying,[1] including insertion of graphics, figures, and tables into the documents, either electronically or manually;
- Coordination of outside printing activities (for example, covers, spines, tabs, and color pages);
- Photoreproduction and in-house color printing;
- Collation;
- Assembly and binding (such as three-ring vubinders, Velobinding, GBC, and plasticoil), including placing covers, spines, and tabs into binders;
- Preparation of required, virus-free CD-ROMs or Zip disk copies;
- Conversion to PDF format for upload to the client's Web site, as appropriate[2];

---

1   With a flatbed scanner, both black-and-white and color prints can be scanned directly into such applications as Microsoft Word and Adobe PhotoShop. Note, however, that flatbed scanning technology has become so inexpensive (less than $100) that your company should strongly consider acquiring a scanner for proposals as well as for contract deliverables. Scanner manufacturers include Hewlett-Packard, Canon, Epson, MicroTek, UMAX, and Agfa.

2   Allow adequate time for uploading to your client's Web site. Be certain to retain the time and date report that verifies when your files were accepted by the client's server. Remember, it is your client's time that is the one that counts, not the one on your company's computer system. Being late by 1/100th of a second is still late, and can result in your proposal being rejected.

- Final quality checking;
- Packaging;
- Completion of specific bid forms and packaging labels; labeling;
- Shipping and delivery (delivery is more personal in the international and commercial arenas; a senior person from your company should perform this function);
- Obtaining time- and date-stamped receipt.

It is recommended that the proposal publication be a centralized corporate function. Continuity of publication staff promotes uniformity of the image of the proposal documents from proposal to proposal. A core group of staff should become intimately familiar with, and cross trained in, the document preparation process. Disciplined policies and procedures are much more easily implemented and repeated in a centralized working environment.

12.1 Internal documentation standards

To ensure that your company's proposal and contract deliverable documents have a similar look and feel, it would be highly beneficial to develop a corporate handbook of format and style, publications guide, or analogous guidebook. The publications group, in conjunction with the business development staff and senior project management staff within your company, should be involved in developing and maintaining the format and style standards that will help to ensure the presentation of a uniform company image for both proposals and contract deliverables. This guidebook should address some or all of the following topic areas:

- Editorial style;
- Technical writing guidance;
- Technical report formats;
- Correspondence (memos, transmittal letters, and so forth);
- Proposals, including résumés and project summaries;
- Marketing presentations and brochures.

Always follow the formatting guidance provided in Section L (instructions to offerors) of the RFP or RFS when preparing your federal proposals. However, in-house written documentation standards will assist technical writers, proposal reviewers, and publication staff to work toward a consistent style and structure of presentation.

A useful technique for each proposal effort is to have someone in your company with editorial capability develop a listing of RFP-specific terms that will appear in your proposal so that writers and publication staff will capitalize, hyphenate, and abbreviate those terms in accordance with the particular government agency's parlance.

12.2 Document configuration management and version control

Once sections of a proposal are written, it is critical to control the internal release of and changes to those sections throughout the proposal life cycle. The centralized publications group should consider maintaining a master hardcopy proposal book for each volume in three-ring binders. Such binders allow for quick page replacements. One set of master proposal books should be made available to the proposal manager, and another should be under the direct control of the publications group. The pages from a third and final master copy may be hung on the cork-lined walls of your secure "war room" facility to allow for comment and review.[3] Then, as new proposal sections are created and existing ones modified with the authorization of the proposal manager, the publications group should be responsible for generating the change pages and inserting them in the master proposal books and hanging them on the walls of the war room. Establishing and maintaining a "living" master proposal document that iterates during the course of the proposal response life cycle will be of valuable assistance to all proposal contributors.

Your company's publications group should maintain electronic and hardcopies of the proposal documents at each review stage. For example, when the publications group produces the Red Team draft, all electronic files should be copied to a storage medium appropriate for the data volume of the proposal. Mass storage devices and removable cartridge drives and media include compact disk (CD), USB flash drives (up to 2 GB), IOmega Zip drives (100, 250, and 750 MB) and SYQT, Inc's EZFlyer, SparQ, and SyJet data cartridges. A universal serial bus (USB) flash drive is a compact USB flash memory drive that acts like a portable hard drive, letting you store and transport computer data. These drives hold large amounts of

---

3   The technique of affixing documents to the walls of the war room is slowly being phased out, in part due to file sharing made possible through computer networks and the Internet. It is still valuable, however, particularly for reviewing oral presentations.

information and are about the size of a pack of chewing gum. Despite the different brands and names—JumpDrives, Pocket drives, Pen drives, and Thumb drives—they are all pluggable and portable. USB flash drives use the USB protocol to interoperate with PCs and Macs with a USB port.

Currently, USB flash drives are available with up to 2 GB of storage capacity. Additionally, USB flash drives act like portable hard drives to which you can add and delete files as many times as you need to do so. With CD-Rs and DVD-Rs, you can only add data once, and you cannot remove data once burned onto the disk. Your firm might also consider tape backup systems, or rewritable Sony or Hewlett-Packard magneto-optical cartridges.

One hardcopy should be retained in the exact form and format of the Red Team draft. Then, if a computer hard drive fails, the network server crashes, or electronic files are corrupted in some way due to power surges or electrical storms, the last major draft version will be available, unmodified, in both electronic and hardcopy forms. It is advisable to back up all electronic proposal files to secure storage media at least once each work day.

A key aspect of proposal document configuration management is the use of time, date, path and file name on the hard drive, and diskette/ CD-ROM/magneto-optical rewritable disk number headers on all draft proposal pages. Headers can be in a small font and, of course, must be removed prior to submittal to the client. The following is an example of a useful header for a page in the technical volume of a proposal to the Federal Aviation Administration:

14:25 Thursday, September 9, 2004 c:\proposal\A-1-3.FAA *CD: Tech-A.1*

Such headers allow easy identification of when a particular page was last modified and on which storage medium the backup file for A-1-3.FAA resides.

Another beneficial configuration technique is the use of colored paper for various proposal drafts—for example, pale blue paper for the Blue Team and pale red for the Red Team.

And finally, do not discard draft proposal sections until after the proposal has been submitted. You can never be sure that you will not want to refer to an earlier version of the proposal. For security of competition-sensitive information, keep all such materials in locked storage and shred them when the proposal is submitted. A document destruction firm might also be used to destroy all interim draft proposal materials. Many will come to your facility and pick up the paper waste and shred it on your premises. Remember to retain a complete hardcopy and electronic copy of the final version of the proposal submitted to the government for use during FPR and orals preparation.

12.3 Freelance and temporary publication staff

Many companies, including small businesses, respond to fluctuating publication workloads with outside support from freelance, part-time on-call, or temporary staff. Your company's publications group manager will be well served to develop and maintain an active listing of prequalified local freelance graphics artists, word processing operators, desktop publishing specialists, editors, and proofreaders to call upon during peak periods. Freelance staff, if called upon consistently, will be able to develop an understanding of your company's internal documentation standards, proposal development processes, and technical and support staff. In the case of freelance staff, make certain that they sign a nondisclosure agreement so that they cannot legally share proposal-sensitive information with any other company or individual.

12.4 Incorporating technical brilliance up to the last minute

One of the most challenging aspects of responding to an RFP or RFS is incorporating the best materials into the final document within a very limited time frame, and still delivering the proposal on time. The balance between process and content is delicate but very important to consider and constantly revisit. Solution sets, technical and programmatic input, tailored boilerplate plans, marketing intelligence, nuances of corporate image, résumés, project descriptions (summaries) of past performance, cost and pricing data, and legal opinion—all of these elements must be brought together quickly and effectively. However, precisely because the proposal preparation cycle must be a controlled, choreographed process of planning, analysis, writing, multiple review, and publication, potentially brilliant technical or programmatic ideas and knowledge might not be incorporated into the final document. There simply may not be enough hours to integrate the change(s) into the text from a publication standpoint. Technical, management, and cost volumes of a proposal are ecological in the sense that a change in a system design concept affects cost, and an alteration in the WBS and bid task list affects the program plan and perhaps the cost as well. These wide-ranging changes take time to identify and make consistent across the proposal volumes both in the text and graphics.

Unfortunately, the publication staff invariably are caught in the crossfire of last-minute changes. Their task is to generate a well-presented, smooth-reading, consistent, and integrated set of documents, which in final form always takes more time to produce than in draft form. Deciding between structure (generating a polished, final document) and chaos (ongoing free-thinking that can result in significant technical and programmatic enhancements) is a challenging pivot point for the proposal manager, but one that can actually work to the benefit of your company's proposal.

The balancing act for the proposal manager is to allow the entire proposal process to remain fluid enough to accommodate evolutionary change (content) while simultaneously maintaining a firm commitment to milestone schedules and completion of action items (process). Regular brainstorming sessions can prove valuable, and building in extra space in the proposal documents will facilitate the easy incorporation of new ideas late into the production process. Human and organizational dynamics come into play in graphic relief during the proposal process. Management commitment of sufficient human and material resources, bonus and incentive programs, effective cross training, and the infusion of a winning attitude all can be brought to bear on the often arduous schedule of responding to federal procurements.

It seems imperative to me for the proposal manager to direct the publications group to generate as complete a document for each volume as possible as early in the proposal response life cycle as they can. Not that final formatting needs to be accomplished, because this can be counterproductive given today's high-end desktop publishing software applications. What is important is seeing and working with a complete rough draft so that technical gaps, RFP compliance issues, and program management inconsistencies are identified early.

12.5 Graphics are an integral part of your proposal

Appropriate use of graphics, photographs, line art, and clip art (noncopyrighted art available on the Web or on CD-ROM, for example) of all kinds will increase the evaluators' interest in and positive response to your company's proposal. It is essential not to submit a boring, lackluster proposal to any potential client organization. Well-designed graphics can convey complex information in an easily understood format. See Figure 12.1 for an example of a graphic appropriate for an executive summary. And in page- or word-limited proposals, graphics can present significant quantities of information in very limited space. Most proposals benefit from and

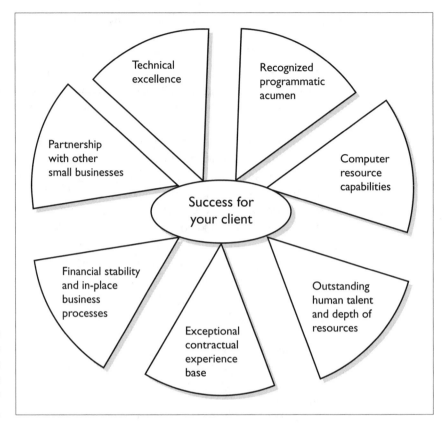

Figure 12.1
Sample graphic that introduces your company's primary thematic story line to your client.

many, in fact, require graphical presentations of staff skills, contractual experience, client-contractor interfaces, company organization, project organization, and project milestone schedules. Photographs might be added to personalize your company's résumés, particularly if your client knows your proposed key staff. Appropriate photographs of your company's facilities, specialized computer equipment, engineering and manufacturing centers, COTS products, and design prototypes can also greatly enhance your proposal's sales value. Photodocumentation brings projects and products to life, and also serves to break up the text for easier reading and fuller comprehension.

A word of caution about graphics found on the Web. The majority of photos and graphics on the Web are designed to look good on your monitor, which can display a maximum resolution of 72 pixels per inch (ppi). Printed material requires at least 150 dpi, with 300 dpi preferred for high-quality hardcopies. Do not assume that you can capture noncopyrighted photos and graphic images from the Internet and use them in your proposal. Plan ahead and purchase photographic images and clip art from such sources as Eyewire (http://www.eyewire.com/products/photo/

photodisc/new.html); Ablestock, which offers high-resolution, royalty-free images (http://www.ablestock.com); and Jupiterimages, Inc.'s http://www.clipart.com, http://www.comstock.com/web/default.asp, and http://www.photos.com/en/index. Two other excellent photo image sites available online are Seattle-based Getty Images (http://www.gettyimages.com) and Creatas Stock Resources (http://www.creatas.com). Make certain that you only purchase *royalty-free* images, that is, an image that does not require your company to invest additional funds every time the image is used in proposals, marketing materials, trade show signage, annual reports, and so on.

One powerful software tool that can convert, capture, compress, organize, enhance, view, and print graphics files is International Microcomputer Software, Inc.'s HijaakPro Version 5. This application supports more than 115 graphics formats, including raster (e.g., .gif, .jpeg, .bmp, .pcx, .psd, and .tiff), vector (e.g., .dxf, .txt, and .pic), and metafile (e.g., .eps, .cdr, .pict, .wmf, and .wpg) types. See http://www.imsisoft.com.

12.5.1 Action captions

Fundamentally, proposals are sales documents. In any sales document, whether it is a brochure or mailer or proposal, there are significant messages conveyed through photographs, artwork, and words. In proposals, these messages are called themes. Think of a theme as a statement that you make in a proposal that explains exactly why the client should select your company for contract award and not select your competition. In short, Why you? themes should always be substantiated and authenticated with facts and metrics. One great place to use theme statements is in the captions for figures, tables, photographs, and other graphic presentations. Let's look at the difference between simply identifying what a graphic or photograph is and using the caption to convey important sales messages tailored to the specific procurement.

- Instead of "Project Organizational Chart," your caption might read *"Our company's project team is configured to manage multiple USACE work assignments efficiently."*

- Instead of "Map of Local Company Locations," your caption might read *"The proximity of our company's offices to Fort Campbell will afford rapid response to changing requirements."*

- Instead of "Capabilities Matrix," your caption might read *"Each one of our eight proposed senior technical staff holds an accredited degree in a relevant engineering and scientific discipline."*

- Instead of "Laboratory Staff," your caption might read *"Our well-trained laboratory staff are dedicated to and trained in quality-*

> *oriented USEPA procedures that ensure appropriate chain-of-custody handling."*

- Instead of "Appropriate H/S Procedures" your caption might read, *"We actively practice a comprehensive Health and Safety program designed to meet the requirements of 29 CFR 1926.65 for hazardous waste site work."*

- Instead of "Scope of Our Support Services," your caption might read *"We provide cost-effective, innovative support services across a broad spectrum of environmental regulatory acts."*

- Instead of "Technical Approach Overview," your caption might read *"Our technical approach embraces documented metrics to ensure performance success today and into the future."*

- Instead of "Understanding of SOW," your caption might read *"Our understanding of Air Force Space Command SOW requirements allows us to apply tailored practices to optimize performance efficiencies."*

In each of these examples, we have taken a flat statement and added positive, dynamic sales messages. We have introduced benefits of our approach and capabilities to the client. These sales messages, or themes, should be developed early in the proposal response cycle so that the entire proposal writing team can incorporate them into their narrative and into the captions they develop for their graphics. If a particular RFP stipulates that figure and tables shall be named in a particular manner, you must adhere to that guidance.

12.5.2 Configuration control of graphics

Configuration control of the graphics in your proposal is made difficult because many graphics are submitted to your publications group without captions (for figures) or legends (for tables) and their number within the proposal volumes may change several times depending upon placement within a specific section or subsection. The proposal manager can assist with this configuration management issue by insisting that writers submit graphics concepts with appropriate numbering and clear linkage to specific outline sections. Adding a file path name in a very small font directly into the graphic file (so that the path specifications print with the graphic image) will also help when these graphics iterate or shift within the document. And retain that identifying file name in the final proposal as well. That way, 6 months later, when one of your staff is building a new proposal response, she can quickly reference the specific graphic for ease of electronic retrieval.

12.6 Role and structure of your publications group

A small company's publications group is often tasked with multiple responsibilities, such as proposals, contract deliverables, presentations, and marketing brochures. The core staff of this group should be cross-trained in a variety of operations, software applications, communications protocols, and hardware platforms. And it is critical that the publications group be able to accept, incorporate, and manage outside assistance. The group's policies and procedures, software, and hardware should not be tailored so that temporary or freelance employees or other in-house staff cannot support the group effectively and efficiently during "crunch" (crisis) documentation periods. In addition to an active list of local freelance staff, the manager of your company's publications group should consider maintaining lists of local photography and visual imaging vendors and photore-prographics houses that offer pickup and delivery services as well as rapid turnaround.

The continuity of publications group core personnel is very important for smoothly operating proposal efforts. Proposal managers benefit from seasoned, competent documentation staffers. Having established successful proposal production departments from the ground up for five federal contracting firms, I can testify to the benefits of human continuity within the publications group. Continuity, cross training, and a positive attitude are salient elements in proposal publication success. Loss of skilled and cross-trained staff can cut deeply into productivity and inflate overall proposal publication costs significantly.

In order to facilitate a smooth proposal operation, your company will need to determine the document throughput capacity of the publications group. How many new proposals of average size (for your firm) per month can be handled adequately with the core staff? If your bid–no bid process causes additional proposals to enter the publications "pipeline" each month, will additional staff, computer equipment, and floor space be needed? Keep in mind that an informed, aggressive bid–no bid or down-select process is the checkpoint that controls proposal document flow-through. Your company should avoid overloading your publications group and your business development infrastructure with proposal efforts that have low-win probabilities. That practice wastes B&P money and can be very detrimental to morale. Preparing and submitting proposals on a law-of-averages basis—the more we submit the better the chance of winning—is a devastating and debilitating practice in both human and financial terms.

Keep in mind that editing and proofreading must be integral parts of the proposal publications process, not merely nice-to-have services. The publications group can be a good source of editorial support.

Software and hardware compatibility, standards, and recommendations

In today's era of desktop publishing—integrating sophisticated text configurations and complex graphics and creating multicolumn page layout designs on a personal workstation—computer platforms, peripherals, and software applications are important considerations for any company that prepares proposals. The ongoing explosive proliferation of third-party hardware compatibles and multiple variations of software applications argues strongly for your company adopting a uniform standard for hardware platforms, software applications and versions, storage media, and peripherals.[4] This standard is best adopted as early in your company's corporate life as possible. The same version of operating software, such as Microsoft Windows 2000 or Windows NT, should be installed on all your company's computers. As much as possible, the default settings should be consistent from machine to machine, and the printer drivers and fonts should be the same.

In addition to software and hardware compatibility, computer and peripheral redundancy and maintenance should also be considered. A limiting factor in the publications process is document printing. Complex graphics, for example, can take many minutes to print just one copy. Your publications group should be adequately equipped with printers that have expanded random access memory (RAM) to print complex graphic images, Pentium-class computers with high-storage hard drives (minimum of 30.0 GB) and expanded RAM (minimum of 256 MB and 1.024 GB desired), and color video graphics adapter (VGA) monitors (preferably 21 inches). A 4-hour response time maintenance agreement for all your proposal publication computer equipment is highly recommended.

When establishing hardware and software standards for your company, keep in mind the various external organizations, vendors, and freelance support with which your system at the hardware, software, and storage media levels must interact on a regular basis. For example, perhaps several of your major federal clients require contract documents be delivered in Microsoft Word 2000 rather than WordPerfect 11.0 for Windows,

---

4   Printers manufactured by the same company, for example, can vary in the number of words printed per page. The Hewlett-Packard LaserJet 6P and 8100 black-and-white printers are cases in point.

or your top two graphics service bureaus[5] use SYQT 200-MB cartridges for transferring large files rather than CD-ROM or magneto-optical disk, or three of the firms with which you team most frequently operate in a Macintosh Power PC environment rather than with IBM-compatible PCs, or the majority of the known freelance staff in your immediate area are most competent and proficient in WordPerfect.

Significant computer-based experience during the past 17 years suggests strongly that whatever system platforms your company selects should be open architecture—designed for and fully capable of responding to the exponential pace of software and peripheral advancements. Your central processing units (CPUs) should be equipped with as much RAM as is necessary to operate and execute all of your software applications and printing tasks in an efficient manner. Printers should have expanded RAM to accommodate graphics files, as well as integrated text and graphics files. Storage media must be capable of handling ever-increasing data volumes and of rapid random retrieval. Today, an average-sized proposal—replete with full-color cover and integrated text and graphics—can extend to more than 200 MB. One proposal alone could fill a USB flash drive, and only three would fit on a CD-ROM. A single graphics file that contains several scanned photographs can be more than 30 MB, the equivalent of more than twenty 3.5-inch high-density diskettes [1]. One final note: Back up files regularly. Your proposal process probably cannot sustain major data losses—ever!

From a software applications standpoint, proposals do not generally lend themselves to being prepared with high-end, full-scale desktop publishing packages such as Ventura and PageMaker. This is because of the very tight production time frames. Ventura, for example, is best suited for the publication of books and other lengthy documents that are frozen (approved with no changes) and then have styles (page layout parameters) applied to them. In the IBM-compatible world, WordPerfect for text and CorelDRAW! for graphics [imported into WordPerfect as Windows metafiles (.wmf)] have proven to be appropriate over the long term as have MS Word and Adobe Illustrator. By keeping text in WordPerfect or MS Word, which many technical and administrative staff can manipulate, the documentation process can receive more in the way of ad hoc support from throughout your company and the local freelance community. In addition, technical changes can be accommodated until a much later stage in the production life cycle. The high-end packages also have an associated steep learning and proficiency curve.

---

5   Service bureaus are for-profit organizations that provide such support as typesetting, electronic file conversions and raster image processing (RIP), 35-mm slide generation, and prepress film generation.

12.8 Electronic proposal submittal and evaluation

In an effort to streamline the evaluation and source selection processes, certain federal agencies are now requiring both hardcopy and electronic copy (such as electronic upload to the Web, CD-ROM, and Zip disk) proposal submittals. It is already commonplace that cost data are submitted on diskette in Lotus or Excel, for example. The technical and management narrative, as well as oral presentations, are migrating in that direction as well. Be certain to provide your proposal in the specific version of the particular software application requested on the correct size and storage capacity diskette. Increasingly, proposals are also uploaded to government acquisition Web sites and no hardcopies are involved whatsoever.

Electronic source selection (ESS) integrates communication technology, database applications, and information management systems to reduce risk and increase the productivity and efficiency of source selection evaluations (see Figure 12.2). ESS uses standard, as well as proprietary, software applications and hardware to produce a paperless environment for source selections and to streamline RFP preparation. ESS has already been used in major DoD programs such as Joint Strike Fighter (JSF), Landing Platform Dock (LPD-17), and Global Broadcast System (GBS).

In the electronic source evaluation process, the government harnesses the Internet to distribute and receive RFPs and RFSs. The productivity of the selected ESS tool reduces risk, cost, and time associated with the government's evaluation of proposals. Sensitive information is controlled,

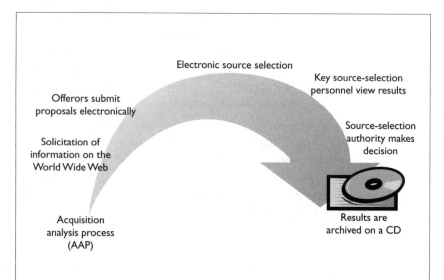

Figure 12.2 Representative electronic source selection process.

Electronic source selection

Offerors submit proposals electronically

Key source-selection personnel view results

Solicitation of information on the World Wide Web

Source-selection authority makes decision

Acquisition analysis process (AAP)

Results are archived on a CD

while simultaneously being available to key decision makers in diverse locations for streamlined consensus building. Electronic source evaluation facilitates an understanding among all government participants of how each part of the procurement affects the whole. This is key to providing a united front during higher-level review. The electronic process also provides flexibility in linking with other analyses and in quantifying the decisions.

The U.S. Air Force Materiel Command (AFMC) has migrated to electronic evaluation for all major system procurements [2]. Electronic evaluations are already being used for many smaller procurements as well. Each AFMC center is in the process of defining and developing its own methods for electronic evaluation.

CACI's FedSelect Windows-based proposal evaluation software (http://www.caci.com/bmd/FedSelect/fedselect.shtml), which is available on the GSA Schedule, is designed to provide a best value source selection process among competing proposals. Version 3.3 of this application uses evaluation factors/areas along with criteria/thresholds. FedSelect provides fully traceable and automatically documented proposal evaluations focused on capturing rating and risk rationales behind the business decision. It is supported by links to proposal strengths and weaknesses. It also has pre-formatted interim and final reports that can be used to brief decision-making officials. FedSelect is in use at the Missile Defense Agency (MDA), Defense Information Systems Agency (DISA), each military service, and NASA, among other agencies.

FedSelect supports source selection/proposal evaluation with Administration, Evaluation, Caucus, and Reports modules. The software uses existing solicitation information and is completely tailorable. FedSelect also does not require the use of a full-time LAN administrator. The software is managed easily by the administrator (contract specialist).

12.9 Important documentation tips

Proposal documents are presented very attractively in double-column format, which provides the added bonus of 10% to 15% more text per page, depending on the font selected. Conserving the white space at the end of lines of text is part of the reason why double-column formatting provides for more words per unit of space. This space savings is very important for page-limited proposals. See Figure 12.3 for an example of a double-column proposal page format.

When compiling a list of acronyms, perform a global search for a beginning parenthesis "(". If acronyms have been defined in the text, and they should be, this technique should identify most of them. The acronym list for your proposal can, like the compliance matrix, be put on a foldout

1.0   TECHNICAL APPROACH

Proposal documents are presented very attractively in double-column format, which provides the added bonus of 10-15% more text per page, depending upon the font selected. Conserving the "white space" at the end of lines of text is part of the reason why double-column formatting provides for more words per unit of space. This space savings is very important for page-limited proposals.

Proposal documents are presented very attractively in double-column format, which provides the added bonus of 10-15% more text per page, depending upon the font selected. Conserving the "white space" at the

> **This space savings is very important for page-limited proposals.**

end of lines of text is part of the reason why double-column formatting provides for more words per unit of space. This space savings is very important for page-limited proposals.

1.1   Understanding the Technical Requirements

Proposal documents are presented very attractively in double-column format, which provides the added bonus of 10-15% more text per page, depending upon the font selected. Conserving the "white space" at the end of lines of text is part of the reason why double-column formatting provides for more words per unit of space. This space savings is very important for page-limited proposals.

Proposal documents are presented very attractively in double-column format, which

provides the added bonus of 10-15% more text per page, depending upon the font selected. Conserving the "white space" at the end of lines of text is part of the reason why double-column formatting provides for more words per unit of space. This space savings is very important for page-limited proposals.

Proposal documents are presented very attractively in double-column format, which provides the added bonus of 10-15% more text per page, depending upon the font selected, as shown in Figure 1-1. Conserving the "white space" at the end of lines of text is part of the reason why double-column formatting provides for more words per unit of space. This space savings is very important for page-limited proposals.

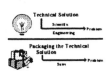

Figure 1-1.

Proposal documents are presented very attractively in double-column format, which provides the added bonus of 10-15% more text per page, depending upon the font selected. Conserving the "white space" at the end of lines of text is part of the reason why double-column formatting provides for more words per unit of space. This space savings is very important for page-limited proposals.

1.1.1   Software Applications

Proposal documents are presented very attractively in double-column format, which provides the added bonus of 10-15% more text per page, depending upon the font selected. Conserving the "white space" at the end of lines of text is part of the reason why double-column formatting provides for more

Figure 12.3   Sample double-column proposal page format.

page so that the government evaluators can have constant and easy access to the acronym definitions.

For lengthy proposals, customized tab pages should be prepared to serve as page dividers for ease in locating important sections. Tabs are ordered by the set and by the cut. Ten sets of seven-cut tabs means 10 complete sets of tabs each of which has seven tabbed pages. Many higher end photocopier machines can now produce tabs, even laminated ones, right in your office.

The human and organizational dynamics of the proposal process are probably the most important elements of all. Chapter 13 will highlight these very real issues.

12.10 Virtual proposal centers, intranets, and extranets

An *intranet* is an internal information distribution system that uses Web technology to allow an organization to access specific information in a form and format designed just like an Internet home page. However, the general Internet community cannot access your company's intranet site because it is firewalled. In effect, an intranet is a separate system with limited access[6] using a Web browser identical to Internet Web browsers such as Netscape Navigator and Microsoft Internet Explorer. The organization develops its own internal home page, with links to various sites or locations containing information within the site. On the other hand, an *extranet* incorporates encryption technology to allow secure access to an intranet over the Internet (see Figure 12.4). The Federal Express do-it-yourself tracking system represents an example of a sophisticated extranet. Now how are these systems applicable to, and useful for, proposal development?

An intranet can be used to create a virtual proposal center (VPC) within your company, which in turn can be used to manage, support, and simplify the proposal development process for the firm. A VPC enables your entire proposal team—even when geographically separated—to have a common electronic work center with access to an expandable repository of important automated proposal tools, project- and marketing-related information and knowledge base, and proposal templates required to help develop a successful proposal. A VPC facilitates real-time status checks for every proposal participant, regardless of physical location. Intranet and extranet technologies become invaluable time and B&P cost-savers with

---

6   As opposed to a file server, electronic mail, or groupware such as Lotus Notes or Microsoft Exchange messaging software.

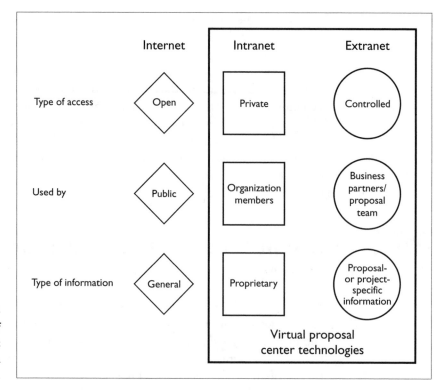

Figure 12.4 Comparison of electronic networked data warehouses.

geographically dispersed, complex teams. Both travel and administrative time and costs can be reduced. Schedules and action item lists can be shared. And appropriate levels of security and access control can be built into your system. Companies such as Intravation, Inc., in Belmont, California (http://www.intravation.com), offer off-the-shelf software applications designed to build virtual proposal centers within your firm.

Intravation's VPC tool is a collaborative, browser-based intranet and Internet application that enables organizations to plan, assemble, review, store, and disseminate proposal information. By reducing the amount of time spent on nonvalue-added tasks, VPC permits teams to focus on proposal content and quality.

Another enterprise document and content management tool of direct benefit to proposal development is infoRouter (http://www.inforouter.com), produced by Active Innovations, Inc., in Bohemia, New York. Having personally leveraged infoRouter to build the Proposal Knowledge Base as well as to establish multiple VPCs for my company, I can attest firsthand to its robust capacity and intuitive architecture and results. With the infoRouter Web-based solution, organizations can save money and time by lowering the cost managing documents. Your company's documents can be accessed securely from anywhere in the world. The only things required are a connection to the infoRouter server and a Web browser. Integrated full-text searching and powerful advanced search features make it simple to locate exactly what you need.

12.10.1 Useful document management systems (DMS)

In general, document management systems (DMSs) are used by individuals, office work groups, and enterprises to store organize, locate, and track the documents being produced as a part of their work. Document management is the automated control of electronic documents—images, spreadsheets, word processing documents, and complex documents—through their entire life cycle within an organization, from initial creation to final archiving. The document management process allows organizations to exert greater control over the production, storage, and distribution of documents. This yields greater efficiencies in the ability to reuse information, reduce duplication of effort, enhance organizational awareness, control a document through a workflow process, reduce document cycle times, and collaborate among geographically dispersed groups and individuals. The full range of functions that a DMS may perform includes document identification, storage and retrieval, tracking, version control, work-flow management, and presentation—across different computer platforms and in asynchronous mode (i.e., at no prescribed time), as well as over a variety of electronic networks, including the Internet.

Document Locator (ColumbiaSoft Corp., Portland, Oregon; http://www.documentlocator.com/productinfo/); DocuXplorer Software (Archive Power Sytsems, Inc., New York; http://www.docuxplorer.com); and Documents@Work (Critical Path Technical Services, Mukilteo, Washington; http://www.cpts.com/) are but several examples of document management systems on the market today.

In the architectural and construction industries, Autodesk's Buzzsaw product suite (http://usa.autodesk.com) allows documents and drawings to be shared with members of your team while being maintained in one secure place. With collaboration and print management benefits, Buzzsaw applications such as ProjectPoint centralizes all project documents and communications in one secure, online location. Document revisions and markups are shared in real time among architects, engineers, reprographers, developers, and contractors.

Novell's GroupWise 6.5 (http://www.novell.com/products/groupwise/) also offers full document management through the same interface that users already employ for messaging and scheduling.

As with any system for information management, there are a number of complementary and competing standards that describe the architecture of DMSs. Some of the more prominent standards include the following:

1. *Open Document Management (ODMA) application programming interface (API).* ODMA is an industry-standard interface for managing documents that allows users to store, retrieve, and share documents with security and version control. ODMA enables desktop applications to interface with document management sytems in a flexible manner without the need of a hard-coded link between the application and the DMS. Thus, ODMA allows multiple applications to access the same DMS. ODMA simplifies integration and interoperability of standard desktop applications with DMSs. Using ODMA, desktop applications access and manipulate documents carried in DMSs as easily as if they are residing in the locally accessible file system.

2. *Object linking and embedding (OLE).* This proprietary standard from Microsoft allows objects in one application to be linked to objects in another. For example, a graph in a word processing document can be linked to the original data in a spreadsheet application. When the data in the spreadsheet change, the graph in the document is automatically updated to reflect those changes. OLE allows applications to share data as well as the functionality of the originating applications.

3. *Object Management Group (OMG).* OMG was founded in April 1989 by 11 companies, including 3Com Corporation, American Airlines, Canon, Data General, Hewlett-Packard, Philips Telecommunications N.V., Sun Microsystems, and Unisys Corporation. The OMG is moving forward in establishing Common Object Request Broker Architecture (CORBA) as the "middleware that's everywhere" through its worldwide standard specifications.

Open Text Corporation's Livelink (http://www.opentext.com) is a highly scalable collaborative application that delivers Web-based intranet and extranet solutions. Livelink's functionality can be accessed using any standard Web browser. Deployable out-of-the-box, Livelink delivers a fully integrated set of robust enterprise services directly to your desktop. A specific Livelink module, called Livelink Activator for Lotus Notes, leverages companies' existing investments in IBM Lotus Notes/Domino. While tapping into Livelink's advanced document management and search capabilities in a Web-based environment, this module allows you to index multiple Lotus Notes databases.

Groupware is technology designed to facilitate the work of groups. This technology may be used very effectively in collaboration across your proposal team. IBM Lotus Notes is a groupware system that provides users with e-mail, work flow, calendars, task lists, and document sharing. Its infrastructure allows convenient flexibility in delivering these features. Lotus Notes is primarily restricted to asynchronous groupware features and not real-time communication capabilities.

Why use document management systems for proposal development? The benefits of document management for small businesses include (1) leveraging intellectual capital such that knowledge is created once, then reused many times; (2) managing workflow, controlling the flow of information through all phases of a proposal process; (3) fostering more effective teamwork among a distributed workforce; and (4) allowing rapid response to events such as RFP amendments and extensions to the proposal due date.

12.11 Using freelance proposal writers to maintain technical productivity

Carefully managed freelance writers and interviewers can be of invaluable assistance to your proposal manager at select phases of the proposal response life cycle (see Figure 12.5). If deployed on a just-in-time, as-needed basis, individual assignments should last from 2 or 3 days to 2 or 3 weeks. The point here is not to add to B&P costs, but rather to maintain your company's technical staff at maximum productivity and billability. The writers and interviewers would be trained by your proposal manager to extract appropriate technical and programmatic information from your firm's professional and managerial staff. Following the interview process, the freelance writers would then translate "technical-ese" into marketing-oriented, high-impact narrative that conveys the benefits and value of your approach and credentials to the client. Interviews can be conducted in person, by telephone, or via carefully constructed e-mail queries, and can be worked in before and after core business hours or during lunch breaks. The goal is to minimize the time that any billable staff person is diverted from direct client support, while simultaneously controlling the proposal schedule and generating integrated, well-crafted proposal text.

Effective freelance writers and interviewers bring an understanding of the documentation process and possess technical or professional writing or English-language and communications skills. Real-world experience with well-managed and trained freelance specialists serving in this capacity demonstrates an astounding 40% reduction in overall B&P costs!

Sources of freelance writers and interviewers include local colleges and universities as well as the following organizations:

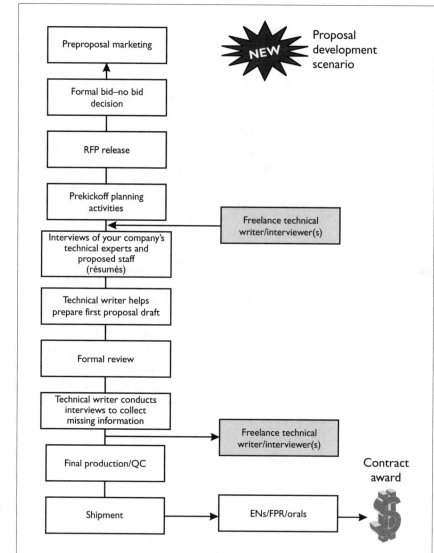

Figure 12.5
Carefully managed freelance writers can help you to control B&P costs.

- Association of Proposal Management Professionals (APMP)
  http://www.apmp.org/
- Washington Independent Writers (WIW)
  733 15th Street NW, Suite 220
  Washington, D.C. 20005
  Tel. (202) 737-9500
  http://www.washwriter.org

- Independent Writers of Chicago (IWOC)
  PMB 119
  5465 W. Grand Avenue, Suite 100
  Gurnee, IL 60031
  Tel. (847) 855-6670
  http://www.iwoc.org
- Independent Writers of Southern California (IWOSC)
  Main Street
  P.O. Box 34279
  Los Angeles, CA 90034
  Toll-free phone: (877) 79-WRITE
  http://www.iwosc.org
- Cassell Network of Writers (CNW)/Florida Freelance Writers
  Association (FFWA)
  Main Street
  P.O. Box A
  North Stratford, NH 03590
  Tel. (603) 922-8338
  http://www.writers-editors.com
- World Wide Freelance Writer
  http://www.worldwidefreelance.com
  (comprehensive listing of independent consultants and firms in all
  areas of expertise)
- Society for Editors and Proofreaders (United Kingdom)
  http://www.sfep.org.uk

ENDNOTES

1. In 1994, the average file size was approximately 18 MB. By late 1995, that had grown to an average of 73 MB. See Hevenor, Keith, "Storage Product Buyers Guide," *Electronic Publishing,* August 1, 1996, p. 10.

2. Information about the AFMC's electronic proposal evaluation initiative was obtained from Davidson, Paul, "U.S. Air Force SMC Innovations in Electronic Procurement," *APMP Perspective I,* Vol. I, July/August 1996, pp. 7, 12, 13, 15.

Chapter 13

Human and organizational dynamics of the proposal process

The classic mechanistic model ... fails to engage to the fullest extent an organization's most important asset—its people ...
—The PeopleWise Organization, Kepner-Tregoe

CRITICAL THINKING IN THE WORKPLACE. In 1996, Kepner-Tregoe, a Princeton, New Jersey–based organizational-management consulting firm, conducted its fourth in-depth, audited survey of more than 1,400 managers and hourly employees. What was being examined in this important study was how much an organization's collective thinking is being tapped to resolve the big strategic and operational issues, as well as the day-to-day impediments to competitiveness and profitability. The results? Sixty-two percent of workers surveyed believed that their organizations are operating at half—or less—of the brainpower available to them. These perceptions were reflected in responses by the managers as well. More than 40% of workers said they do not feel valued by their organization. And yet delighting clients and pleasing stakeholders "are the hallmarks of organizations in which people can—and want to—contribute their very best" [1].

13.1 Modifying our thinking to win

Your company's entry into the arena of competitive procurements brought with it the requirement for a fundamental shift in thinking and business-related behavior throughout the ranks of management as well as the professional and support staff. Senior management, for example, must appreciate the significance of proposals and proactively support the development and enhancement of the processes associated with the proposal response infrastructure. In addition, each division of your company must make full use of the collective human and contractual expertise resident throughout the entire firm. Divisions cannot operate effectively in isolation from each other or from corporate direction. The dynamics of genuine teamwork and knowledge sharing take on accentuated meaning in the competitive marketplace. People at all levels within the company must work together with clearly defined objectives, for example, to collect and assess marketing intelligence, write proposals, review proposals, and publish proposals. In too many companies, a small cadre of technical and support staff produce the significant bulk of the proposals. That practice does not encourage team building, knowledge sharing, and a wider sense of ownership for the company's success.

Proposals do not follow a fully democratic process. Companies must recognize that effective proposal management follows from informed, albeit authoritative decision making at the appropriate time. A proposal effort can become mired if people are not instructed to heed the guidance of the proposal manager. On the other hand, the proposal manager must listen to team members and make informed and balanced judgments. And senior management must clearly and repeatedly reinforce the role and authority of the proposal manager, the criticality of sharing knowledge internally, and the importance of meeting the schedule milestones.

There is no more important corporate activity than proposal development for a contracting firm.

An entrepreneur—the very person(s) who founded the company—can often be an unwitting impediment to the development, implementation, and enforcement of rational, formal, repeatable business development processes. Not because they do not want their company to succeed. Quite the contrary. However, they know that they built their company to its current revenue base and position within the small business contracting community on savvy, intuition, risk taking, and shoot-from-the-hip decision making. Successful marketing and proposal development in the competitive arena are, on the other hand, built upon a methodically applied and continuously enhanced set of disciplined business practices and processes. And that is an entirely different mind set than that which founded and nurtured the fledgling company.

13.2 Building a competitive work ethic

It is likely that your company's very existence and future growth depends upon winning 25% to 40% of the formal proposals you submit. Proposal development is an absolutely essential corporate activity within your organization. Every resource in your company must be made available to the successful completion of each proposal. There are many details that must be attended to in the course of preparing a proposal that do not require the direct, hands-on attention of the proposal manager. Such activities include electronic searches of the résumé and project description subdirectories or data bases to determine appropriate staff to meet position descriptions and to ascertain relevant company contractual experience. Technical editing and tailoring of résumés, project descriptions, and the technical and management narrative would be of invaluable assistance. In addition, administrative support in the form of photocopying, meal arrangements for evening and weekend work, and text-entry word processing are each vital to the overall success of the proposal preparation process.

No one person can address all of the details, for example, of technical content quality, writing consistency, and compliance with the RFP. Therefore, your company would benefit from more people being trained and available on an as-needed basis to assist in the preparation of the proposal in addition to the technical authors, proposal manager, and publications group. The burden of preparing winning proposals must be shared among all professional and support staff. Instill this work ethic among all your staff through regular briefings on new business prospects, competitors, and new technologies. A rotating schedule of administrative support being on call to support proposal activities after core business hours is one mechanism for ensuring broader involvement and appropriate levels of support for the proposal team and publications group. Everyone's future depends upon winning! The acquisition/capture team is an excellent vehicle through which to build a winning attitude for a given proposal effort. And when your company expands, you should hire technical and programmatic staff who possess strong proposal-related backgrounds, either from successful competitors or good business colleges and universities. Successful proposal experience and the mindset that fosters proposal success can be highly synergistic. I have personally witnessed a small company, which had very few people trained in the art and science of proposal development and management, become far more sophisticated and successful with the addition of key technical and programmatic staff who had proposal backgrounds. This knowledge influx helped to underwrite and authenticate the ongoing proposal development training programs that had been initiated in the company previously.

13.3 Strong link between project performance and proposal success

The integral connection between superior project performance and proposal success cannot be overemphasized. Past performance and key staff are two of the most significant evaluation factors in most support services proposals. Your client or potential client wants to ensure that your company has performed to expectation on contracts of similar type, size, technical scope, staffing level, complexity, and geographic area. Solid references from COTRs and COs are critical to your company being awarded the next contract. Project schedule slippages, low-quality interim and final deliverables, and cost overruns on your current or past contract work translate into hurdles that have to be overcome during the bid–no bid, marketing, and proposal processes. Verify all of your project-related references before your proposal is submitted to ensure that you will receive a positive reference from the government points of contact. Also ensure that telephone numbers, mailing addresses, and e-mail addresses are current.

Client staff have demonstrated on numerous occasions that they have enduring memories—for superior as well as poor performance. Perceptions are paramount in contracting. Just as your company networks and shares information, so do your client counterparts. If your company has allowed significant schedule slippages or incomplete deliverables to be submitted, you will be remembered for the wrong reasons. Those negative perceptions can take several contract cycles to overcome. They can also seep throughout the client's organization; so that if your performance was below average at a particular Air Force installation in Texas, a potential client at an Air Force location in Colorado Springs may have heard negative comments. The key is to leverage outstanding contractual performance into exceptional project experience in your proposals. Project and proposal processes and successes are closely linked, as shown in Figure 13.1.

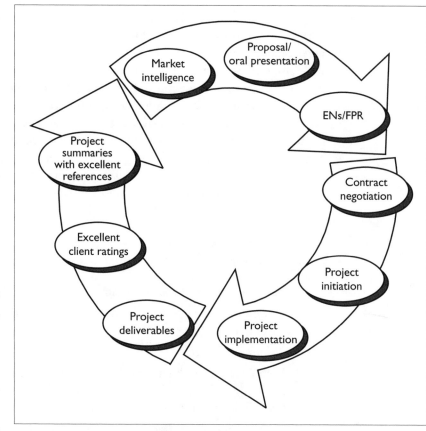

Figure 13.1
Proposals and
project
deliverables each
require planning,
technical effort,
senior review,
contractual
involvement, and
production
expertise.

13.4 Past performance—it's more important than you think!

Confidence in a prospective contractor's (in effect, your company) ability
to perform satisfactorily—or better—is an increasingly important factor for
the federal government in making a *best value* source selection decision.
For more than 15 years under the Federal Acquisition Regulations (FARs),
federal government acquisition officers have been allowed to consider
contractor past performance in their selection criteria for contract award.
According to the FAR, "past performance shall be evaluated in all source
selections for negotiated competitive acquisitions expected to exceed
$1,000,000" [FAR 15.304(3)(i)].

In recent years, however, changes fostered by the Federal Acquisition
Streamlining Act (FASA) and the Federal Acquisition Reform Act
(FARA), also known as the Clinger-Cohen Act, have expanded the role of
past performance in the federal acquisition process. In fact, the

government's desire to select high-quality contractors has resulted in past performance representing up to 50% of the source selection criteria in certain instances. On a high-profile $800 million Air Force procurement let in 2003, for example, past performance accounted for approximately 30% of the score under the evaluation factors for award (Section M of the RFP).

Past Performance Information (PPI) is the *relevant* data regarding a contractor's actions under previously awarded contracts. It includes the contractor's record of (1) conforming to specifications and standards of good workmanship, (2) containing and forecasting costs on any previously performed cost reimbursable contracts, (3) administrative aspects of performance, (4) history for reasonable and cooperative behavior, and (5) commitment to customer satisfaction and business-like concern for interests of the customer.

The federal government uses a variety of sources to obtain past performance information regarding contractors, including the following:

- Past Performance Proposal Volumes supplied by the contractor;
- Questionnaires completed by the contractor's federal government customers;
- Telephone interviews;
- Web-enabled Contractor Performance Assessment Reporting System (CPARS) via the Past Performance Information Retrieval System (PPIRS) (http://www.ppirs.gov/);
- Central Contractor Registration (CCR) Program;
- U.S. Army Corps of Engineers Architect/Engineer Contract Administration Support System (ACASS);
- Corporate Web sites;
- Internet.

Let's take a closer look at the questionnaires that are completed by a contractor's federal government customers. Frequently, the questionnaire will include requests for specific information about the contractor's quality of service, project management effectiveness, timeliness of performance, and cost-effectiveness. From the contractor's perspective, these questionnaires should be a vital part of the overall proposal response. *Why?* First, because the government's response to these questionnaires carries far more weight than the contractor's own self-appraisal about its past performance excellence. Second, because the questionnaire provides valuable and detailed insight into the performance, schedule, and cost parameters that the government agency letting the new contract deems to be important.

Therefore, the key elements found in the past performance questionnaire should be built into your proposal responses, as appropriate.

For instance, a questionnaire may include queries related to accurate and timely deliverables, resolution of contract performance problems, ability to fill technical vacancies rapidly, and responsiveness to end-user requirements. Other queries relate to phase-in/transition, management of multiple projects simultaneously, responsiveness to government requirements, ability to work independently of the government's guidance, and compliance with quality control plans. These elements can then be woven into the contractor's proposal volumes in such a manner as to demonstrate the processes, tools, and staff resources that will be applied and deployed to meet these subtle requirements that may or may not be stated explicitly elsewhere in the RFP or RFS.

Additionally and very importantly, certain federal government agencies will go beyond the past performance citations of the prime contractor and its subcontractors included in a prime contractor's Past Performance Proposal Volume. For example, the Air Force's Performance Risk Assessment Groups (PRAGs), which are chartered to assess the *performance confidence*[1] each contractor, will also unilaterally consider additional projects and programs not included in a contractor's Past Performance Volume. Performance confidence evaluates a contractor's present and past performance to establish a confidence rating of that company's ability to successfully perform as proposed. What this new practice means in very practical terms is that past performance "blemishes" can and will be uncovered by government evaluation teams. For the prime contractor and its subcontractors, this practice on the part of the government underlines further the close connection between *project performance success* and *new business success.* In effect, the business development and proposal development success of a contractor is directly dependent on its operational and contractual success.

Past Performance Information Retrieval System (PPIRS)

The PPIRS database mentioned earlier is available to all source selection officials across the entire federal government. Its purpose is to assist federal acquisition officials in purchasing goods and services that represent the best value for the federal government. Contractor access to PPIRS is gained through the Central Contractor Registration (CCR) (http://www.

---

1   Rather than *performance confidence,* some federal government organizations use *performance risk* in their assessment of contractor past performance proposal submissions.

ccr.gov) process. A contractor must be registered with the CCR and must have created a Marketing Partner Identification Number (MPIN) in the CCR profile to access PPIRS information.

Formerly called the Past Performance Automated Information System (PPAIS), PPIRS is currently sponsored by the DoD E-Business Office and administered by the Naval Sea Logistics Center Detachment Portsmouth (NAVSEA). The federal PPIRS database is a Web-enabled application that allows the retrieval of contractor past performance information. It is also a central warehouse used to retrieve performance assessment reports received from the recognized federal report card collection systems, including the following:

- *Contractor Performance Assessment Reporting System* (CPARS) (http://www.cpars.navy.mil/) used by the Navy, USMC, Air Force, Defense Logistics Agency (DLA), and other defense agencies;
- National Institutes of Health (NIH) *Contractor Performance System* (CPS) (http://cps.od.nih.gov/);
- National Aeronautics and Space Administration (NASA) *Past Performance Data Base*;
- Army's *Past Performance Information Management System* (PPIMS)

Contractor Performance Assessment Reporting System (CPARS)

CPARS is a Web-enabled application that collects and manages the library of automated CPARs. CPARS is for unclassified use only. A CPAR assesses a contractor's performance and provides a record, both positive and negative, on a given contract during a specific period of time. Each assessment is based on objective facts and supported by program and contract management data, such as cost performance reports, customer comments, quality reviews, technical interchange meetings, financial solvency assessments, construction/production management reviews, contractor operations reviews, functional performance evaluations, and earned contract incentives. Contracts of $1 million and up in the areas of services and information technology, for example, are required to have an associated CPARS report on the contractor. Because contractors are vital partners in the performance assessment process, contractor senior management representatives may request access to all records pertinent to their corporate entity by completing a CPARS Corporate Senior Management Access Request Form (http://www.cpars.navy.mil/accessforms/csmarf.htm).

DoD policy states that past performance evaluation information (e.g., CPARs) is privileged source selection information and is not releasable under the Freedom of Information Act (FOIA). Further, FAR Subpart 42.1503 (b) states: "The completed evaluation shall not be released to other than Government personnel and the contractor whose performance is being evaluated during the period the information may be used to provide source selection information." What this means is that your company cannot make a FOIA request for a competitor's CPARS information.

National Institutes of Health (NIH) Contractor Performance System (CPS)

Contractor ratings under the CPS system (http://cps.od.nih.gov/) range from outstanding to unsatisfactory for (1) quality of product or service, (2) cost control, (3) timeliness of performance, and (4) business relations (level of response to inquiries and/or technical, service, and administrative issues). CPS also captures government assessment information about a contractor's key personnel on a given project, subcontracting plan, and customer satisfaction level.

The NIH CPS was designed, developed, and implemented by the NIH Center for Information Technology (CIT) under the guidance and sponsorship of the Office of Contracts Management, Office of the Director, NIH. This system supports acquisition activities in all 50 states and more than 100 foreign countries. CPS contains evaluations from the following federal departments/agencies: Department of Health and Human Services (DHHS), Department of Agriculture (USDA), Department of Treasury, Department of Commerce (DOC), Department of Justice (DOJ), Department of Energy (DOE), Department of the Interior (DOI), Department of Labor (DOL), Department of Veterans Affairs (VA), Social Security Administration (SSA), Agency for International Development (USAID), Environmental Protection Agency (EPA), Federal Emergency Management Agency (FEMA), General Services Administration (GSA), Department of Transportation (DOT), Department of Education (DOEd), Department of State (DOS), Export-Import Bank (EX-IM), and the Architect of the Capital.

U.S. Army's Past Performance Information Management System (PPIMS)

The Past Performance Information Management System (PPIMS) is the Army's central repository for the collection and utilization of Army-wide contractor Past Performance Information (PPI). Available to authorized government personnel, PPIMS is used to support both the contracting performance review process and future award decisions.

Within DoD, there are also specialized past performance databases for the architect-engineering (ACASS) and construction (CCASS) and business sectors. If you are doing business in those specialized business sectors, your past performance information would be in either ACASS or CCASS. ACASS and CCASS data are not yet in PPIRS, but plans are under way to make CCASS and ACASS information available in PPIRS.

Architect-Engineer Contract Administration Support System (ACASS)

ACASS is a database of selected information on architect-engineering (A-E) firms. ACASS provides required information for DoD and federal government selection committees to aid them in their process of awarding A-E contracts. It is maintained by the Portland, Oregon, office of the U.S. Army Corps of Engineers (USACE). It contains detailed Standard Form 254 and now SF330 data on contractors, DoD awards, and past performance evaluations for the last 6 years of A-E services contracts. For more information on ACASS, go to https://www.nwp.usace.army.mil/ct/I/welcome.htm.

ACASS provides an efficient and economical automated database system by which to adhere to these requirements. The use of ACASS for past performance history is mandated for all DoD agencies in the Defense Federal Acquisition Regulation Supplement (DFARS). For the Corps of Engineers, ACASS eliminated the need for 43 district offices to disseminate performance evaluations among each other and to maintain copies of SF254s. ACASS is a catalyst for overall improvement in performance of the A-E community. If the performance evaluation is to be rated "below average" or "poor," the A-E firm will receive notification of such prior to the evaluation being signed by the reviewing official. The firm then has 15 days to appeal the proposed rating.

Construction Contractor Appraisal Support System (CCASS)

CCASS is an automated database of performance evaluations on construction contractors. CCASS provides PPI for DoD and federal government contracting officers to aid them in their process of evaluating construction contractors' past performance. As with ACASS, it is maintained by the Portland, Oregon, office of the USACE. For more information on CCASS, go to https://www.nwp.usace.army.mil/ct/I/welcome.htm.

Recent acquisition reform legislation, policies, and guides are intended to move source selections into increasingly greater use of "best value" procedures, in which considerations of noncost factors (such as technical, management, and PPI) might justify award to other than the apparent low-cost offeror. In particular, these reforms require source selection officials to consider relevant past performance as a measure of performance risk for future contract awards. *Make certain you know what the grades are on your company's report card!*

13.5 Proposals can be fun!

Successful proposal efforts can be fun, in addition to being a lot of hard work. The level of fun is directly dependent upon how well a variety of people can interact during the arduous proposal response life cycle. Publications staff, for example, must interact effectively and courteously with technical contributors. A graphics artist might need to spend some time talking with one of the technical writers in order to understand fully what that author wanted to convey through a particular graphic concept or to help that author conceptualize a graphic. That interaction need not be confrontational in either direction. The writer should want to ensure that his graphic is well presented and be pleased that another professional is attending to that detail. And the artist should want every graphic to be accurate and convincing and not become overly frustrated with the back-of-the-envelope stick figures that the writer may have submitted originally. Communication and cross-fertilization of ideas within the Proposal Team should be actively encouraged by the proposal manager.

Fun can take the form of some office-appropriate humor injected into the proposal process in liberal proportion. And a special meal involving the proposal team members is a nice perk, too. The proposal manager should help make the proposal team members feel appreciated and good about what they are doing for the company. As the days and weeks of the proposal response life cycle unfold and the proposal team becomes more tired and worn, pleasant humor, nice meals, meaningful compliments, and effective communication assume even larger importance. And gift certificates distributed to the team on a late night or weekend can lift everyone's spirits and contribute to even more productivity.

13.6 Maximizing human intellect

[U]nlike traditional raw material—which is inspected, warehoused, bar-coded, and audited—corporate knowledge is scattered, hard to find, and prone to disappear without a trace.
—Thomas A. Stewart, *Fortune* magazine

Converting human intellect into useful products and services, including proposals, is becoming increasingly critical to any business's success [2]. "We now know where productivity—real and limitless productivity—comes from. It comes from challenged, empowered, excited, rewarded teams of people. It comes from engaging every mind in the organization" [3]. There is an accelerating recognition among corporations that their future profitability depends upon intangible assets such as "creativity, flexibility and the speed with which they can share new ideas and information. In a literal sense, modern corporations *are what they know*" [4]. Major international corporations such as Skandia AFS (Sweden), Xerox, Hewlett-Packard, British Petroleum, and Dow Chemical are spending significant resources on knowledge identification and management. And the U.S. Securities and Exchange Commission (SEC) hosted a conference that focused on accounting for and reporting intangible assets. Alan Webber, a founding editor of the Boston-based business magazine *Fast Company*, has noted that "[t]he world of business is realizing that ultimately, what matters is the quality of the people in the organization, and how they treat their customers and how creative they are in coming up with new products and new services" [5].

Another significant transition among such major corporations such as Microsoft, Compaq, Sun Microsystems, and Hewlett-Packard is the paradigm shift away from business as a battlefield to that of business as a "complex ecosystem in which the company that wins is the one that is quickest to adapt. Employee individuality and diversity are honored and encouraged" [6]. To adapt to new market circumstances and translate them quickly into meaningful opportunities, your employees simply cannot be viewed and treated as mere mindless troops, crank-turners, or line items on an accounting spreadsheet. The originators of business process reengineering (BPR)—Michael Hammer, James Champy, and Thomas Davenport—have come to recognize the real-world failure of their concept of "fundamental rethinking and radical redesign of business processes to achieve dramatic improvements in critical, contemporary measures of performance, such as cost, quality, service, and speed." In essence, they forgot the people [7]. As Michael Verespej cautions, "We must remember that knowledge workers keep their tool kits in their heads. We must learn to lead people, not contain them. We must learn to listen"[8].

With small companies in particular, leveraging the collective intellect and knowledge is integral to business acquisition achievement. Providing work environments that are conducive to the generation, exchange, respect, and appropriate application of ideas will pay dividends in morale, employee retention, and revenue. Pleasant physical facilities, progressive human resources policies, an open-door management culture, and concrete incentive programs, for example, all participate in inspiring and harnessing the best in and from your employees.

Dr. Douglas Soat, a psychologist who consults to businesses regarding human resources issues, has identified two critical motivators for both technical and nontechnical staff that result in excellence on the job. The first is "setting challenging but realistic expectations," and the second is "demonstrating a true concern for employees" [9]. I can certainly attest to the validity of the latter. I have seen my role in proposal development during the past 17 years to be that of facilitator—a person to grease the wheels of the proposal response process as well as the management decision-making mechanism. How can I help my staff do their jobs more effectively and efficiently? What training and additional equipment do they require? Do they need a ride to public transportation after working very late at night? I have arranged for and purchased meals for my staff and have had special breakfasts at which I have brought in the food from home. I have talked with my employees, at their request, about personal issues, not offering any solutions, but simply providing an attentive ear. In addition, I have also pursued and received out-of-cycle raises, spot bonuses, and gift certificates for my staff in an effort to demonstrate in a tangible way their value to the proposal process.

Your company will be well served to recognize the value of competent, dedicated, team-oriented, and solution-oriented support and technical staff in maximizing performance and engineering business development success.

On a regular basis, I have worked hand-in-hand with my staff when we had to stay 26 hours straight (with breaks and meals) to finish a proposal or provide support over a weekend or on a holiday. My staff knows that I will be there with them to the end to get the job done. The concern for quality is lived each day. They also know that I will share their accomplishments on and contributions to proposals and contract deliverables with senior management. Their successes are documented and reappear in detail in the narrative of their performance appraisals. I have extremely high standards for quality and productivity, and these are communicated to my staff regularly and in a variety of forms. I ask a lot of them and attempt to give a lot back to them as well. In every employee relationship, I have attempted to foster trust in the person by trusting him or her and treating the person with dignity and respect. I recognize the value of competent, dedicated, team- and solution-oriented support and technical staff in maximizing performance and fostering business development success.

13.7 Proposal professionals as change agents

Professional proposal staff can become integral participants in the positive transformation of an organization and its business processes [10]. The detailed analysis and articulation of a company's identity as well as its business processes, operational functions, relationships with clients and vendors, success stories and best practices, and internal resource- and knowledge-sharing mechanisms—the very activities that proposal staff do on a day-to-day basis—can be of direct benefit to the company in ways that extend far beyond proposal development and production. Your professional proposal staff can be among the most effective facilitators of knowledge exchange across your organization's divisional, geographical, and hierarchical boundaries. The "excellent ideas, streamlined operations, cost efficiencies, and technical superiority" [11] that are shared internally with and by your proposal staff can be of immense benefit, not only for the next contractual competition, but also to each element of the company.

13.8 Wellness in your proposal process

Fundamental health and wellness involve nutrition, exercise, sleep, and mind-body interrelationships. Employees who manage stress well tend to be more healthy, productive, cooperative, and creative. By supporting employee stress resilience, proposal managers may improve the quality of the work environment and reduce their own stress.

Senior management should investigate ways to apply flex time realistically in order to allow employees to participate in wellness programs that help them recharge during the work day. Proposal managers should actively encourage staff to take breaks away from their workstations, during which they might read, walk, or exercise in some other way. Supply fresh fruits, vegetables, other low-fat, low-sodium nutritious foods, and spring water to the proposal team on a regular basis. Ensure that staff are given some dedicated time during the height of the proposal life cycle to get out of the office physically and mentally and share time with family and friends. The proverbial all-nighter is not only counterproductive to the overall proposal process and product, but physically and mentally detrimental to the people who are the proposal team.

Proposal managers must make every effort to support the proposal team in a proactive manner with words and actions of encouragement and reliable guidance. Provide a private forum for staff to express their feelings in appropriate ways rather than letting those feelings fester and build to the point of an inappropriate emotional outburst sometime later. Help everyone know and feel that they are appreciated and respected, and are an integral part of an exciting proposal process.

During the marketing life cycle, your company will want to control costs very carefully. Monitoring and managing bid and proposal costs are discussed in Chapter 14.

ENDNOTES

1. *Minds at Work: How Much Brainpower Are We Really Using? A Research Report,* Princeton, NJ: Kepner-Tregoe, 1997.

2. Quinn, James Brian, Philip Anderson, and Sydney Finkelstein, "Managing Professional Intellect: Making the Most of the Best," *Harvard Business Review*, March/April 1996.

3. Quinn, Judy, "The Welch Way: General Electric CEO Jack Welch Brings Employee Empowerment to Light," *Incentive*, September 1994, p. 50.

4. Hamilton, Martha A., "Managing the Company Mind: Firms Try New Ways to Tap Intangible Assets Such as Creativity, Knowledge," *Washington Post,* August 18, 1996, p. H1.

5. Hamilton, Martha A., "Managing the Company Mind: Firms Try New Ways to Tap Intangible Assets Such as Creativity, Knowledge," *Washington Post,* August 18, 1996, p. H5.

6. James, Geoffrey, "It's Time to Free Dilbert," *New York Times*, September 1, 1996, p. F-11.

7. Mariotti, John, "Nursery-Rhyme Management," *Industry Week*, May 5, 1997, p. 19.

8. Verespej, Michael A., "Only the CEO Can Make Employees King," *Industry Week*, November 16, 1998, p. 22.

9. Soat, Douglas M., *Managing Engineers and Technical Employees: How to Attract, Motivate, and Retain Excellent People*, Norwood, MA: Artech House, 1996, p. 119.

10. McVey, Thomas W., "The Proposal Specialist as Change Agent," *APMP Perspective*, May/June 1997, p. 13.

11. McVey, Thomas W., "The Proposal Specialist as Change Agent," *APMP Perspective*, May/June 1997, p. 3.

Chapter **14**

Controlling bid and proposal costs

S MALL BUSINESSES SPEND on the order of $1,000 to $2,000 for every $1 million of total contract value. This expands to a range of $30,000 to $50,000 for a $5 million procurement, or approximately 1%. Proposal consultant Hyman Silver and many others have suggested that 5% of total contract value be spent on marketing, B&P, and IR&D. Decision drivers regarding marketing, B&P, and IR&D expenditures will, of course, vary according to strategic importance, contract type (cost-reimbursement or fixed-price), fee and profit margins, and your company's incumbency status regarding the particular business opportunity. Small businesses often cannot spend 5% of total contract value due to cash flow, line of credit, and contract backlog parameters. It certainly does require money to win proposals, but the careful application and control of that money can be leveraged into contract award.

14.1 What does it cost to get new business, and how are those costs recovered?

If a company wants to capture $5 million of new business during fiscal year (FY) 2005 and it has a win percentage on proposals of 30%, then it has to estimate the costs associated with pursuing $15 million of new business.

All companies, including those competing in the federal marketplace, need to develop strategies for becoming knowledgeable about and then capturing new business. Within the small business contracting community that serves the federal government, a relatively narrow subset of professional staff is involved in identifying, tracking, and capturing new business. This staff may include business development and marketing, proposal development, and division-level management professionals. Coordinating short- and long-range opportunity tracking can be challenging in small businesses because of staffing limitations and also because centralized control of business development in such firms is often an evolutionary development rather than a fact of life present from the point of the company's founding. Many times, division managers are primarily responsible for "growing" their respective divisions and have been accustomed to overseeing their own business acquisition activities. Corporate marketing staff may then assist division managers with particular marketing activities.

Marketing and B&P costs can only be projected, monitored, and controlled once a company has determined the extent to which it wants to expand each operating division in accordance with an agreed upon mission statement and strategic plan. Extending the mission statement and strategic plan down to the competitive procurement level is the key to success. Procurements within the appropriate dollar range and contractual type must be ascertained to be available before growth projections can be made realistically. In-depth knowledge of specific client organization planning, programming, budgeting, and procuring processes, cycles, and decision-making pathways is imperative to make this determination. If, for example, a company wants to capture $5 million of new business during its FY2005 and it has a win percentage on proposals of 30%, then it has to estimate the costs associated with pursuing $15 million of new business. (Start by listing the potential marketing opportunities, identifying the best opportunities, calculating associated B&P costs, and totaling the numbers.) Fifteen million dollars of viable marketing opportunities in this company's specific LOBs must then be available during FY2005. This small company should plan to spend $150,000 during FY2005 in marketing and B&P costs. These cost estimates should be tracked, monitored, evaluated, and modified on an ongoing basis.

Marketing, B&P, and IR&D costs are recoverable through G&A charges to the federal government. Upon invoicing the government for services rendered, a contractor will typically collect G&A and other indirect costs associated with operating a business.

14.2 Tracking B&P expenditures

Analyzing direct labor and ODCs associated with preparing a proposal as well as the associated postproposal activities in advance of a bid–no bid decision and then tracking these costs carefully during the entire proposal life cycle will be beneficial to your company for several reasons. First, accumulating real B&P data over time will allow for increasingly accurate B&P projections as new estimates are calibrated against these real data. Early B&P analysis and estimation will help ensure that adequate resources are made available to prepare an outstanding proposal. Figure 14.1 provides a representative B&P initiation form. Tables 14.1 and 14.2 present templates that can be used to capture and track B&P data. These data should be entered electronically into a spreadsheet or data base software application such as Microsoft Access to access easily and prepare comparative graphics for trend evaluation and future projections.

14.3 Business development bonus policy

B&P expenditures increase the indirect rates that your company must submit on competitive procurement solicitations. Proposal managers are the individuals charged with developing, defending, monitoring, and controlling B&P budgets. In effect, they are fully accountable. Success in the highly cost-competitive federal marketplace of the new millennium demands that acquisition and proposal development costs be kept to a minimum. Therefore, to the fullest extent possible and keeping within Department of Labor guidelines, proposal development efforts must be accomplished during nights and weekends. And the reality, as I have observed it and lived it on both coasts, is that many major U.S. contracting and consulting firms nationwide also minimize B&P costs by unpaid night and weekend work. As a small company, your professional, exempt employees should keep B&P charges to an absolute minimum. If they do not, you will have difficulty in expanding your company's revenue base; to do so you must incentivize them and communicate regularly with them about the good job they are doing.

B&P Initiation Form

B&P initiation date: _____   B&P internal tracking number: _____

Client: **FAA**
Procurement title: **Software engineering**
Division responsible for contract (if awarded): _____

Acquisition manager: _____
Proposal manager: **M. Hendrick**

Due date of proposal: _____
Prime: _____   Sub: _____
Teaming partners: _____
_____
_____

Location of performance contract: _____
Type of contract: _____
Expected period of performance: _____ (base and option years)
Persons/year staffing level: _____
Estimated contract dollar value: _____

Estimated B&P costs (please attach the direct labor analysis and ODC forms)
    Hours required: **175**
    Direct labor: **$4,925**
    Overhead: _____
    ODCs: **$825**
    Total: _____

Start date of B&P effort: **April 2005**   Ending date of B&P effort: **June 2005**

Signatures below indicate that the staff have (1) full awareness of this procurement, (2) have reviewed the items above for accuracy and completeness, and (3) accept the B&P hours estimate as reasonable for a procurement of this dollar and level of effort.

Signature lines: _____

Figure 14.1 B&P initiation form.

In recognition of the extended effort of your professional-level and support staff, your company should strongly consider developing and implementing a clearly defined business development bonus policy. The policy should be designed to provide incentives and extra monetary compensation for those employees who have provided outstanding support for proposals or business development. You certainly want to retain your technical and support talent base.

| Date prepared: March 11, 2005 | | | | | | | |
| Analyzed by: M. Hendrick, proposal manager | | | | | | | |

Projected Spending

| Month | April 2005 | | May 2005 | | June 2005 | | Total | Total |
Employee Name	Hours	Dollars	Hours	Dollars	Hours	Dollars	Hours	Dollars
B. Frank	15	$525	35	$1,225	25	$875	75	$2,625
G. Sinclair	5	$125	45	$1,125	10	$250	60	$1,500
J. Landis	10	$200	25	$500	5	$100	40	$800
Monthly Total	30	$850	105	$2,850	40	$1,225	175	$4,925

Table 14.1   Direct Labor Analysis—Software Engineering Procurement for the FAA

Date prepared: March 11, 2005							
Analyzed by: M. Hendrick, proposal manager							
Include types of expenditures, such as documentation, travel, training, advertising, equipment purchases, lease agreements, supplies, software.							

Month Projected

	Amount	Amount	Amount	Amount	Amount	Amount	Total
Type of expenditure	April	May	June				
Local travel	$75	$150	$50				$275
Position advertising	$550	0	0				$550
	$625	$150	$50				$825

Table 14.2   ODC Analysis—Software Engineering Procurement for the FAA

In a reward and motivational survey conducted by Ms. Jo W. Manson, then of the Andrulis Research Corporation in Arlington, Virginia, it was determined that during the actual proposal response life cycle, special meals were used as a motivator by 25% of the respondents. After a proposal had been submitted, expressions of appreciation, bonuses, and time off were used as motivational tools.[1] Ms. Manson reminds us that "we are in the people business and, no matter how automated and process driven we may become, we must still rely on well-motivated staff to produce our winning proposals."

According to research by Shari Caudron that appeared in *Industry Week* magazine, "[C]ompanies are starting to extend their incentive programs to nonsales employees with the hope of generating renewed enthusiasm and commitment" [1]. Incentives have begun to take the form of select merchandise, travel, catalog merchandise, cash, jewelry, and honorary titles.

14.4 Stretching limited marketing funds

Most small companies face limited staff power for marketing and business development activities. One mechanism to augment and supplement your company's marketing staff is to use the services of any of a wide variety of marketing consulting firms, such as Federal Sources, in McLean, Virginia (http://www.fedsources.com); INPUT, which has offices in Reston, Virginia, London, and Bangalore, India (http://www.input.com); and Environmental Information Ltd., headquartered in Edina, Minnesota (http://www.envirobiz.com). Consulting companies can provide market research and trend analyses, competitor analysis, marketing opportunity identification, procurement informational databases, intercompany information sharing, strategic planning support, and proposal planning and guidance. Spending dollars for specific, well-directed marketing and planning support from an outside source can help to stretch your company's internal resources and actually help them to be more efficient. Taking the marketing information from outside firms and assessing, distributing, and archiving it in an easily retrieved manner is critical in order to maximize its value to your company.

Let us consider several specific ways in which outside marketing consulting and research firms can assist your company. They can be used to identify and conduct initial screening of potential teaming partners for an important procurement. Such firms can also help validate the funding for a particular procurement. They can be employed to collect competitor newsletters and annual reports. Often, these marketing research firms offer a basic suite of services such as a regularly updated automated database

---

1   The survey was distributed to 400 members of the Association of Proposal Management Professionals (APMP) and resulted in a 25% response rate. Expressions of appreciation noted included letters, verbal praise, win parties, and publicity within the company.

and hardcopy notebooks of marketing information. This standard suite is, in turn, augmented for additional costs with such services as on-demand telephone hotline support, Web-based information access, news release support and access, and senior-level consulting on a one-on-one, personal level.

Direct experience has shown that a company is well served to judiciously employ the services of several outside marketing consulting firms. Although there will be overlapping information provided, each firm contributes potentially important analysis and data that another may not.

Accurate, current marketing intelligence must be built into your company's proposal. We will now examine proposal writing techniques that are effective in incorporating that intelligence into a solutions-oriented, compliant set of knowledge-based sales documents that convey a compelling, client-focused story.

ENDNOTE

1. Caudron, Shari, "Spreading Out the Carrots," *Industry Week,* May 19, 1997, p. 20.

Chapter 15

Tried-and-true proposal writing and editing techniques

15.1 Proposals are knowledge-based sales documents

Accuracy, brevity, and clarity

The essence of effective proposal writing is responding to the RFP or RFS requirements while completing the process begun during marketing, namely, convincing the client that your team and approach are the most appropriate and cost-effective. Remember that proposals are, first and foremost, sales documents. They are not technical monographs, white or position papers, or user's manuals. Conceptualizing and developing the technical solution is an engineering or scientific issue. Packaging that solution in a combination of crisp, convincing narrative and high-impact graphics is a sales issue (see Figure 15.1).

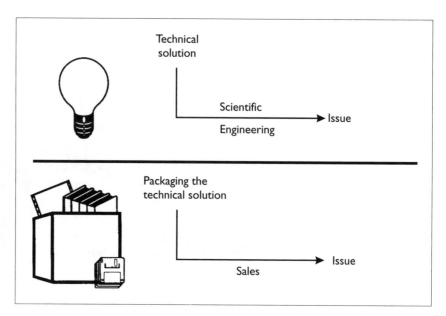

Figure 15.1
Packaging the
technical solution
is a sales problem.

Your company's proposal manager must articulate clearly to his or her technical writing team exactly what the expectations and acceptance criteria are. He or she must foster ongoing communication and feedback not only to himself but also among all of the writers. This will help to ensure a consistent approach and minimize rewriting.

Most proposals for a given procurement look and read essentially the same. The challenge is to incorporate well-substantiated information in your proposal that only your company can say. Identify precisely what will separate, or discriminate, your company from the competition. If your company is the incumbent contractor for a particular project and has performed favorably, use the names (and photos) of your incumbent staff people throughout your proposal. Reassure the client that they will be working again with the same competent technical staff who will be supervised by the same responsive management team they know and trust. Help the client understand that their investment in training and building intellectual capital will be retained. Write to a level of technical detail that exceeds what nonincumbents could gain from your project monthly reports (obtained through the FOIA),[1] published articles, and marketing conversations with government staff. Demonstrate that your company understands the technical and programmatic risks as well as the success criteria. Strive to have your proposals not be boring.

Contributors should keep in mind the following ABCs of proposal writing:

- *Accuracy:* It is not enough, for example, to tell your client that your company has saved time and money on projects of similar size, contract type, technical scope, and geographic area. Tell the client exactly how much time you have saved, how you accomplished this, precisely how much money, and the reasons why. Authenticate your claims with concrete, accurate information. Quantify your claims whenever possible with reference to your company's ACASS, PPIRS, CPARs, or PPIMS (see Appendix D for full discussion of these terms) ratings, contractual evaluation reports prepared by your clients, or FASA-driven[2] client "report cards" for contracts in excess of $1 million.

- *Brevity:* Keep your writing crisp. Do not assume that volume will make up for quality. Too many times I have heard technical staff say, "Give them 100 pages, and they'll just assume the correct responses are in there." Guess what? No, they (the evaluators and their support staffs) will not. Unless your company demonstrates through subheadings, graphical pointers, chapter-level or sectional tables of contents, and cross-reference matrices that it has responded completely and appropriately to all elements in Sections L, M, and C, you will not score the maximum points available.

- *Clarity:* Assist the client's evaluators to understand, for example, exactly how your company proposes to manage this new project, how you will integrate your subcontractor staff on the job, how your proposed project manager can access corporate resources quickly, and how you will manage tasks in geographically diverse locations. Amplify your crisp, direct narrative descriptions with clear, easily interpretable graphics.

Ten pages of thematically integrated, well-constructed, and compliant prose are much preferred to 50 pages that contain most every technical detail the writers happened to know. Technical "data dumps" glued

---

1   FOIA requests should be made to the appropriate agencies in written form. Letters should contain only one specific request. The reason for this practice is that an agency could decline to process one item of a multi-item request, and then the entire request would be returned to the contractor and the process would have to begin again. It is suggested that you begin your request with, "Pursuant to the Freedom of Information Act, 5 U.S.C., Section 552 as amended, Company XYZ hereby requests a copy…." Keep an electronic log of each FOIA request, and place follow-up calls to the agencies after 2 weeks have elapsed.

2   As required by the FASA, a contractor's past performance in government contracts is now considered a relevant factor in the award of future contracts.

together produce a very uneven and most likely noncompliant proposal. Keep your ideas focused on the client's requirements as stated in the RFP, and keep your sentences short. This author has seen sentences in proposal narrative that have extended more than 80 words!

Focus on translating the features of your technical and management approaches and into benefits to the client and his project, his budget, his schedule, his career, and his organization's mission (see Figure 15.2). Demonstrate in clear terms that your company will help him succeed in his mission within his organization. This is an aspect of proposal writing that demands concrete, current, and in-depth marketing intelligence and sharing of this knowledge internally. Without this, your proposal will describe your staff, facilities, contractual experience, and technical capabilities in a vacuum. There will be little or no direct linkage of your company's capabilities with your client's perceived requirements.

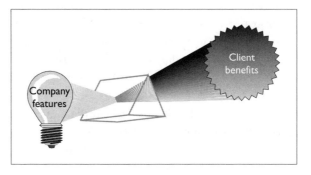

Figure 15.2
Translate features into benefits for your clients.

15.2 Active voice adds strength and saves space

Proposal writers should make every effort to employ the active voice: "We accomplished XYZ," not "XYZ was accomplished by us." The active voice adds strength to the proposal presentation and can save as much as 14% in terms of space as compared with passive voice depending upon the font that is used.

Contributors should attempt to vary sentence and paragraph structure. If you examine a proposal and most of the paragraphs on a page begin with "The" or your company's name, the narrative requires editing to infuse variety.

Writers should also try to employ strong and descriptive verbs, adjectives, and adverbs (see Figure 15.3), such as the following:

- Our incumbent staff of six seasoned engineering professionals all exceed the stipulated experience level requirements.

High-impact descriptors

- Rigorous development approach
- Principled management
- Seasoned field technicians
- Innovative networking solutions
- Comprehensive community-relations plan
- Results-oriented project team
- Industry-standard sampling methodologies
- Demonstrated success in providing superior-quality engineering solutions
- Responsive organizational structure and seasoned program leadership
- Proven and externally certified business processes
- Technology-driven core competencies

Figure 15.3
High-impact descriptors for proposal text.

- Over the course of 3 years of successful enterprise IT service to the Air Force Space Command, our senior programmers developed and implemented effective and cost-efficient ...

- Our company offers field-tested prototyping experience supported by a local testbed laboratory.

- Our 15 analysts embody a knowledge base and legacy of experience that is unmatched.

- Best practices and lessons learned from award-winning performance-based IT support will be shared through established KM processes.

- We are a people-oriented firm with a very low turnover rate as measured against the industry standard in the Washington metropolitan area.

- We will deliver highly defensible results through the conduct of all activities in strict compliance with our sound metrics-based QA-QC program.

- We will maintain full technical continuity of operations (COOP) and preserve intellectual capital on this important Department of State program through proactive incumbent hiring.

- Our technical understanding is greatly enhanced by our award-winning past performance on two critical contracts for the Office of Personnel Management.

- With an extensive IT portfolio that includes Sun workstations to support off-site scientific data processing, our company ...

- Our company's 4 years of progressive support of the VA validates our commitment to ...

- We are committed to joining a working partnership with NOAA National Weather Service during the life of this contract.

15.3 Guide the client's evaluators through your proposal

Formulate ideas precisely, concretely, and simply. To include every technical detail each writer knows about a certain topic will probably not result in a successful proposal. This is where storyboards and bullet drafts come into play—to guide your writers in preparing a thematically consistent response of uniform technical depth. Proposals must appear as if written by a single, well-organized individual. Your proposal narrative must contribute to fast and easy comprehension by a variety of client evaluators who are reading your document with technical, contractual, and programmatic frames of reference. Use frequent subheadings or graphical icons to break up text and to facilitate the evaluation of your proposal.

Contributors should write from the general to the specific, from the easy-to-understand to the more difficult to comprehend. Proceed from an overview of each major topic to the finer grained technical and programmatic details. Always ask yourself the question: What's next? [1]. Writers should identify and discuss tangible benefits to the client of your company's technical and management approach, such as the following:

1. When we build and implement solutions, we offer value engineering recommendations that result in significant cost savings for our clients, including the DOE and GSA.

2. We have established clear lines of authority and channels of communication to ensure accountability for this important FAA program at all levels within our organization and to expedite decision making.

3. We will incorporate our formalized and widely deployed operating procedures along with automated applications into the operational environment to ensure that state-of-the-art services are being provided that meet the requirements of Department of Treasury end users.

4. Our company has been successful in transitioning five other projects by focusing on the prospective employees and their career goals, and by offering a superior benefits package.

5. Our single-vendor contract management approach will eliminate program accountability complexities, reduce costs, and minimize overall management risks.

6. Our demonstrated record of careful cost controls on Army and Air Force programs will result in best value support for the Navy throughout the life of this important contract.

7. With an established regional presence in Texas since 1998, we have performed 16 projects in the Dallas–Ft. Worth area in the past 4 years.

8. Complementing our presence in Connecticut and our technical knowledge base with Coast Guard programs is the collective technical, programmatic, and human talent of our two teaming partners.

9. To effectively manage and conduct technical work under each task order assigned, we have created the position of task leader. The task leader will serve as the single point of contact with the Department of Homeland Security.

10. Our outstanding and well-established lines of communication among our team and with the Air Combat Command will help to ensure rapid problem resolution.

There is a natural tendency to dwell on familiar ground. Often, this is of the least significant as far as the RFP requirements are concerned. Demonstrate how the features of your approach translate into benefits. Use facts—support all statements with concrete, quantified (if possible) examples. Include the following:

- What your company will do for the client;
- How your staff will do it;
- Why you will do it that way;
- What you did in the past, that is, your previous or current contractual experience, highlighting the relevant lessons learned, best practices discovered, and successes generated from similar contractual experience.

Summarize the content of each proposal section in the first paragraph of discussion. Write for a variety of readers, including the skim or executive-level reader. To be a winner, a proposal must contain concise, understandable, and closely related thoughts woven into a compelling story. Identify the critical technical areas. In discussing them, use a level of detail that exceeds what a nonincumbent could use, but do not drown the reader in jargon or equations. Your proposal should be built on solution-oriented writing. Write-ups should be risk-aware and solution-oriented and demonstrate an understanding of the evolutionary changes likely to occur over the life of the contract.

Make your company's compliance with the requirements and responsiveness to your client's evolving needs apparent to the evaluator over and over again. Use RFP or RFS terminology exactly (or in shortened form) for your proposal headings and subheadings. Employ the RFP terminology as a point of departure for further original writing. Do not simply recite the RFP verbiage in the actual narrative of your proposal. A DoD COTR once told me that he felt personally insulted when contractors, both large and small, merely replaced the words "the contractor shall" that appear in the SOW, with "our company will." Such an approach demonstrates no technical understanding whatsoever.

Define acronyms and abbreviations the first time they are used in each major proposal section. Not all evaluators see the entire technical and management proposal volumes, so redefinition of acronyms is certainly acceptable assuming page-count constraints allow. Avoid sectional or page references in the text because references must be changed each time the section or page numbers change. References such as "See Section 5.1.1 on page 5-34" should not be used. Let the compliance matrix and table of contents assist the evaluators in locating specific pages or sections. In addition, avoid fifth-order headings (for example, Section 4.3.1.3.1).

Proposal writers should attempt to think graphically as well as in written form. Every proposal section should have a figure or table associated with it that is referenced and discussed clearly in the text. Figure captions should appear centered beneath the appropriate figure. Table legends should appear centered above the appropriate table.

15.4 Action captions

A great place to use theme statements is in the captions or legends for figures, tables, photographs, and other graphic presentations. Let's look at the difference between simply identifying what a graphic or photograph is and using the caption as a forum to convey important sales messages that have been tailored to the specific procurement.

- Instead of "Project Organizational Chart," your caption might read "Our project team is proficient in each of the USACE technical requirement areas."

- Instead of "Map of Our Local Locations," your caption might read "The proximity of our offices to Fort McCloud will afford rapid response to changing requirements."

- Instead of "Capabilities Matrix," your caption might read "Each one of our four proposed senior technical staff holds an accredited degree in a relevant engineering and scientific discipline."

- Instead of "Our Laboratory Staff," your caption might read "Our well-trained laboratory staff are dedicated to and trained in quality-oriented procedures that ensure appropriate chain-of-custody handling."

- Instead of "Appropriate H/S Procedures," your caption might read "We actively practice a comprehensive health and safety program designed to meet the requirements of 29 CFR 1926.65 for hazardous waste site work."

- Instead of "Scope of Support Services," your caption might read "We provide cost-effective, innovative support services across a broad spectrum of financial regulatory acts."

In each one of these examples, we have taken a flat statement and added positive, dynamic sales messages. We have introduced benefits of your approach and capabilities to the client. These sales messages, or themes, should be developed early in the proposal response cycle so that the entire proposal writing team can incorporate them into their narrative and into the captions they develop for their graphics. Please note, however, that it is always critical to be totally compliant with the RFP. If a particular RFP stipulates that figure and tables shall be named in a particular manner, you must adhere to that guidance.

15.5 Methods of enhancing your proposal writing and editing

Toward the goal of keeping your proposal writing highly focused, consider the four following stages [2]:

1. Collect the technical and marketing intelligence information you need for your section(s). Focus on the basic features of those materials.

2. Identify the information in those materials that is relevant to the RFP requirements for your particular section and which

supports the strategic win themes that have been noted for that section.

3. Extract, organize, and reduce the relevant information. Begin to compose the relevant information into sentence form in accordance with the proposal outline that was part of the Kickoff Package or proposal directive. Slant your writing toward your audience, that is, the client evaluators of the proposal. Be informative but brief. Be concise, exact, and unambiguous. Use short, complete sentences.

4. Refine, review, and edit the relevant information to ensure completeness, technical accuracy, and the inclusion of theme statements.

One of the best ways to polish the writing in your company's proposals is to read it aloud to a colleague or yourself. Awkward sentence constructions quickly become obvious. The flow of the narrative can be smoothed and refined through this process of reading.

A longer term key to effective proposal writing is reading many different types of materials across a variety of disciplines and media—such as professional and scholarly journal articles, public relations brochures, newspapers, textbooks, novels, and other works of literary fiction. An enhanced functional vocabulary is one result. Another is learning how to use the same word in a broad spectrum of contexts. Reading in this manner can also produce an appreciation of nuance, that is, using precisely the most appropriate word for the given application.

There are many courses available through local colleges and universities that can be of assistance in enhancing your staff's technical and business writing capabilities.

In terms of editing proposal narrative, the following checklist can prove very useful [3]:

1. Check for completeness.

2. Check for accuracy.

3. Check for unity and coherence.

4. Include effective transition statements.

5. Check for consistent point of view.

6. Emphasize main ideas.

7. Subordinate less important ideas.

8. Check for clarity.

9. Eliminate ambiguity.

10. Check for appropriate word choice.

11. Eliminate jargon and buzzwords.

12. Replace abstract words with concrete words.

13. Achieve conciseness.

14. Use the active voice as much as possible.

15. Check for parallel structure (for example, a list begins with all -*ing* or all -*ed* words).

16. Check sentence construction and achieve sentence variety.

17. Eliminate awkwardness.

18. Eliminate grammar problems.

19. Check for subject–verb agreement.

20. Check for proper case (such as upper case and initial caps).

21. Check for clear reference of pronouns.

22. Check for correct punctuation.

23. Check for correct spelling, abbreviations, acronym definitions, contractions (avoid them), italics, numbers, symbols, for example.

24. Check for correctness of format.

15.6 Government-recognized writing standards

There are several writing, editorial, and proofreading standards recognized by the federal government. Among these are the 29th edition of the *United States Government Printing Office Style Manual 2000* (available online at http://www.gpoaccess.gov/stylemanual/browse.html); the Supplement to the *United States Government Printing Office Style Manual, Word Division* (1987); the *Gregg Reference Manual* by William A. Sabin, 8th ed. (1996); and MIL-HDBK-63038-2 (TM) (1 May 1977). The consistent application of one such standard is important to implement within your company's publication or editorial group. Pages 4 and 5 of the *GPO Style Manual 2000* provide a useful reference for proofreader's markings.

15.7 Additional sources of writing guidance

Judith A. Tarutz, *Technical Editing: The Practical Guide for Editors and Writers* (Hewlett-Packard Press, 1992); Philip R. Theibert, *Business Writing for Busy People* (Career Press, 1996); *The Chicago Manual of Style: The Essential Guide for Writers, Editors, and Publishers,* 15th ed. (University of Chicago Press, 2003), Edward T. Cremmins, *The Art of Abstracting,* 2nd ed. (Info Resources Press, 1996); Charles T. Brusaw, Gerald J. Alred, and Walter E. Oliu, *Handbook of Technical Writing,* 6th ed. (St. Martin's Press, 2000); Marjorie E. Skillin and Robert Malcom Gay, *Words Into Type,* 3rd ed. (Pearson, 1974); Robert A. Day, (ed.), *How to Write and Publish a Scientific Paper,* 5th ed. (Oryx Press, 1998); and Herbert B. Michaelson, *How to Write and Publish Engineering Papers and Reports,* 3rd ed. (Oryx Press, 1990) each offer insight and direction for clearer and more concise writing. Edward Cremmins' work is exceptional in helping one to understand how to convey information accurately and robustly while simultaneously conserving the number of words used.

We will look at the critical information management processes that link directly to business acquisition success.

15.8 Storytelling as an art form

Storytelling as an art form—coupled with quantitative, scientific, and technical validation—constitutes an intensely powerful and highly cost-effective framework for developing and conveying proposal solutions, which can in turn lead to winning significantly more new and recompeted federal government contractual business. In addition, organizational performance as it relates to business development and the management of bid and proposal (B&P) dollars can be optimized.

Proposals are, in fact, complex contractual and legal entities. In addition, they can be perceived as "knowledge products," that is, the synthesis of technical, programmatic, past and present contractual performance, and costing knowledge coupled with a detailed understanding of government customers, industry competitors, and current and emerging technologies. Furthermore, they can correctly be considered as "sales products"—focused on *completing* the sale to the government that began during the marketing phase of your small company's overall business development life cycle for a specific government procurement opportunity.

This is precisely where the art of fact-based storytelling becomes so important as the tool and framework by which to convey solution sets, in effect, knowledge and "sense." It is also precisely where the majority of companies—both small and large—fail in their capture management and proposal development strategies and efforts. Compliance-oriented and process-driven capture management and proposal development approaches alone cannot and will not produce winning proposal documents and stellar oral presentations on a sustainable basis. The oft-cited "answer the mail" model for responding to government RFPs and RFSs is necessary, but certainly not sufficient for long-term success in the federal marketplace.

Fact-based storytelling is the knowledge transfer and communications mechanism that constitutes the core of proposal solution development. Research suggests that sharing experiences through narrative (i.e., storytelling) "builds trust," "transfers tacit knowledge," and "generates emotional connections" [4]. Solution development, in turn, refers to the critical activity of building and articulating fact-based *storylines* or *story arcs* that convey your small business's specific approaches to providing meaningful, measurable, achievable, and risk-aware outcomes for your federal government customer. Robust solution sets must encompass such proposal areas as technical, program management, staffing, contract phase-in, and past/present contractual performance. The storylines must draw together such major elements as "Understanding the Customer Environment," "Technical and Programmatic Approach," and "Measurable Outcomes/Value Proposition" in a manner that ensures continuity of major sales messages or themes. More to the point, these storylines convey both sense and sensibility to the federal customer.

People buy from people, and people buy emotionally. Transactions are very personal. For these critical reasons, fact-based storytelling should be considered the central cornerstone of proposal development. To the detriment of countless small and large businesses, however, it is not afforded that keystone role at present.

The real-world implications of this approach for today's and tomorrow's high-velocity business environment are substantive and tangible. Small companies can optimize their valuable and perennially limited B&P dollars by employing the storytelling framework for successful capture management and proposal development. In today's extraordinarily competitive performance-based federal marketplace, the *delta*, or difference, between winning and coming in second borders on the microscopic. Storytelling in the proposal process can help articulate and deliver meaningful and positive differences between one company and its competitors. Through well-articulated stories augmented by appropriate and quickly comprehensible graphics and photographic images, federal proposal evaluators as well as federal customers and stakeholders can come to understand "possible futures" [4] that include a given small company's suite of solutions being deployed to meet the government's set of requirements and solve the government's challenges. This positive sentiment ("customer sentiment" is a term employed by proposal expert Mr. Hyman Silver) on the government's part will, in turn, contribute to increased proposal win rates and revenue streams for the proposing small company.

The validity of this innovative fact-based storytelling approach for capture management and proposal development has been established empirically during the past 6 years with a small, minority-owned federal contracting company. Staffing levels increased by a factor of *11.3* to nearly 1,600 within that period, and revenues increased by a factor of *16.1* to $250 million. In addition, the scholarly literature review also provides a defensible theoretical basis for the foundational planks of this storytelling approach.

Endnotes

1. Miner, Lynn E., and Jeremy T. Miner, *Proposal Planning and Writing*, 3rd ed., Westport, CT: Greenwood Press, 2003.

2. Adopted from Cremmins, Edward T., *The Art of Abstracting*, Philadelphia, PA: ISI Press, 1982, pp. 17, 73.

3. Adopted from Cremmins, Edward T., *The Art of Abstracting*, Philadelphia, PA: ISI Press, 1982, p. 85.

4. Sole, D., and D. G. Wilson, "Storytelling in Organizations: The Power and Traps of Using Stories to Share Knowledge in Organizations;" http://lila.pz.harvard.edu/_upload/lib/ACF14F3.pdf.

Chapter 16

Packaging and managing proposal information and knowledge effectively

Overview

Proposals are authenticated information and knowledge in carefully packaged form. Organizing a centralized, fully operational, and scalable informational data system or knowledge base is imperative to your company's business development success. Maintaining frequently used proposal materials in compatible electronic formats as well as in easily accessible hardcopy form will help your proposal planning and preparation process significantly. These materials include résumés, project descriptions or summaries, proposal boilerplate or reuse material such as configuration management and health and safety plans, and previously submitted proposals. Proposal managers require ready access to the latest[1] company and marketing intelligence information. But be sure to store all company-sensitive marketing and proposal information in a secure physical area and within a secure partition of an electronic network.

16.2 The all-important résumés

Your company's human, technical, and programmatic talent is the basis for your success to date. Staff are a particularly important consideration in the case of support services contracts, because the client is, in effect, purchasing human expertise in, for example, pollution prevention, electrical engineering, systems design, network engineering, facilities management, applications development, or integrated logistics support (ILS). Indeed, a company's staff being presented in the best possible light is valid for most every federal government procurement. This means that résumés have to be customized or tailored for each proposal. Tailoring is not altering or misrepresenting a person's experience or education in an unethical or illegal manner but rather involves highlighting relevant experience, publications, certifications, and education to the exclusion of other information. Some RFPs or RFSs require that each person proposed sign a statement that the information contained in his or her résumé is accurate. Adequate time within the proposal response schedule needs to be built in to secure those requisite signatures. Smaller companies need stellar résumés; they function like name brands.

At times, the RFP will provide a very detailed résumé format, or indicate in narrative form how the issuing government agency wants to receive the résumés. When résumé format, content, or page count are stipulated, you must comply fully. However, many times you will have a degree of autonomy to structure the résumé.

A technique for helping evaluators quickly understand your staff's capabilities is to include a box entitled "Experience Summary," "Benefits to the Client," "Basis of Team Selection," or "Relevance to the Project" on the first page of each résumé required by the RFP. Within that summary box, highlight the particular individual's expertise and accomplishments relevant to

---

1 Ensuring that proposal managers are provided with the latest company information is an ongoing effort in configuration management. For example, a comprehensive written overview about your company that is used in most every proposal needs to contain the most up-to-date information on contract awards, company commendations and success stories, annual revenue, staff level, funded contract backlog, and corporate organizational structure and leadership. If this business overview is updated for a given proposal, the corporate library boilerplate file that contains this same overview must be revised as well.

the specific position description or technical requirements of the procurement—this might include education, management experience on specific contract types, number of staff managed, publications, certifications, professional society memberships, work on similar projects, innovations, relevant training, and years of successful experience in a given operational environment. Other items to consider include award/honors, knowledge of federal agency policies, and skill sets mapped to relevant task areas. Three to four bulleted items are sufficient. If you do not use an "Experience Summary" box, you might consider including a "Benefits to the Project" section in each résumé. Do not assume that the evaluators will take the time to carefully review each résumé and extract the relevant details that correspond to the position description. If you make their job difficult, you simply cannot expect to receive the maximum points possible.

Résumés should focus on relevant technical and programmatic accomplishments and excellence. They should accentuate results—increased production rates, improved quality, on-schedule performance, cost savings, meeting performance thresholds, and the implementation of innovative techniques and technologies. As appropriate, they should be client specific, site specific, and geography specific. Résumés should use transferable, action-oriented lead words to describe an individual's activities and contributions (see Figure 16.1). These types of words help the evaluators to understand the relevance of your staff's current and past experience to the new project at hand. They facilitate the transfer of talent and capability from one project to another.

Figure 16.1
Use action-oriented words to transfer capabilities from past contracts to the current opportunity.

Action-oriented words for staff résumés

• Manages	• Negotiates	• Leverages	• Brings	• Field-tests
• Performs	• Reviews	• Translates	• Provides	• Establishes
• Implements	• Writes	• Offers	• Facilitates	• Cocreates
• Conducts	• Designs	• Serves	• Enforces	• Assists
• Directs	• Supports	• Oversees	• Selects	• Plans
• Coordinates	• Builds	• Delivers	• Trains	• Employs
• Prepares	• Drives	• Ensures	• Operates	• Studies
• Initiates	• Leads	• Participates	• Troubleshoots	• Examines
• Develops	• Determines	• Maintains	• Audits	• Assesses
• Investigates	• Applies	• Practices	• Analyzes	• Consults
• Inspects	• Characterizes	• Deploys	• Compiles	• Surveys

Review the following example.

- *Wrong way:* "Mr. Jones managed an investigation at Ft. Baltimore, Maryland."
- *Appropriate way:* "Mr. Jones managed a lead contamination groundwater investigation at Ft. Baltimore, Maryland. He negotiated technology-based cleanup criteria with the Maryland Department of the Environment in the decision document."

One of the most important information-management activities your company can undertake is to create, update, archive, and make available for retrieval a résumé for each member of your technical, management, and support staff. Use the same software application and version that you use to produce your proposals, and format each résumé in a style that will be compatible aesthetically with the rest of your proposal text. Store hardcopies of all résumés in alphabetical order in three-ring notebooks for ease of access. Establish an electronic subdirectory of all company résumés so that they might be searched (using infoRouter from Active Innovations, for example) for a combination of appropriate keywords based upon the position descriptions in the RFP. And you would be well served to generate and maintain a genuine database (using database software such as Microsoft Access) of résumé information to facilitate rapid searches to meet position description parameters or prepare tables and graphics providing information about your technical and professional staff. Many times RFPs will request the number of staff with specific degrees, certifications, or levels of experience. Appendix B of this book illustrates a form that can be used to capture the employee information used to populate the database.

Creating, word processing, editing, proofreading, and updating résumés is a time-consuming and ongoing task. But without résumés readily accessible, much of your proposal response time will be consumed preparing them. And having more information than is required by the average RFP—such as grade point average, name and address of high school, previous supervisors and telephones, security clearances, years of supervisory experience, and professional references (including telephone numbers and e-mail addresses)[2]—is always preferable to having to search for that information under the pressure of preparing a proposal.

16.3 Project descriptions (project summaries)

A cumulative, regularly updated hardcopy and electronic file copy of your company's contracts, both as prime and sub, is a vital building block in the knowledge foundation.

Create project descriptions at the time of contract award and update them at the completion of specific tasks, or semiannually, and then again at contract completion. As with résumés, storing hardcopies of each project description in a three-ring notebook allows ease of access. In the front of the binder, include a current list or matrix that includes all project description titles, contract and purchase order numbers, and client agencies in some logical order. It would be advisable to put your boilerplate project descriptions in the format required by your major client(s). One such format is presented in Figure 16.2. Specific information can always be added or removed in the tailoring process. Project or division managers should oversee and approve the updating process to ensure completeness and accuracy. As with the proposal response process, the updating of résumés and project summaries will require top-down management support, otherwise it will rarely be given the attention it deserves.

Project descriptions or summaries should focus on project successes—cost savings to the client, schedule adherence, awards (including repeat business from the same client), value-added service provided, conformance with performance metrics and application of innovative technologies. Be specific, and be quantitative in your narrative descriptions. Highlight applicable best practices and lessons learned, such as the development of appropriate timesaving techniques or the application of certain automated project management tools.

As client evaluation factors for award are weighted increasingly toward past performance, well-written and photodocumented project summaries become more and more valuable. Clients buy from companies that have performed similar work and can demonstrate and validate that fact effectively and appropriately in their proposals. You might consider investing

---

2   To this listing might be added date of hire, current approved job title, total years of professional experience, maximum number and type of staff supervised, technical publications written, knowledge of ISO and SEI CMMI® standards as well as industry-specific regulatory guidelines such as technology export compliance rules, experience working for or supporting specific federal agencies, general technical areas of expertise, and computer-related experience [subdivided to include such items as hardware; hardware operating systems; computer-aided software engineering (CASE) methodologies and products; application development tools and languages; communications protocols; communication controllers, hubs, and hardware; network operating systems; databases; system tools; and software].

Title of project:
Name of client:
Address of client:
Contract number:
Contract type:
Contract value (with options):
Period of performance (with options):
Place of performance:

COTR:
Telephone:
E-mail:

Contracting officer:
Telephone:
E-mail:

Company point of contact (POC):
Subcontractors:
Brief description:
Detailed description:

Method of acquisition (competitive/noncompetitive):
Nature of award (initial/follow-on):
Overview of customer's mission and operational environment:
Number of users served
Technical performance/accomplishments/innovations:
Programmatic performance:
    Contract transition
    Business management
    Personnel management/staffing
    Reporting
    Schedule control
    Program management tools
    Application of ISO 9001:2000, SEI CMMI®, and ITIL structured processes
    On-time rating and quality rating of deliverables
Cost-price management history:

Risk mitigation approach:
Performance metrics and service-level agreements (SLAs):
Quality assurance/process improvement:
Lessons learned:
Best practices:

Client performance evaluation highlights (award fees, CPARS, customer-approval rating):
Average number of personnel utilized per contract year:
Percentage turnover:
Incumbent capture:
Incumbent retention rate:
Add-on/follow-on work:
Alliances with local colleges and universities:
Mentor-protégé programs:
Termination history:

Figure 16.2
Effective project
summary format.

in a digital camera to take client-approved photographs on site for your company's projects.

Sometimes in small businesses, because of the corporate culture or the pathway of company growth, one division may be reluctant to use staff and contractual experience from other divisions when responding to a procurement. This is one example of the change of thinking that needs to occur as a small business enters the competitive arena. Each division of your company must harness fully the collective human and contractual expertise and knowledge base resident in the entire firm.

One important caveat: Be certain to obtain your client's permission to prepare an in-depth project summary about your company's support for that client organization. In particular, private-sector clients, as well as federal law enforcement and intelligence agencies, can be extremely sensitive about having the scope of work and specific points of contact made public. It is always the best policy to confirm what you intend to do with your client point of contact.

16.4 Proposal boilerplate (canned or reuse material) as knowledge assets

Translate boilerplate into client-focused text.

As your company prepares more and more proposals, it is important to extract certain boilerplate sections from past proposal documents, copy the text and graphics (see Figure 16.3) from these sections, and archive them in a central corporate library, proposal data center, or knowledge base. Boilerplate must still be tailored for each application, but it represents a rich collection of fodder for the writers and planners of future proposals. Table 16.1 presents examples of boilerplate files, which are actually knowledge assets.

Proposal writers should not have to start from scratch on every proposal effort. You should create an electronic proposal toolbox or knowledge base so that proposal writers can consult the latest reuse material under one electronic umbrella. Invest the time and resources early in your corporate life to develop these knowledge assets. They will be used again and again to develop your proposals.

16.5 Marketing targets

As your company grows, it will reach a point at which it needs a systematic, formalized mechanism to track its business opportunities. A variety of excellent data base software packages, such as Microsoft Access or IBM Lotus Notes [1], can be employed for this purpose. Database software allows for the generation of customized business reports and provides a quickly retrievable information source, for example, for planning and

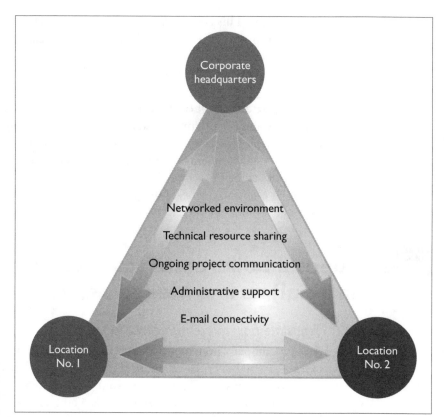

Figure 16.3
Integrated staffing
and resource
management.

Company Assets
Company core competencies (technical discussion)
Company bonding, registrations, and certifications
Licenses
Maps (office locations, project locations, number of projects by state or region)
Customer base
Patents
Trademarks
Company organizational charts
Company best practices
Industry certifications and accreditations [e.g., Software Engineering Institute (SEI) Capability Maturity Model Integration® (CMMI®), ISO 9001:2000, and IEEE]
Facility and personnel security clearances

Table 16.1
Examples of
Boilerplate Files

Industry and business community awards, commendations, and recognition (e.g., SBA Entrepreneur of the Year, Ernst & Young)

Customer endorsements and commendations

Small-minority-business involvement; HUBZone involvement

Corporate-sponsored community service and outreach activities

Company videos

Facilities diagrams

Best management practices (BMPs)

Technical and business articles and books authored by company staff

Interview guides

Vendor manuals

Contractual Resources

Project and product descriptions or summaries

Project performance metrics

Award fee statistics per project

Project success stories

Human Resources

Résumés

Personnel diversity data (minorities, women, veterans, handicapped individuals)

Incumbent capture rates

Employee turnover rates

Employee recruitment plan and metrics

Employee retention metrics

Staff by project, office, technical discipline, degree

Business Development and Proposal Development Assets

Strategic plans and benchmarking data

Market segmentation analyses

Competitor information

B&P spending patterns

Table 16.1
(Continued)

Winning proposals

Multimedia presentations to clients

Teaming agreements

Teaming agreement SOWs

Proposal templates (e.g., executive summaries, cover letters)

Freedom of Information Act (FOIA) documents

Proposal lessons learned (from client debriefings)

Sales volumes by client sector, line of business (LOB), and geographic area

Tactical business plans

Corporate Infrastructure and Process Assets

Documentation capabilities

Business processes (procurement systems and mechanisms, financial systems, invoicing procedures, cost accounting, scheduling, automated tracking tools)

Project-Specific Assets

Customer evaluation reports [A/E Contract Appraisal System Support (ACASS), CPARS, PPIRS, and PPIMS ratings]

Contract deliverables

Earned value analysis (EVA) results by project

Work breakdown structures (WBSs)

Project management plans

Computer-Related Assets

Software development capabilities and practices

Software source code

Nonoperational hardware and software test-bed environments

Computer equipment and resources

Financial Assets

Budget forecasts

Capital expenses

Annual revenues

Table 16.1
(Continued)

Relationship Assets
Cooperative agreements with colleges and universities, particularly HBCUs
Strategic business alliances (with organizations such as Microsoft Solutions, IBM, Dell, and Oracle)
Public Relations Assets
Ads placed in professional and trade journals
Trade show modules (text and graphics)
External news releases
Engineering and Manufacturing Assets
Engineering white papers (position papers)
Manufacturing defect rate data
Engineering drawings
Engineering change proposals (ECPs)
Test procedures
Requirements data
Policies and Plans
Make-or-buy policy
Management plans
Health and safety plans
Technology transfer plans
Professional compensation plans
Design control plans
Software quality assurance (SQA) plans
Phase in/phase out and transition plans
Subcontractor management plans
Configuration management (CM) plans

Table 16.1
(Continued)

bid–no bid decisions. General access to the network drive on which marketing targets reside should be controlled carefully through password protection.

16.6 Corporate library

An important step in the transition from the 8(a) environment to the competitive world is to establish a centralized library, proposal data center, or knowledge base (electronic and hardcopy) in which previously submitted proposals and presentations, RFPs, proposal debriefings files, résumés, project summaries, technical materials, FOIA-requested materials (competitor proposals and monthly progress reports for specific projects), professional and trade journals, potential teaming partner information, public relations materials, and photographs reside. The library is best administered by the business development group.

Cataloguing materials can be done using a variety of alphanumeric schemes that best fit your needs and uniquely identify each informational entity in the library. For example, all proposal documents could be catalogued with the prefix 1, followed by a two-digit calendar year designator, a three-digit federal agency code of your choosing, and so forth. Be prepared to dedicate at least one large office with lateral and vertical file cabinets to your company's library. The volume of hardcopy information multiplies very rapidly. Using magnetic-backed drawer labels in the library is particularly advantageous. As your company's collection of materials increases in certain areas, the labels can be switched easily from drawer to drawer.

Personal experience has shown that storing a duplicate set of CD-ROM or tape copies in appropriate containers along with the paper copies of proposals and other library documents is very beneficial. The original storage medium for a given document can reside with the publications group, but the duplicate medium can be used for electronic searches to locate specific verbiage, numbers, for example, that would be difficult to find manually. And having a backup of electronic files is always a good idea. An alternative to a duplicate set of stored CD-ROMs or tapes is network storage, such as an intranet. Intranets allow companies to connect their own people as well as their vendors. "Participants using browsers inside and outside the company share a common interface and communications environment" [2]. All of a company's business processes can be executed online in a secure manner. Important intranet standards include TCP/IP and HTML. "[T]hink of the intranet's design just as you do the physical organization of your company. If there's a locked door in front of an office, lock that door to the virtual office. If you have a secretary acting as a gatekeeper in front of certain offices, require passwords to access those files in the virtual office. Model the virtual world on your real world" [3]. A significant advantage of an intranet is the "increased productivity that results from quick, companywide communications and data sharing....

Some of the many functions that intranets provide include repositories for volatile company information; easy access to company handbooks, guidelines, and forms of all kinds; and real-time calendaring and scheduling services" [4].

As data volumes of proposal-related files increase due to high-end desktop publishing software and the use of integrated text and graphics, most proposals can no longer be stored effectively and efficiently on high-density (1.44 MB) floppy diskettes. With files such as style files, graphics files, text files, and mirror files, a midsized set of proposal volumes produced with desktop publishing software can range from 40 MB to more than 200 MB. Storage media are now more appropriately CDs, Zip disks, USB drives, magnetic tape, and multigigabyte hard drives.

Effective information management is a critical factor in achieving proposal and marketing success. It is essential that this activity within your company receives the attention and support of senior management. Although it appears administrative and mundane, nothing could be farther from reality. Even very large companies have not done a good job in documenting and tracking their human and contractual talent and experience. Your company, being small at present, has the opportunity to start on the right footing. Build an appreciation for the integrated nature of externally directed business development, internal sales support, information management, and corporate image.

To be effective, the information management activity within your company must receive the support of senior management.

16.7 Proposal lessons-learned database

To ensure that your company will derive both short- and long-term benefits from both proposal successes as well as losses, it is advisable to develop and maintain an automated proposal lessons-learned database (see Table 16.2) using commercially available software applications such as IBM Lotus Notes or Microsoft Access. Capture managers should be responsible for providing the specific information for the lessons-learned database when proposals are won or lost. Once the data have been captured, they should be subject to senior-level review to help ensure impartiality.

By carefully tracking the reasons for wins and losses, you can begin to discern patterns in your business development performance with regard to specific client organizations, particular lines of business, certain types of contract vehicle, and proposal costing strategies. Are you consistently receiving low scores on ID-IQ contracts on which you submit proposals?

Client name

Brief scope of work

Place of project performance

Contract type (ID/IQ, CPFF, FFP, T&M)

Proposal due date

Internal proposal B&P number

Contract value to our company

Status of proposal (won or lost)

Capture manager/proposal manager and other key staff present

Was a debriefing conducted with the client organization?

Date of debriefing

Has our company worked for this client before? (yes or no)

What was our relationship with the client before the RFP was released?

Was our proposal determined by the client to be in the competitive range (shortlisted)?

Did we conduct advanced marketing with this client?

Winning contracting firms

Winning bid amounts ($)

Our bid ($)

B&P budget estimate versus actual direct labor and ODCs

Specific reasons why we were not selected (check all that apply):
❑ Technical approach
❑ Lack of understanding of client mission and requirements
❑ Past performance
❑ Contractual experience
❑ Staff qualifications
❑ Management approach
❑ Company size
❑ Financial condition
❑ Direct labor costs
❑ ODCs
❑ Subcontracting plan
❑ Teaming arrangements
❑ Geographic location of staff, projects, or offices
❑ Missing information
❑ Late delivery of proposal
❑ Noncompliance with RFP-driven proposal structure, format, or page count

Additional relevant information

Client's source of feedback

Table 16.2
Sample Information for a Proposal Lessons-Learned Database

Is your win percentage on Navy proposals particularly low in the Pacific Northwest? Does the lack of an established office near Wilmington, Delaware, seem to impact your chances of being selected for contract award for particular DuPont projects? Are your management plans missing the mark far too consistently? Those patterns must be discerned and communicated to appropriate business development and operations staff within your company, and then, most importantly, translated into prudent shifts in marketing, intelligence gathering, proposal development, and operational strategies and processes. In addition, through this database you can begin to generate a solid basis for estimating B&P costs for specific types and sizes of proposal efforts. This is critical for efficient staff and equipment resource planning and allocation, as well as more accurate B&P fiscal and resource projections.

It is worth noting that your company must strive—from the top downward—to develop a business culture that allows for appropriate and constructive introspection. The lessons-learned database will be most valuable when internal company politics do not skew the interpretation of the information that the database contains. For example, let's say that your company received a low score under the key staff evaluation factor because your proposed project manager had allowed a previous contract for the same client to run significantly over budget and behind schedule. Although this fact does not place this PM candidate in the best possible light, it is important to record the causes for your loss as completely and accurately as possible, and then act upon the patterns discerned in the causes. For example, management plans in your future proposals may have to emphasize specific strategies and tools that your firm has implemented and deployed to monitor and control costs and schedule effectively and successfully.

16.8 Applying IT solutions: scalable informational data systems

As your company grows, staff in various office locations need easily accessible and usable proposal-related information that can be modified to reflect local and regional requirements. Ongoing advances in information and communications technologies continue to facilitate data and information sharing across distributed locations and networks, and include scenarios such as virtual proposal centers.

16.8.1 **IBM Lotus Notes scenarios**

One powerful automated solution to this ever-growing requirement is an IBM Lotus Notes–based domain of information we will refer to as *proposal building blocks.* Lotus Notes groupware provides a searchable, sortable document storage and management system for group use. As illustrated in Figure 16.4, the proposal building blocks are represented by icons on a Lotus Notes screen on a PC. From their desktops and with no third-party

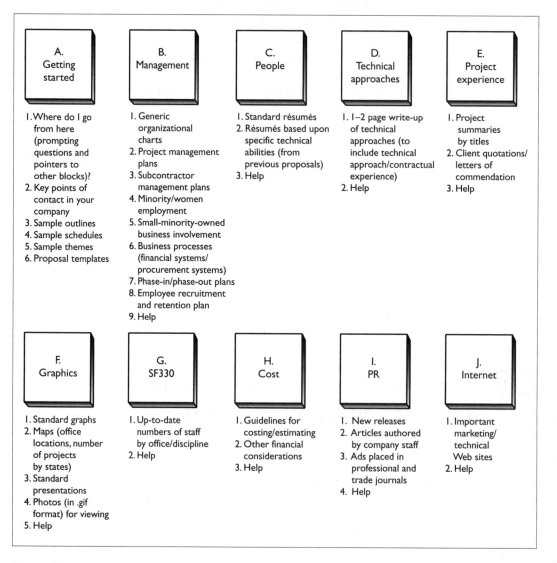

Figure 16.4 Guiding geographically distributed staff through the proposal building process—beginning your knowledge base.

interaction, your field office or corporate staff can click on these icons to access proposal guidelines and key internal points of contact for support along with sample outlines, résumés and project summaries, technical narrative, and standard graphics. Each icon should have a HELP function associated with it to further enhance this system's self-containment. The building blocks should be designed to guide proposal managers and technical staff through the proposal-building process.

Drawing on Lotus Notes' replication feature, electronic files can be updated, amended, or deleted from the proposal building blocks domain by a corporate-based systems administrator in a controlled manner and made available to your staff nationwide in real time. Everyone in the company will be able to access the latest materials. Such a system facilitates localized customization of proposals that are built upon a common, corporate-approved source domain of information. Distributed office locations can have access to the latest narrative on your company's business processes, résumés of recent new hires, key success stories from ongoing projects, and the most current letters of commendation from your clients.

16.8.2 **CD-ROM scenarios**

To address these same information requirements in very small companies, critical proposal files, which are password protected or encrypted, can be loaded onto a CD-ROM or removable drive (e.g., Zip or USB drive). This technique can also be useful for staff who travel frequently and need rapid access to proposal information. Data security is the primary concern with this scenario of information sharing.

16.8.3 **Intranet scenarios**

An *intranet* is an internal information distribution system supported by a server or multiple servers that uses Web technology and graphical user interfaces. An intranet can be used to make the information depicted in Figure 16.4 available to your staff at multiple geographic and office locations. Search engine software such as infoRouter can be employed to find and filter relevant information on your company's intranet.

16.9   Small business KM success story—this stuff really works!

16.9.1   Small-scale, pilot KM initiatives applied to proposal development

Small-scale pilot KM initiatives are critical to launch and sustain, and then to leverage and propagate their success throughout your organization. Successful pilots become proofs of concept that can be transplanted and adopted in other parts of your organization. Most important, knowledge initiatives focused on measurable and achievable business benefits will have a higher probability of acceptance and sustained success.

My employer, RS Information Systems, Inc. (RSIS), in McLean, Virginia, funded and proactively supports the RSIS Business Development Knowledge Base as an integral part of an enterprisewide KM initiative. This knowledge repository is a robust Web-based, password-protected interactive tool that RSIS staff professionals can access remotely to qualify marketing opportunities and develop proposals and presentations. Approximately 50 RSIS staff currently access the knowledge base on PC workstations and laptop computers, although the system can serve more than 2,000 end users. Using the infoRouter search engine manufactured by Active Innovations in a secure networked Windows NT 4.0 environment, company employees can browse thousands of indexed files or perform keyword searches to identify relevant files in multiple application formats, including MS Word, Corel WordPerfect, HTML, XML, RTF, and PowerPoint, as well as PDF (Adobe Acrobat), Excel, and MS Outlook e-mail files. Queries result in lists of files prioritized by relevance and presented to users in the familiar format of a Google or Yahoo search result. The RSIS knowledge base captures the staff's domain-specific technical knowledge, as well as its programmatic and client-specific expertise. In addition, the repository archives current best-of-breed proposal narrative, graphics, and oral presentation slides. Currency is a critical parameter of the knowledge base's validity and value.

The direct business development benefits of the Business Development Knowledge Base have been rapid proposal prototyping, as well as informed bid–no bid decision making. By leveraging the knowledge base, RSIS proposal managers can generate first-draft proposal documents quickly with minimal B&P expenditure and little impact on billable technical staff. The proposal response life cycle can then be used to enhance and polish proposal documents, rather than expend time and resources tracking and locating résumés, project summaries, and other company information assets. By 2000, we had progressed to the point where three people could and did develop a 50-page technical and management proposal, including résumés and project descriptions, to the National Oceanic and Atmospheric Administration's (NOAA) Geophysical Fluid Dynamics Laboratory (GFDL) at Princeton University within 24 hours from RFP release to client delivery! Without the knowledge base, it would have taken those same people at least a week to produce the proposal. RSIS marketing staff and capture managers also use the Business Development knowledge base to make informed decisions on the feasibility of pursuing a given procurement opportunity by assessing past and present contractual experience and technical skill sets of professional staff.

16.9.2 Balance of tools, disciplined methodologies, and a supportive business culture

Structured processes for knowledge transfer and capture ensure the sustained value and validity of the RSIS knowledge base. Best practices, lessons learned, client commendations and awards, performance metrics and standards, emerging technologies, staffing success stories, incumbent capture statistics, and information regarding rapid contract transitions are routinely funneled into the RSIS Proposal Development Department. Sources of this information span executive management, monthly technical progress reports for projects nationwide, division-level leadership, and direct interviews of select subject-matter experts and other technical staff. The knowledge transfer and sharing expertise and experience of RSIS professional staff are also captured annually as an integral part of the corporate résumé update process. All levels of management reinforce these knowledge transfer processes regularly. Successful KM at RSIS results from the dynamic integration of supportive, forward-looking executive leadership; a business culture of and organizational commitment to mutual trust and sharing; the passion of people at all levels of the organization; and technology-based, user-driven solutions and tools.

16.9.3 Development drivers and challenges

The RSIS Business Development Knowledge Base, and indeed the company's entire KM initiative, began late in 1998. Given the reality of 120 total company staff in 1998 and only two full-time professionals in proposal development at the time, the need to do more with limited resources was immediate and critical. We had no centralized, electronic repository of information. There were no institutionalized or even ad hoc processes in place to identify, audit, collect, archive, and leverage key knowledge within the company. On the other hand, we had a significant number of electronic directories and files of proposals, presentations, résumés, and project summaries. One individual prior to 1998 had actually attempted to extract, catalog, and organize proposal-related information into some semblance of order, but that initiative had never been completed. After having developed and received management buy-in for the architecture of the initial proposal system in mid-1999, one of my staff (a full-time proposal manager) and I began the arduous task of manually sorting through scores of proposal directories and hundreds of related files. This was done in addition to a full-time proposal development workload. Within 7 months, 5 years' worth of electronic files for proposals, presentations, résumés, project summaries, awards and commendations, white papers, public relations materials, and other information had been reviewed, purged as appropriate, and organized into the preapproved architecture. The initial search engine selected was dtSearch, manufactured by dtSearch Corporation, which is incorporated in Virginia. Unfortunately in late 1999, dtSearch required frequent, time-consuming, and manual indexing and reindexing. Server space demands were significant as the number of proposal modules increased. Search results were not as user-friendly as they needed to be for our requirements.

By mid-2000, one of my proposal managers, who was also our knowledge engineer, along with one Web developer and I migrated the prototype knowledge base from dtSearch to Microsoft Index Server. With this application, which was fed continuously by established knowledge transfer processes, the current RSIS Business Development Knowledge Base was fielded in late 2000. Our KM specialist and one Web developer now maintain the repository, which was migrated to infoRouter in November 2001. System maintenance requirements are modest at present. Incoming information is evaluated during the course of several days and then rapidly indexed into the knowledge base. As vice president of knowledge management for the firm, I provide both oversight for the KM initiative enterprise-wide, as well as hands-on collection and analysis of various knowledge assets.

Concurrent with our efforts to architect and populate our fledging knowledge base, I worked closely with RSIS executive management to secure their support to institute knowledge transfer and collection processes in a disciplined and repeatable manner nationwide. Monthly technical progress reports are now sent to me, as well as to the technical managers. Those reports are then combed for relevant technical innovations, lessons learned, staffing successes, best practices, and so forth. Résumés are generated for every new hire, and annual résumé updates are linked to the performance appraisal process. Eventually, knowledge-sharing experience was a key element incorporated into each RSIS résumé for all staff nationwide. Project summaries are now created according to a structured, consistent, and comprehensive template when contracts phase in. Summaries are updated when major new tasks are added and then again upon contract completion. On a regular basis during business development review meetings and strategic planning meetings, critical knowledge is shared from across the many defense, civilian, and law enforcement contracts that RSIS supports. The culture of knowledge sharing within the company has extended to the establishment of technical Centers for Excellence (CFEs) and our Birds-of-a-Feather program. Through these institutionalized programs, domain-specific technical knowledge and best practices can be applied rapidly to provide our clients with proven solutions in near-real time. Recently, an important, relevant development in corporate communications and public relations was integrated directly into a civilian agency proposal within 1 hour of notification.

16.9.4 Sustainment and future enhancements

As part of our KM continuous process improvement (CPI) program, we envision the following enhancements to the maturing RSIS Business Development Knowledge Base by 2005:

- Additional structured categories to facilitate rapid retrieval of select full-text proposal modules;
- Linkage with photographic lightboxes for browsing the current inventory of stock and company photographs;
- Monitoring and reporting of performance metrics (time and B&P cost savings) associated with the KM processes at RSIS for use in future planning and funding decisions.

16.9.5 Transferable lessons learned

There are four key ingredients in the RSIS KM success story. The first was a KM champion, an individual who understood and articulated the tangible benefits of knowledge management to executive management, as well as business development and technical staff. The second comprised the executive leadership, support, and vision necessary to grasp the value of KM and then fund the processes and clear the internal impediments to knowledge sharing. The third entailed the disciplined and repeatable processes put in place enterprisewide within the company to funnel knowledge and information into one central point in near-real time. And the fourth included the Web-based KM tools.

Experience has demonstrated the value of starting your own company's business development KM initiative as soon as possible in the corporate life of your firm. This approach does two things. One, there will be less raw information and data to review, purge, and categorize. And two, the sooner your staff develops and hones their knowledge-sharing skills and behaviors, the more quickly your firm will emerge as a learning organization—one that adapts and prospers in a business environment of fast-paced and unpredictable change.

16.10 Leveraging federal performance appraisal systems to your company's benefit

An important element of your company's proposal information repository is the evaluations and appraisals conducted by your clients about your project support. Make certain to request and archive the most recent appraisal results from the federal evaluation systems, as appropriate. Your company can certainly trumpet your excellent scores accrued under these federal performance appraisal systems in your proposals, as shown in Figure 16.5.

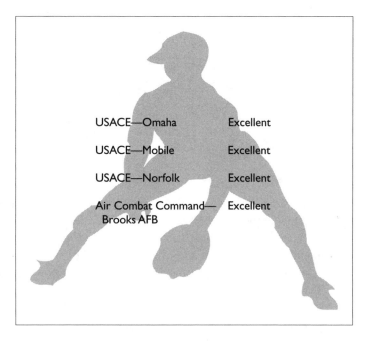

USACE—Omaha	Excellent
USACE—Mobile	Excellent
USACE—Norfolk	Excellent
Air Combat Command—Brooks AFB	Excellent

Figure 16.5
Project prowess—our ACASS ratings are consistently in the strike zone.

16.11 ISO-driven proposal and business development excellence

The International Organization for Standardization, or ISO, is a network of national standards institutes from 147 countries working in partnership with international organizations, governments, industry, and business and consumer representatives. ISO 9001:2000 is one of the standards in the ISO 9000 family's core series on quality management systems (QMSs). *How do ISO 9001:2000 standards apply to the actual processes of proposal development and business development? And why do these disciplined processes contribute to the generation of excellent proposal end products?* To be clear, we will not be focusing on companywide or program-specific ISO certifications that are discussed and illustrated within management and technical proposals, as important as these are to a company's success. Rather, our emphasis will be on how ISO-driven standards are applied successfully to define, clarify, and enhance proposal development as well as business development processes.

Small business success story

Under forward-looking executive leadership, RS Information Systems, Inc. (RSIS), established a corporate-level Quality Improvement Office (QIO) to initiate and manage quality initiatives and certifications across the company and its federal contracts. The QIO implemented a Quality Management System (QMS) that is now certified to the ISO 9001:2000 standard by an auditing firm approved by the American National Standards Institute (ANSI)/Registrar Accreditation Board. RSIS joins only 10% of U.S. service providers that have registered their QMS to ISO 9001:2000 (Certificate # US2195).

The QMS manual details the specific controls that are implemented at RSIS to ensure quality service, customer satisfaction, and continual improvement of all of the company's systems and processes from end to end. The QMS manual is divided into five sections modeled on the sectional organization of the ISO 9001:2000 standard. Importantly, RSIS exceeded ISO standards with the development of its QMS. The company proactively included all of its corporate infrastructure and program support processes such as proposal development, business development, contract management, human resources and recruiting, and finance and accounting. In addition, RSIS developed a methodology that enables it to apply its QMS and ISO 9001:2000-certified processes to any contract for certification.

As part of RSIS' continuous improvement initiative, the company added a variety of programs into the scope of its QMS and had them audited for certification to the ISO 9001:2000 standard. Recently, a high-profile program with the Air Force Space Command (AFSPC) received a recommendation for ISO 9001:2000 certification. Of note is that the independent auditor determined that "RSIS' QMS has best in class processes for software design and development activities." RSIS' structured quality approach provides definitive quality assurance (QA) responsibility and accountability, coupled with an independent reporting tool to top management.

Associated costs

Up-front costs are associated with implementing an internal QMS. In RSIS' case, it was approximately $3,000. Once implemented and if working well, however, the QMS will *reduce* costs by:

- Making work processes more efficient and streamlined, thereby reducing complexity (without documented processes, most staff professionals perform work in their own unique way, which increases complexity and the potential for rework);
- Reducing rework and work product correction;
- Having core business processes documented, which aids in bringing new employees up to speed faster;
- Providing a management system that aids in disaster recovery should an unfortunate event occur;
- Providing a management system that enables management decisions based on validated performance data.

ISO certification of business processes helps a corporation meet customer requirements, achieve business goals and objectives, keep within budget guidelines, provide consistent services and products, train new employees on the appropriate way to perform tasks, and improve internal and external communications.

Methodologies, tools, and training

RSIS' Quality Operational Procedure (QOP) for "New Business Identification, Qualification, and Capture" stands among 26 other QOPs. The Business Development QOP is further decomposed into seven detailed work instructions (WIs). The entire body of QOPs, associated WIs, and standard forms is fully documented and available electronically to all company staff professionals via a secure intranet. In fact, staff understand that the most current version of all QOPs and WIs resides solely on the intranet. This site was built using Active Innovations Inc.'s commercial off-the-shelf (COTS) infoRouter tool, a powerful Web-based document management system (DMS) and search engine.

Of note is that all RSIS Business Development and Proposal Development Department staff have completed a minimum of two internal training courses. These courses are entitled "Introduction to the ISO 9000 Process" and "Process Management and Procedure Development." In addition, as an entire department, mandatory interactive meetings are held every 6 months to provide an overview of the Business Development QOP and Business Development and Proposal Development WIs, to review forms and any changes to those forms, and to ensure that hand-off procedures are well understood. For example, proper hand-off procedures of proposal documents to the Finance and Contracts Department for cost reviews are emphasized. Appropriate, timely, and quality-conscious interaction at various connection points in the proposal and business development processes is an essential concern in the ISO paradigm.

As part of the ISO process, annual *internal* customer satisfaction audits are conducted of Proposal Development and Business Development processes and level of support. Internal customers include, for example, representative leaders from the three RSIS operating divisions as well as from the Contracts and Finance Department. The audit scale ranges from 1 to 4, with 4 being a superior metric. When the audit data are collected and synthesized, an action plan is developed and implemented to ensure that any issues identified are addressed and fixed promptly.

Proposal development work instruction

The parent (or master) Business Development QOP presents a detailed and comprehensive flow diagram of all Business Development and Proposal Development processes. Decision points, end points, and connection points are identified in that flow diagram. Under this specific QOP is a WI entitled "Proposal Development and Review." The overarching *purpose* of this WI is to establish the processes for proposal development and review, including solution development; review team selection, scheduling, and conduct; and reporting of findings relating to business proposal content. The *scope* of this WI identifies the dollar thresholds to which these processes apply under prime and subcontractor conditions. *Inputs,* in effect, triggers, for this WI include a "go" bid decision, B&P charge number assignment, and release of an RFP orRFS by the federal government.

Roles and responsibilities are delineated clearly in tabular form for all cognizant staff, including business development directors, capture managers, proposal managers, production managers, and proposal coordinators. *Outputs* of the WI span (1) completed and delivered proposal documents and/or electronic files, (2) completed proposal approval forms, and (3) completed proposal delivery forms.

Case study insights

RSIS' ISO implementation process got under way in March 2003. Registration was achieved in December 2003—less than 9 months. Typically, ISO implementation takes 12 to 18 months within the industry. Visionary *strategic leadership* enabled implementation throughout all areas of the company from business development through service delivery. Having no gaps in process certification (in effect, no exclusions) ensured that service delivery is managed from end to end consistently with customer requirements always in the foreground.

Another key is to find an *auditing firm* that understands your corporation's primary business. RSIS interviewed many auditing firms, and determined that Intertek Testing Services, NA, Inc., had the closest understanding of our core business areas.

Finally, having an *executive champion* aided in RSIS' achievement of registration. In the case of RSIS, this was the company's executive vice president and chief operating officer. This executive champion helped to remove internal barriers, find alternate solutions, and speed the certification. It is also important to have an executive champion who is very cognizant of QMS implementation and certification activities.

Benefits of the ISO-driven processes

RSIS' ISO-driven proposal and business development approach has resulted in and continues to generate important, tangible benefits:

- *Repeatable* proposal processes that leverage lessons learned;
- *Documented* processes that serve as excellent training tools for new staff;
- Built-in focus on *continuous process improvement,* a recent example of which was the incorporation of a Blue Team Review process for storyboard, storyline, and solution set assessment prior to any proposal text development. This has proven to be tremendously beneficial in terms of launching the proposal in the proper direction early in the process. It was directly related to RSIS winning a major $65 million procurement with NASA in 2003.

- Generation of a living *knowledge base* to retain intellectual capital as proposal staff transition to other positions, leave the company, or retire.

ENDNOTES

1. Elbert, Bruce, and Bobby Martyna, *Client/Server Computing: Architecture, Applications, and Distributed Systems Management*, Norwood, MA: Artech House, 1994.

2. Blankenhorn, Dana, "11 Lessons from Intranet Webmasters," *Netguide*, October 1996, p. 82.

3. Blankenhorn, Dana, "11 Lessons from Intranet Webmasters," *Netguide*, October 1996, p. 89.

4. Tittel, Ed, and James Michael Stewart, "The Intranet Means Business," *Netguide*, July 1996, p. 122.

Chapter 17

Leveraging business complexity in a knowledge-based economy

Knowledge = Productivity

17.1 Turbulent transition toward knowledge-based business

In merely 50 years, "the transistor, whose modest role is to amplify electrical signals, has redefined the meaning of power, which today is based as much on the control and exchange of information as it is on iron or coal. The throbbing heart of this sweeping global transformation is in the tiny solid-state amplifier invented by [John] Bardeen, [Walter] Brattain, and [William] Schockley. The crystal fire they ignited during those anxious postwar years has radically reshaped the world and the way in which its inhabitants now go about their daily lives" [1]. When processed individually and collectively, information is interpreted, congealed, repackaged, and produced into personal and organizational knowledge. And it is knowledge that rose to prominence as the currency of the global economy late in the twentieth century and continues today. The timeworn aphorism that cash is king is undergoing emendation, and employees are being reconceptualized as knowledge or gold-collar workers who provide and possess valuable, strategically advantageous, and renewable intellectual capital. As British research associate Michael Lissack asserts, "In the future information will be the load-bearing structure of organizations." Economies of scale—bigness—are being replaced by economies of speed in this digital age. And Swedish Professor Johan Roos, formerly of the International Institute for Management Development (IMD) in Switzerland, speaks of the "knowledge economy" and "knowledge landscapes."

The transition from the money-capital and management skill-capital eras to the intellectual-capital era has been concretized by a host of discrete decisions and events, including the creation of careers based upon knowledge management, measurement of (metrics for) knowledge-creating processes and intangible assets, and implementation of knowledge-sharing business cultures. Noteworthy among these events was when Skandia Assurance & Financial Services (AFS)—a major Stockholm-based insurance company—appointed Lief Edvinsson as the corporate world's first director of intellectual capital in 1991. Other companies and organizations at the vanguard of recognizing and acting upon the reality, and the value, of intellectual capital and intangible assets (e.g., corporate brain power, organizational knowledge, client relationships, innovation ability, and employee morale) include the Canadian Imperial Bank of Commerce; Booz-Allen & Hamilton; DuPont; and Hughes Space & Telecommunications, which have launched considerable efforts to understand and enhance intellectual capital management. In addition, Buckman Labs, a U.S.-based biotechnology firm, has created a Knowledge Sharing Department. Buckman Labs incentivizes those employees who contribute to its knowledge-sharing culture through financial rewards and management positions. Chevron has engineered a best-practice database to capture and make available the company's collective experience with drilling conditions and innovative solutions to technical problems on site. Sweden's Celemì company published the world's first audit of intangible assets in its 1995 Annual Report. Celemì's Intangible Assets Monitor focuses on its customers (image-enhancing customers, from whom testimonials are valuable; brand names; trademarks); their internal organization (patents, computer systems, management infrastructure); and their staff (competencies, flexibility). Skandia published the first ever annual report supplement on intellectual capital. Pfizer of Switzerland has created competence models for recruiting executives that include knowledge building and sharing as important criteria. WM-data of Sweden, a fast growing IT company, links nonfinancial indicators to strategy, and considers financial ratios of little use for management. The Swedish telecommunications company Telia has published its annual Statement of Human Resources for 10 years. This statement includes a profit-and-loss account that visualizes human resources costs and a balance sheet that shows investments in human resources [2].

During a speech before the American Accounting Association's annual meeting held in Dallas, Texas, in August 1997, Michael Sutton, Chief Accountant of the United States Securities and Exchange Commission (SEC), noted that "historically, accounting has been strongly influenced by the reporting needs of a manufacturing-based economy" [3]. In April 1996, the SEC had convened a symposium on intangible assets in Washington, D.C., during which invited participants from prestigious business, academic, and government organizations discussed issues related to the measurement of intangible assets by preparers of financial reports, concerns about disclosures related to intangible assets, and the experience of U.S. and overseas trendsetters with regard to the accounting and disclosure of intangible assets. As of 2004, however, the SEC had not provided any guidelines or issued any directives vis-à-vis intangible assets for direct application in American corporations.

Converting human intellect into proposals, and ultimately into useful products and services, is becoming increasingly critical to any business' success [4]. "We now know where productivity—real and limitless productivity—comes from. It comes from challenged, empowered, excited, rewarded teams of people. It comes from engaging every mind in the organization" [5]. There is an accelerating recognition among corporations that their future profitability depends on intangible assets such as "creativity, flexibility and the speed with which they can share new ideas and information. In a literal sense, modern corporations *are what they know*" [6]. Major firms like Xerox, Hewlett-Packard, and Dow Chemical are spending significant resources on knowledge identification and management. Alan Webber, a founding editor of the Boston-based business magazine *Fast Company,* has noted that "[t]he world of business is realizing that ultimately, what matters is the quality of the people in the organization, and how they treat their customers and how creative they are in coming up with new products and new services"[7].

Another significant transition among such major corporations as Microsoft, Compaq, Sun Microsystems, and Hewlett-Packard is the paradigm shift away from business as a battlefield to that of business as a "complex ecosystem in which the company that wins is the one that is quickest to adapt. Employee individuality and diversity are honored and encouraged" [8]. To adapt to new market circumstances and translate them quickly into meaningful opportunities, employees simply cannot be viewed and treated as mere mindless troops, crank-turners, or line items on a traditional Excel or Lotus accounting spreadsheet. With small companies in particular, leveraging the collective intellect is integral to business development achievement and superlative, long-term client support. Providing work environments that are conducive to the generation, exchange, and respect of ideas will pay dividends in morale, employee retention, and financial revenue. Pleasant and secure physical facilities; progressive human resources policies; an open-door management culture; articulated, achievable proposal win and performance incentive programs; and so forth all participate in inspiring and harnessing the best in and from employees.

How to communicate effectively on your knowledge landscape

The command-and-control model ... remains part of the baggage carried by many of our best executives.
—The PeopleWise Organization, Kepner-Tregoe

Profound redefinition of the images we hold and the language we use to define ourselves as well as our companies, product lines, colleagues, competitors, proposals, business processes, and the entire economic spectrum is required to generate and sustain business health within our increasingly knowledge-based economy. Metaphors and language directly influence the manners in which corporate and project managers perceive the world and, consequently, the approaches by which they manage their companies, projects, and proposals. Transformed iconography, or "word pictures," can expand the domain of possible and realizable interactions and approaches. Business metaphors that include concepts of war, a race, survival of the fittest, or a jungle and machine- and clock-like images (e.g., go full steam on a project; crank up the pace of activity; C^3 or command, control, and coordinate; and reengineering) will do little to enhance the fitness of American companies on a worldwide knowledge landscape that is subject to nondeterministic change, plateaus of relative stability, and coevolution. Open, complex systems such as company organizations demand concomitant open, flexible architectures of understanding and interaction that have matured far beyond the linear, top-down paradigms that served to generate business success from the Industrial Revolution to the mid-twentieth century. Frederick Taylor's enduring principles of scientific management no longer provide valuable enhancement of human productivity.[1]

"Metaphors allow the transfer of bands of information where other means only transfer smaller bits" [9]. In effect, new words and metaphors constitute new linguistic domains. These new domains promote the conceptualization of avant-garde models and discourse that in turn can guide business decision making on an economic landscape contoured functionally and in real time by human thoughts and ideas. Johan Roos and David Oliver assert that contemporary management issues will increasingly be hindered by the lack of appropriate language to describe emerging organizational phenomena. This is precisely the point at which complexity theory may, in fact, offer senior managers and project managers a robust, multidimensional mode of understanding and decision making.

Within a complexity framework, a company might be conceptualized as a complex adaptive organization, a shifting constellation (à la Henry Mintzberg) with independent and semiautonomous organizations. Machine-like, linear, cause-and-effect must be transformed to the nonlinear ecological interconnectedness of living organisms. Rather than being fiscal liabilities, employees might be perceived to provide valuable intellectual capital and intangible assets within a structured organization and environment. The complex role of management then shifts to visualizing, articulating, measuring, leveraging, expanding, and creating options for the knowledge resources of the firm. Dr. Touraj Nasseri, a consultant and strategist for industry and government, suggests that mapping a company's intellectual capital and employing it productively to achieve strategic objectives are valuable initiatives for senior management to pursue [10]. Nasseri goes on to say that "[t]he potential inherent in well managed intellectual capital extends its impact well into the future as it adapts, renews, and replaces capabilities so that strategies remain responsive to rapid change and much uncertainty" [11]. "Management needs to develop systematic processes that stimulate languaging throughout the company so that, over time, an internal lexicon is formed that can provide the foundation for an effective corporate identity" [12]. *Languaging*, according to Professor Johan Roos, is the art of word choice. And Karl Sveiby introduces the term

---

1   Frederick Winslow Taylor, M.E., ScD. (1856–1915), conceptualized a system that he termed *scientific management* to address "the great loss which the whole country is suffering through inefficiency in almost all of our daily acts." In his work *The Principles of Scientific Management* (1911), Taylor asserted that "the remedy for this inefficiency lies in systematic management, rather than in searching for some unusual or extraordinary man." Taylor's form of industrial engineering established the organization of work on Henry Ford's assembly lines in Michigan, which were the pride of the industrialized world at that time. However, many of Frederick Taylor's principles were misinterpreted or misapplied, and too often were translated into time-and-motion studies to extract more work from people for less pay. The migration away from labor-intensive mass production operations in American business and manufacturing has neutralized the effectiveness and applicability of Taylor's principles.

knowledge management, "the art of creating value from an organization's Intangible Assets." Traditional command-and-control management functions must now give way to mapping and acquiring intellectual capital, communicating a clear vision for the firm, developing and implementing a business lexicon with shared meanings and nuances to foster generalized understanding of strategic goals, and facilitating the rapid assessment and multidirectional flow of knowledge throughout the organization. Caveats and pronouncements issued from senior management will not be internalized throughout the knowledge organization.

In the knowledge economy, companies should focus on adapting, recognizing patterns, and building ecological webs to amplify positive feedback rather than trying to achieve optimal performance. Managers should attempt to discern interrelationships and recognize patterns rather than conduct forecasting based upon rationalist (read: Tayloresque) causes and effects. Hierarchical management schemas that rely heavily upon the paternalistic authority that marked the Industrial Era are best transformed into partnership models for the Knowledge Era that incorporate values such as cooperation, caring, creativity, empathy, connectedness, mutualism, and compassion. A sustainable competitive advantage is now maintained on a nonlinear, unpredictable economic landscape (i.e., a rugged knowledge terrain) through leveraging the collective knowledge base of a company and fostering a learning organization rather than simply effecting financial capital investment in infrastructure and institutional legacy systems.

At a practical level, sustainability in this new century follows from nurturing a business culture, infrastructure, and managerial approaches that encourage *self-organization*. (Biologists use the term self-organization to describe organisms that continually adapt to their environment without losing their basic identity.) Fundamentally, there are no accurate predictions that pertain to a business landscape that is subject to multidimensional, emergent change and on which companies coevolve, that is, mutually affect the development and direction of each other.

Michael Lissack correctly and profoundly observes that "[t]he descriptive metaphor that everything is changing and thus an organization must be poised to adapt to change says nothing about what to do next or about how to convert conflict into cooperation" [9]. What does the "everything is changing" metaphor mean at the level of hour-by-hour, day-to-day managerial decision making? Functionally, very little, according to this author's observation during the course of the past 17 years in the aerospace, defense, IT, and environmental marketplaces. Many American companies continue to operate according to a cluster of timeworn nineteenth-century paradigms, superimposed with a smattering of late twentieth-century management metaphors and buzzwords such as value innovation, total quality management (TQM), business process reengineering (BPR), continuous process improvement (CPI), and quality circles. The parental, autocratic, or mechanistic models of management, inextricably coupled with and often driven by manufacturing-related paradigms of accounting and asset evaluation, remain deeply entrenched and closely guarded. Business performance is still measured largely in financial terms, rather than collectively from a knowledge development perspective, infrastructure perspective, and customer perspective, as well as a financial perspective [13]. However, this situation presents the opportunity to think, and therefore act, in fundamentally different ways. Collectively, American companies are fully capable of inspired vision.

17.3 Envisioning supple business models

Creating open-architecture business models that can function effectively (i.e., converge on reality rather than diverge from real-world data and experience) and evolve within a knowledge-based economy will involve a shift in approach that spans several orders of magnitude. With information and knowledge becoming the currency of American business, the organizations that can share this new currency "ecologically" in an effort to arrive at effective and appropriate solutions for both their external and internal clients in the shortest possible time and with the least amount of resources expended will be those firms that occupy positions, or *optima*, above the knowledge landscape. From these vantages, companies will be less likely to be displaced by new technologies, reconfigured marketing paradigms, social and demographic shifts, and political and legislative climate changes.

Let us reconsider the notion of clients, for example. Within a traditional company, clients are perceived as the buyers or procurers of products or services being offered. In a knowledge-oriented company, clients include both external elements such as buyers, key subcontractors, regulatory agencies, and government institutions, as well as internal clients. The latter might include professional, project management, and support staff; operational units; sales and marketing as well as proposal development elements; financial and accounting modules; senior management; and so forth. The currency of knowledge must flow in meaningful, multidirectional pathways among all of these clients. Each client is critical to the sustainability and performance (i.e., positive energetics) of the company, and each requires, and in turn provides, vital information and knowledge to the other much as enzymes are shared within a living cell to perform a host of vital maintenance, replicative, and growth functions.

The traditional organizational diagram that most American companies use to illustrate internal structure will have little applicability in the knowledge-based economy. Organizational charts are by nature highly linear, pyramidal, and top-down in direction. As Figure 17.1 depicts, however, businesses in 2005 and beyond should strongly consider an ecological model with knowledge constituting the currency or fuel that is transferred among clusters of entities we will call clients, business process and metaprocesses, employees, and finances. Table 17.1 suggests how traditional line and staff organizations can be reconceptualized according to this new model.

In this model, every cluster and module within the learning organization is afforded a pivotal complex of roles in ensuring the sustainability of the company. Everyone is involved with and contributes to the new currency—knowledge, which is quite unlike the traditional paradigm wherein any element not involved in the most linear manner with generating cash is perceived as a "leech on the corporate purse." In fact, within a knowledge-based and forward-looking organization, intangible assets are considered quantitatively along with such tangible assets as cash, accounts receivable, computers and other amortizable equipment, facilities, and inventories.

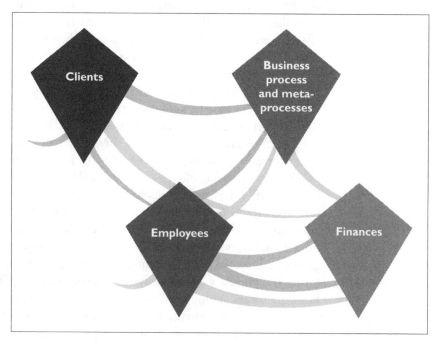

Figure 17.1
Knowledge fuels complex, ecological business interactions.

Traditional Organizational Element/Function	Reconceptualized Cluster in a Complex, Learning Organization	Reconceptualized Functions/Comments
Chairman of the Board Chief Executive Officer President	Metaprocesses	Map intellectual capital, communicate a clear vision for the firm, develop and implement a shared business lexicon to foster generalized understanding of strategic goals, and facilitate the rapid assessment and multidirectional diffusion of knowledge throughout the organization. In effect, senior management must leverage and manage the intellectual capital of the company, as well as conduct strategic planning activities.
Risk Management	Business Processes	Develop business continuation strategies that permit the transfer of core business-specific knowledge to a new management team.
Operations Manufacturing	Business Processes (now called Process Owners)	Leverage and empower the intellectual capital of knowledge workers to generate localized, real-time solutions that can then be cataloged and transferred to other operational modules within your company.
Finance, Procurement, Travel, and Facilities	Finances	Assess and report intellectual and structural capital in monetary terms. As Dr. Nasseri notes, "physical assets owe most of their value to intellectual capital" [14].
Human Resources	Business Processes (now called Human Capital Development)	Acquire, train, retain, foster continuing education for, and incentivize human intellectual capital. Develop compensation policies appropriate for knowledge workers. Ensure that employees are always climbing a learning curve.*

* Far from being a detriment, a learning curve within a learning organization is a prerequisite to continued sustainability.

Table 17.1 Reconceptualizing Organizational Elements Toward a Knowledge-Based Framework

Traditional Organizational Element/Function	Reconceptualized Cluster in a Complex, Learning Organization	Reconceptualized Functions/Comments
Information Services (IS)	Business Processes	Facilitate the rapid assessment and multidirectional flow of knowledge throughout the organization via data warehousing, data mining, and data filtering supplemented by ongoing training.
Sales and Marketing Secondary Marketing Business Development Advanced Planning	Business Processes	Assess and communicate the economic landscape, develop and share knowledge of clients, conduct benchmarking[†] and competitor assessments, define client and product lines (congealed knowledge), and design and develop proposals.
Legal Support (contracts, insurance, litigation, compliance and regulatory support) Auditing	Business Processes	Congeal intellectual capital as patents, software source code, published papers and books, marketable user manuals, research papers, productizable business processes and protocols.
Project Management[‡]	Business Processes (now called Knowledge Integrators or Knowledge Managers rather than Project Managers)	Serve external and internal clients; implement client-focused vision; empower staff for success; and foster accountability, commitment, and ownership. Serving external clients extends to assembling appropriate knowledge sources from within the organization and matching those sources to external client requirements on budget and within the required time frame. Internal client servicing includes reporting requirements.

[†] Benchmarking is a rigorous process for linking competitive analysis to your company's strategy development. Benchmarking is a method that measures the performance of your best-in-class competitors relative to your industry's key success factors. It also is a mechanism for determining how the best-in-class achieve those performance levels and competitor assessments, define client and product lines (congealed knowledge), and design and develop proposals.

[‡] Too often in American business, the functional reality is that project managers are technical staff who also happen to sign time sheets. Their focus is on technical nuts-and-bolts, frequently to the exclusion of client interaction, staff empowerment, knowledge sharing, and so forth.

Table 17.1 (Continued)

Traditional Organizational Element/Function	Reconceptualized Cluster in a Complex, Learning Organization	Reconceptualized Functions/Comments
Administrative Support	Business Processes	Produce congealed knowledge products in the form of proposals, contract deliverable reports, marketing presentations, annual and other SEC-required reports, client correspondence.
Internal Research and Development (IR&D) Product Development	Business Processes	Innovation in a company is nourished and driven by knowledge-based capabilities and by management systems that leverage the capabilities [11].
Health and Safety Quality Control/Quality Assurance	Business Processes	

Table 17.1 (Continued)

17.4 Sample application: tracing complexity and KM through the proposal development process

Technically, a proposal is an offer prepared by a contractor (such as your company) to perform a service, supply a product, or a combination of both in response to an RFP or RFS document by a procuring organization within government or private industry. Proposals are legal documents, often incorporated by reference into the final contract. That means that they must be factual and accurate—including information provided in résumés and project summaries, which summarize applicable contractual experience. In actuality, résumés and project summaries reflect a significant percentage of intellectual capital of a service organization. Relevant knowledge resident in the employees of the company as gained through education, training, and professional experience coupled with initiative, dedication, and innovation should be highlighted in each résumé. Similarly, to be effective, project summaries must reflect the organizational knowledge of the firm from technical, managerial, and geographic perspectives, to name but several.

Proposals are, first and foremost, knowledge products that include a host of marketing, technical, programmatic, institutional, pricing, and certification information. Through a choreographed process of knowledge generation, transfer, and congealment, a proposal is designed to sell both technical and managerial capabilities of a company to accomplish all required activities on time and at a reasonable cost. Your company's proposal document(s) is scored, literally, by the client's evaluators against formalized, specific standards in a process analogous to pattern recognition. A proposal is the tangible result of knowledge-building processes, supported in turn by hard and directed work and buoyed by a positive collective attitude within your company's proposal team of knowledge workers.

Crucial to proposal success both as a process and a product is the transfer of relevant marketing information to appropriate members of the capture and proposal team. From there, that information must be interpreted, assigned value, and translated into knowledge that will ultimately be congealed into the actual proposal documents—into narrative, theme statements, graphics, captions, compliance matrices, tabs, oral presentation slides, and so forth. If marketing data and information do not find their way into the knowledge product called a proposal, then a company's proposal will not receive the maximal point score. Therefore, the marketing data are valueless. In effect, if intellectual capital measured in client, project, competitor, and political knowledge is not transferred to appropriate staff (knowledge workers) and converted efficiently into hardcopy or electronic proposal documents via established, articulated business processes, your company's proposal win ratios in terms of dollars (x dollars of contracts awarded for every \$100 of contract value pursued) and number (x number of proposals won) will be suboptimal.

17.5 Summation

Evidence "suggests that the business world may be a complex system poised at the edge of chaos" [15]. If we grant that such is the case, then we who are an integral part of this world must leverage the complexity with which we are faced every day into optima (that is, business processes, business thinking, and the like) that lift us above the unpredictable knowledge landscape and ensure business sustainability. Implementing business processes that actively promote knowledge valuation, development, transfer, management, and congealment will help American companies maximize both intellectual and structural capital as we enter the twenty-first century.

We now turn our attention to a specific type of contracting with the federal government—the important SF330 solicitation process. These solicitations are presented in the Federal Acquisition Regulations (FAR) at Part 53.236-2, Architect-engineer services. SF330 documents are prepared for business opportunities with the federal government as well as states, municipalities, and quasigovernmental agencies such as port and transit authorities.

ENDNOTES

1. Riordan, John, and Lillian Hoddeson, *Crystal Fire: The Birth of the Information Age*, New York: W. W. Norton & Co., 1997, p. 10.

2. Information in this paragraph was drawn from Karl E. Svieby, "What Is Knowledge Management?" pp. 5–6, http://www.sveiby.com/articles/ KnowledgeManagement.html, April 2001, as well as Sacha Cohen, "Knowledge Management's Killer App," http://www.astd.org/magazine/current/ cohen.htm, and Laton McCartney, "Getting Smart About Knowledge Management," *Industry Week*, May 4, 1998, p. 32.

   See also Karl E. Svieby, "The Intangible Assets Monitor," http://www.sveiby.com/articles/IntangAss/CompanyMonitor.html, December 20, 1997. Celemì employs metrics such as efficiency (sales per customer, staff turnover), stability (repeat orders, growth in sales per administrative staff), and growth and renewal (average years of professional competence). Celemì is headquartered in Malmö, Sweden, and is dedicated to creating processes that help companies leverage the power of learning.

   Sacha Cohen notes that Hewlett-Packard employs a Web-based system called Connex to help identify subject experts in specific geographic locations. Also, Booz Allen uses a best-practice application on its intranet, Knowledge On Line (KOL), an automated knowledge repository in which that firm's collective knowledge and expertise are captured, classified, and quantified.

3. Sutton, Michael H., "Dangerous Ideas: A Sequel," Remarks delivered during the American Accounting Association 1997 Annual Meeting in Dallas, Texas, on August 18, 1997, p. 3; http://www.corpforms99. com/83.html.

4. Quinn, James Brian, Philip Anderson, and Sydney Finkelstein, "Managing Professional Intellect: Making the Most of the Best," *Harvard Business Review*, March/April 1996.

5. Quinn, Judy, "The Welch Way: General Electric CEO Jack Welch Brings Employee Empowerment to Light," *Incentive*, September 1994, p. 50.

6. Hamilton, Martha A., "Managing the Company Mind: Firms Try New Ways to Tap Intangible Assets Such as Creativity, Knowledge," *Washington Post*, August 18, 1996, p. H1.

7. Hamilton, Martha A., "Managing the Company Mind: Firms Try New Ways to Tap Intangible Assets Such as Creativity, Knowledge," *Washington Post*, August 18, 1996, p. H5.

8. James, Geoffrey, "It's Time to Free Dilbert," *New York Times*, September 1, 1996, p. F-11.

9. Lissack, Michael R., "Complexity Metaphors and the Management of a Knowledge Based Enterprise: An Exploration of Discovery," http://www.lissack.com/writings/proposal.htm.

10. Nasseri, Touraj, "Knowledge Leverage: The Ultimate Advantage," *Kognos: The E-Journal of Knowledge Issues*, Summer 1996, p. 4.

11. Nasseri, Touraj, "Knowledge Leverage: The Ultimate Advantage," *Kognos: The E-Journal of Knowledge Issues*, Summer 1996, p. 2.

12. Roos, Johan, and Georg von Krogh, "What You See Depends on Who You Are," *Perspectives for Managers*, No. 7, September 1995, p. 2.

13. Roos, Johan, "Intellectual Performance: Exploring an Intellectual Capital System in Small Companies," October 30, 1996, pp. 4, 9.

14. Nasseri, Touraj, "Knowledge Leverage: The Ultimate Advantage," *Kognos: The E-Journal of Knowledge Issues*, Summer 1996, p. 1.

15. Phelan, Steven E., "From Chaos to Complexity in Strategic Planning," Presented at the 55th Annual Meeting of the Academy of Management, Vancouver, British Columbia, August 6–9, 1995, p. 7; http://www.aom.pace.edu/bps/Papers/chaos.html.

Chapter **18**

Planning and producing SF330 responses for architect-engineer services

18.1 **SF330 and the FAR**

The FAR at Part 53.236-2(b), Architect-engineer (A-E) services, describes a special type of solicitation that has broad-based applicability at the federal, state, municipality, and quasigovernmental levels such as port authorities and transit authorities. This type of solicitation is called an SF330. Architectural and engineering firms of all sizes submit this document routinely to establish their credentials with client organizations as diverse as the U.S. Coast Guard (USCG), U.S. Army Corps of Engineers (USACE), U.S. Navy, U.S. Department of Agriculture (USDA), the National Forest Service, and Harford County, Maryland.

SF330, the Architect-Engineer Qualifications form, was developed to replace the long-standing SF254 and SS255. SF254, Architect-Engineer and Related Services Questionnaire, and SF255, Architect-Engineer and Related Services Questionnaire for Specific Projects, had been largely unchanged since their introduction in 1975. The A-E industry, however, has changed substantially and the requirements of federal agencies have evolved considerably. In response to these changes, the SF330 was developed by an interagency committee, submitted to industry and government for comment through the *Federal Register*, and fielded.

SF330 reflects current A-E practices and provides the essential information agencies need for the selection of A-E firms in a streamlined format. SF330 is a two-part form (see Figure 18.1). Part I, Contract-Specific Qualifications, replaces SF255. Part II, General Qualifications, replaces SF254. SF254 is now essentially used as an appendix to SF255 and is rarely used alone. Hence, SF330 was designed as a single two-part form. Similar to SF254, Part II of SF330 can still be separately submitted to agencies to be maintained in their A-E qualification data files.

The objectives of the SF330 are as follows:

- Merge SF254 and SF255 into a single streamlined form;
- Expand essential information about qualifications and experience;
- Reflect current A-E disciplines, experience types, and technology;
- Eliminate duplicate information;
- Eliminate information of marginal value;
- Facilitate electronic usage.

Presented in Figure 18.1 are the forms and formats in which the government wants to receive the specific information regarding, for example, contractor personnel, office locations, and contract experience. SF330 is available on the Web at http://contacts.gsa.gov/webforms.nsf.

18.2 Understanding the required structure of the response

SF 330 submittals are prepared in response to synopses for A-E services that appear on FedBizOpps. Although generally prepared for federal government agencies such as the USACE, USCG, U.S. Army, or specific Air Force bases, SF330s or modified versions of this form can be required submittals for states and municipalities as well. An example is the Form CEB 02, A-E and Related Services Questionnaire for Specific Project (City

ARCHITECT - ENGINEER QUALIFICATIONS

PART I - CONTRACT-SPECIFIC QUALIFICATIONS

A. CONTRACT INFORMATION

1. TITLE AND LOCATION *(City and State)*

2. PUBLIC NOTICE DATE

3. SOLICITATION OR PROJECT NUMBER

B. ARCHITECT-ENGINEER POINT OF CONTACT

4. NAME AND TITLE

5. NAME OF FIRM

6. TELEPHONE NUMBER

7. FAX NUMBER

8. E-MAIL ADDRESS

C. PROPOSED TEAM
(Complete this section for the prime contractor and all key subcontractors.)

(Check)			9. FIRM NAME	10. ADDRESS	11. ROLE IN THIS CONTRACT
PRIME	J-V PARTNER	SUBCONTRACTOR			
a.			CHECK IF BRANCH OFFICE		
b.			CHECK IF BRANCH OFFICE		
c.			CHECK IF BRANCH OFFICE		
d.			CHECK IF BRANCH OFFICE		
e.			CHECK IF BRANCH OFFICE		
f.			CHECK IF BRANCH OFFICE		

D. ORGANIZATIONAL CHART OF PROPOSED TEAM ☐ *(Attached)*

AUTHORIZED FOR LOCAL REPRODUCTION

STANDARD FORM 330 (6/2004) PAGE 1

Figure 18.1 SF330.

E. RESUMES OF KEY PERSONNEL PROPOSED FOR THIS CONTRACT
(Complete one Section E for each key person.)

12. NAME	13. ROLE IN THIS CONTRACT	14. YEARS EXPERIENCE	
		a. TOTAL	b. WITH CURRENT FIRM

15. FIRM NAME AND LOCATION *(City and State)*

16. EDUCATION *(DEGREE AND SPECIALIZATION)*	17. CURRENT PROFESSIONAL REGISTRATION *(STATE AND DISCIPLINE)*

18. OTHER PROFESSIONAL QUALIFICATIONS *(Publications, Organizations, Training, Awards, etc.)*

19. RELEVANT PROJECTS

	(1) TITLE AND LOCATION *(City and State)*	(2) YEAR COMPLETED	
		PROFESSIONAL SERVICES	CONSTRUCTION *(If applicable)*
a.	(3) BRIEF DESCRIPTION *(Brief scope, size, cost, etc.)* AND SPECIFIC ROLE	☐ Check if project performed with current firm	

	(1) TITLE AND LOCATION *(City and State)*	(2) YEAR COMPLETED	
		PROFESSIONAL SERVICES	CONSTRUCTION *(If applicable)*
b.	(3) BRIEF DESCRIPTION *(Brief scope, size, cost, etc.)* AND SPECIFIC ROLE	☐ Check if project performed with current firm	

	(1) TITLE AND LOCATION *(City and State)*	(2) YEAR COMPLETED	
		PROFESSIONAL SERVICES	CONSTRUCTION *(If applicable)*
c.	(3) BRIEF DESCRIPTION *(Brief scope, size, cost, etc.)* AND SPECIFIC ROLE	☐ Check if project performed with current firm	

	(1) TITLE AND LOCATION *(City and State)*	(2) YEAR COMPLETED	
		PROFESSIONAL SERVICES	CONSTRUCTION *(If applicable)*
d.	(3) BRIEF DESCRIPTION *(Brief scope, size, cost, etc.)* AND SPECIFIC ROLE	☐ Check if project performed with current firm	

	(1) TITLE AND LOCATION *(City and State)*	(2) YEAR COMPLETED	
		PROFESSIONAL SERVICES	CONSTRUCTION *(If applicable)*
e.	(3) BRIEF DESCRIPTION *(Brief scope, size, cost, etc.)* AND SPECIFIC ROLE	☐ Check if project performed with current firm	

STANDARD FORM 330 (6 2004) **PAGE 2**

Figure 18.1 (Continued.)

F. EXAMPLE PROJECTS WHICH BEST ILLUSTRATE PROPOSED TEAM'S QUALIFICATIONS FOR THIS CONTRACT *(Present as many projects as requested by the agency, or 10 projects, if not specified. Complete one Section F for each project.)*	20. EXAMPLE PROJECT KEY NUMBER

21. TITLE AND LOCATION *(City and State)*	22. YEAR COMPLETED	
	PROFESSIONAL SERVICES	CONSTRUCTION *(If applicable)*

23. PROJECT OWNER'S INFORMATION

a. PROJECT OWNER	b. POINT OF CONTACT NAME	c. POINT OF CONTACT TELEPHONE NUMBER

24. BRIEF DESCRIPTION OF PROJECT AND RELEVANCE TO THIS CONTRACT *(Include scope, size, and cost)*

25. FIRMS FROM SECTION C INVOLVED WITH THIS PROJECT

	(1) FIRM NAME	(2) FIRM LOCATION *(City and State)*	(3) ROLE
a.			
b.			
c.			
d.			
e.			
f.			

STANDARD FORM 330 (6/2004) PAGE 3

Figure 18.1 (Continued.)

G. KEY PERSONNEL PARTICIPATION IN EXAMPLE PROJECTS

26. NAMES OF KEY PERSONNEL (From Section E, Block 12)	27. ROLE IN THIS CONTRACT (From Section E, Block 13)	28. EXAMPLE PROJECTS LISTED IN SECTION F (Fill in "Example Projects Key" section below before completing table. Place "X" under project key number for participation in same or similar role.)									
		1	2	3	4	5	6	7	8	9	10

29. EXAMPLE PROJECTS KEY

NO.	TITLE OF EXAMPLE PROJECT (FROM SECTION F)	NO.	TITLE OF EXAMPLE PROJECT (FROM SECTION F)
1		6	
2		7	
3		8	
4		9	
5		10	

STANDARD FORM 330 (6/2004) PAGE 4

Figure 18.1 (Continued.)

H. ADDITIONAL INFORMATION

30. PROVIDE ANY ADDITIONAL INFORMATION REQUESTED BY THE AGENCY. ATTACH ADDITIONAL SHEETS AS NEEDED.

I. AUTHORIZED REPRESENTATIVE

The foregoing is a statement of facts.

31. SIGNATURE

32. DATE

33. NAME AND TITLE

STANDARD FORM 330 (6/2004) PAGE 5

Figure 18.1 (Continued.)

ARCHITECT-ENGINEER QUALIFICATIONS

	1. SOLICITATION NUMBER *(If any)*

PART II - GENERAL QUALIFICATIONS
(If a firm has branch offices, complete for each specific branch office seeking work.)

2a. FIRM (OR BRANCH OFFICE) NAME		3. YEAR ESTABLISHED	4. DUNS NUMBER
2b. STREET		colspan: 5. OWNERSHIP	
		a. TYPE	
2c. CITY	2d. STATE / 2e. ZIP CODE	b. SMALL BUSINESS STATUS	
6a. POINT OF CONTACT NAME AND TITLE		7. NAME OF FIRM *(If block 2a is a branch office)*	
6b. TELEPHONE NUMBER	6c. E-MAIL ADDRESS		

8a. FORMER FIRM NAME(S) *(If any)*	8b. YR. ESTABLISHED	8c. DUNS NUMBER

9. EMPLOYEES BY DISCIPLINE

10. PROFILE OF FIRM'S EXPERIENCE AND ANNUAL AVERAGE REVENUE FOR LAST 5 YEARS

a. Function Code	b. Discipline	c. No. of Employees (1) FIRM	(2) BRANCH	a. Profile Code	b. Experience	c. Revenue Index Number *(see below)*
	Other Employees					
	Total					

11. ANNUAL AVERAGE PROFESSIONAL SERVICES REVENUES OF FIRM FOR LAST 3 YEARS *(Insert revenue index number shown at right)*	PROFESSIONAL SERVICES REVENUE INDEX NUMBER
	1. Less than $100,000 6. $2 million to less than $5 million
a. Federal Work	2. $100,00 to less than $250,000 7. $5 million to less than $10 million
b. Non-Federal Work	3. $250,000 to less than $500,000 8. $10 million to less than $25 million
c. Total Work	4. $500,000 to less than $1 million 9. $25 million to less than $50 million
	5. $1 million to less than $2 million 10. $50 million or greater

12. AUTHORIZED REPRESENTATIVE
The foregoing is a statement of facts.

a. SIGNATURE	b. DATE
c. NAME AND TITLE	

AUTHORIZED FOR LOCAL REPRODUCTION

STANDARD FORM 330 (6/2004) PAGE 6

Figure 18.1 (Continued.)

of Baltimore), which is analogous to SF330. One major difference in contractor response to a government RFP versus an SF330 synopsis is that labor rates and other direct costs are included with a proposal but are not part of an SF330 response.

18.3 Overall strategy of response

SF330, Part I, submittal consists essentially of a series of nine major elements. As with response to RFPs, you should follow the guidance provided in the FedBizOpps synopsis very carefully.

When strategizing your response to an SF330 opportunity, focus on Section E, Résumés of Key Personnel Proposed for This Contract, first. During the prekickoff planning phase, reach a consensus regarding the personnel you will be proposing and exactly what position on the project organizational chart they will occupy. Particularly critical to decide on are the program/project manager and the delivery/task order managers. Attempt to structure the project organizational chart in accordance with the government agency's organization and in compliance with the technical subareas named in the FedBizOpps synopsis. If, for example, the government agency is divided into three primary elements, then your company's project organization might reflect three similarly named elements as well. Your marketing intelligence should indicate the appropriate number of delivery or task order managers to propose. Does the particular government agency want streamlined contractor project management structures? In that case, one or two delivery or task order managers for each major organizational element might be appropriate. When selecting personnel for Section E, keep in mind that these staff should have been integrally involved in the projects that you will select in Section F, Example Projects Which Best Illustrate Proposed Team's Qualifications for This Contract. Section G, Key Personnel Participation in Example Projects, requires this involvement. The government wants to be assured that the staff your company is proposing have actually worked on the contracts that you are presenting as similar technical or programmatic experience in Section F.

Once the project organizational chart has been constructed for Section D, Organizational Chart of Proposed Team, your next step is to identify 10 relevant projects for inclusion in Section F. Most government agencies limit the number of projects to 10, although there are occasional exceptions.[1] When the applicable projects have been identified, your proposal

---

1   In one SF254/255 solicitation, Tinker Air Force Base in Oklahoma allowed up to 50 projects in Block 8, which is analogous to Section F of the SF330!

manager can then turn his or her attention to Section H, Additional Information. Most companies and individual proposal managers make the mistake of beginning their response to an SF330 with Section H because this is the technical heart of the document. However, it is extremely difficult to design a successful SF330 response without first having made the pivotal decisions regarding Sections E and F. Remember that evaluators are reviewing 30 to 50 SF330s for even small procurements. They may spend only 30 minutes on any one proposal during their preliminary evaluation sequence. That is all the time your company will get to make a positive, lasting impression.

18.4 Section F: selling your project experience

In selecting the 10 projects for Section F, you might apply the following criteria to determine their relevancy to the particular procurement.

- Same client (USACE, Portland District);
- Same agency (USACE);
- Same technical work (landfill closure);
- Same geographic location (Pacific Northwest);
- Same type of contract vehicle [Indefinite Delivery/Indefinite Quantity (ID-IQ)].

18.5 Section H: structure according to the evaluation criteria

Your company's response to the technical requirements and evaluation criteria found in the FedBizOpps synopsis is the foundation of Section H. Ensure that each of the technical areas in the synopsis is addressed in your response. Use the wording of the particular synopsis to build the headings and subheadings of Section H. In responses that are not page limited, Section H affords an opportunity to present additional project-specific experience to amplify and expand that which was provided in Section F. There is one caveat, however: As with all proposal writing, keep the narrative in Section H crisp and focused on the requirements and how your company can meet them fully.

Section H also includes your company's project management plan. Link this narrative to the staff proposed in Section E. Highlight the advantages of selecting your company at the conclusion of Section H. Provide authenticated benefits of your technical and management approach to the client. In addition, Section H should include the specific rationales for why your company has teamed with other firms (if this is the case) in order to propose on this procurement.

18.6 Section H outlining

Let us suppose that the USACE, Savannah District, your client, has released a synopsis in FedBizOpps for miscellaneous environmental services. The weighted evaluation criteria are (1) technical excellence, (2) prior experience and past performance, (3) management capabilities, (4) personnel qualifications, and (5) computer capabilities. How should you outline Section H?

Build the Section H outline around these stated evaluation criteria, making sure to include subheadings for each one of the points stated under a given evaluation criterion. For example, you might consider outlining this section as shown in Table 18.1, if the management capabilities criterion (3) reads as follows:

(3) Management capabilities (23%) as demonstrated by internal quality control and quality assurance procedures used to ensure the technical accuracy and the coordination of disciplines, the structure of the firm as it relates to the overall approach to project management, capability to perform work in house, and ability to manage fluctuating workload.

Consider building and maintaining an electronic subdirectory of high-impact client quotations to authenticate your proposals and SF330s.

18.7 Subcontractor participation

During the prekickoff planning phase of the SF330 process, ensure that written teaming agreements are in place with all of your subcontractors. Call on them during the writing phase to provide résumés for Section E, project summaries for Section F, a company overview and technical information for Section H, any appropriate letters of commendation, and a letter of commitment from the president of the subcontracting firm to your company's president or senior vice president. (See Table 18.2 for a comprehensive list of materials required and guidelines for submittal.) Include résumés and project summaries from your subcontractors in accordance with the percentage of the contract that is being subcontracted. In effect, if

H.3	Management Capabilities (Evaluation Criterion 3)	
	H.3.1	Overview
	H.3.2	Management of Projects of Similar Size and Scope
		• Cost-Effective Strategies
		• Time-Efficient Strategies
	H.3.3	Structure of the Firm
	H.3.4	Project Execution Strategies
		H.3.4.1 Project Initiation and Kickoff
		H.3.4.2 Internal QC-QA Procedures
		• QC-QA Metrics for Deliverables
		H.3.4.3 Ability to Manage Fluctuating Workloads
		H.3.4.4 Cost and Schedule Control Approaches
		H.3.4.5 Performance Measurement and Control Systems
	H.3.5	In-House Capabilities
	H.3.6	Project Leadership
		• Project Manager's Role and Span of Authority
		• USACE-Company Interfaces
		• Monthly Reporting
	H.3.7	Subcontractor Management

Table 18.1
Section H:
Management
Section Outline

only 15% of the contract will be performed by the subcontractors, then one or two project summaries should be from the subcontractors.

Your subcontractors can also be used in a review capacity for the SF330s. As with responses to RFPs and RFSs, your SF330s should also undergo internal reviews and quality checking throughout the response life cycle. SF330s have specific due date and times, just like formal RFPs. Ensure that your company receives a time- and date-stamped receipt for your submittal.

18.8 Building teaming agreements

Teaming agreements between your company and other organizations are established for several important reasons. Many times, FedBizOpps synopses for SF330s (and RFPs as well) stipulate that a certain percentage of the contract must be performed, for example, by SDBs, WOBs, and

A. Software Application Requirements

1. Submit all text in Microsoft Word 2000.

2. Submit all line-art graphics as CorelDRAW files or Windows metafiles.

3. Any photos, submit as, tiff files or as 35-mm slides or prints.

4. All text and graphics provided in electronic form MUST be provided on diskette or CD-ROM, not via e-mail.

5. No graphics from the Internet may be submitted. We need 200-dpi-resolution photos or images.

B. Formatting Requirements

1. Use as few formatting commands (e.g., **bold**, underline, *italic*, centering, indenting) as possible in creating your text. Make text flush left, and separate paragraphs with hard returns.

C. Required Materials from Each Teaming Partner

1. Signed, original letters of commitment on company letterhead

2. Tailored résumés, including all relevant professional certifications and registrations, along with associated dates and states

3. Tailored project summaries

4. Overview of company history and relevant technical, programmatic, and contractual capabilities

5. Statement of financial strength and solvency

6. Camera-ready company logos

7. Project-specific photos and identifying captions

8. Relevant client references, along with full title, name, address, telephone, fax, and e-mail

9. Recent Architect/Engineer Contract Administration Support System (ACASS) performance ratings

10. Office locations map

11. Total number of projects performed for the client organization

12. Mileage from nearest office locations to client HQ

13. Specific numbers of personnel by discipline in the relevant geographic area

14. Relevant client commendation letters or awards

Table 18.2
Comprehensive Guidelines for Teaming Partners

historically black colleges and universities (HBCUs). Depending on its size, your company may need to enter into teaming agreements with these organizations to fulfill these stipulated contractual requirements. It is beneficial to build a database of potential teaming partners that fall into these categories and have been prequalified with the particular client agency, if necessary. Your company can and should also consult the Central Contractor Registration (CCR) database (http://www.ccr.gov) for small business partners. You can use the Dynamic Small Business Search feature of CCR.

Another common reason for entering into a teaming agreement is that each company will benefit from the other's technical expertise, contractual experience, programmatic strength, geographic proximity to the government site, and number and diversity of professional staff. For example, perhaps your company offers many important and relevant environmental services, but the FedBizOpps synopsis for a specific USCG project also requires support in the areas of unexploded ordnance and noise reduction. Under these circumstances, teaming is essential in order to meet the USCG's technical needs completely.

Teaming agreements are formal documents that define the specific association of companies and other entities, and which offer the basis for legal action if they are violated. In actual practice, they are not infallible but do articulate the basis of mutual interaction for the purpose of acquiring business. Agreements may be exclusive, or nonexclusive, depending upon the procurement. In effect, a given company may elect to join several teams to heighten its chances of winning a portion of the contract work. When your company serves as the prime contractor, your goal will be to secure exclusive teaming agreements with your subcontractors. This helps to eliminate the possibility of company-sensitive or procurement-sensitive information leaking to your competitors.[2]

The language of teaming agreements varies depending upon the procurement, the exact nature of the working relationship, and the business culture of the companies involved. Articles that might be built into a teaming agreement include (1) the particular purpose of the agreement, (2) the relationship of the parties involved, (3) scope of services to be performed upon contract award, (4) costs incurred, (5) the process for executing contracts, (6) termination procedures, (7) notices (i.e., points of contact in each party), (8) nondisclosure of proprietary information, and (9) grounds for and limits of any liability. Presented in Figure 18.2 are topics

---

2   I have had the experience of coordinating a major IRS proposal effort that involved a team of companies, some of which were important members of competitor proposal teams. The logistics of conducting effective strategy meetings and comprehensive review sessions were quite challenging.

Potential Teaming Agreement Topics

1. Mutual support during the entire proposal and postproposal life cycle, including negotiations.

    Timely submission of all proposal materials.

    Availability of qualified staff to assist in developing the proposal or SF330, as well as in conducting discussions and negotiations with the government.

2. Prime contractor's right to include additional subcontractors on the team.

3. Each party will bear all costs, risk, and liabilities incurred. Prime will be responsible for the layout, printing, binding, and delivery costs of the proposal.

4. Prime will have sole right to decide the form and content of all documents submitted to the government.

5. Prime will make every reasonable effort to subcontract to the sub that portion of the work stipulated in a statement of work (attached to the teaming agreement).

6. Any news releases, public announcements, advertisement, or publicity released by either party concerning the agreement, any proposals, or any resulting contracts or subcontracts will be subject to the prior approval of the other party.

7. Nondisclosure and protection of proprietary data and information.

    Each party will hold the other party's information confidential, unless such information becomes part of the public domain or unless subject to lawful demand (e.g., subpoena).

8. Duration or term of the agreement.

    Termination of the agreement.

9. Noncompete clause.

10. Percentage of the contract to be allocated to the subcontractor.

11. Interfaces with the government.

12. Patentable inventions and software developed pursuant to the work performed as a result of the agreement.

13. Nonproselytizing clause.

    Neither party shall solicit for employment any employee(s) of the other party to work on the project contemplated in the agreement.

14. Conditions and mechanisms for amending, modifying, or extending the agreement.

15. Communication between the two parties.

16. Authorized signatures from both parties with witness.

17. Date of execution.

18. Agreement shall be enforced and interpreted under applicable state laws.

Figure 18.2
Potential topics for teaming agreements.

that your company may want to consider including in your teaming agreements when you serve as the prime contractor. It is advisable to include a specific list of proposal-related information and materials (as presented previously in Table 18.2) as an attachment to your formal teaming agreement. Too often, proposal-related requirements and the specific process of information exchange are poorly defined and result in production challenges and delays throughout the proposal response life cycle. These topics include those for SF330s as well as RFPs and international procurements.

It is strongly suggested that your company's legal counsel review the terms and conditions of all teaming agreements into which you enter.

Epilogue

Thinking to win small-business competitive proposals

The magic is in the people. And so are the results.
—The PeopleWise Organization, Kepner-Tregoe

S MALL FIRMS ARE IN the enviable position of being able to respond rapidly to changing business environments. As a key builder of your company's business future, you should maximize your agility as a small contracting firm and establish as soon as possible in your corporate history appropriate internal and external business development and knowledge management processes and patterns of thinking and behavior that will facilitate winning proposals in the federal, private-sector, and international marketplace. Genuine teamwork is a critical element of success in the contracting arena. Proposalmanship is a capability that needs to be cultivated throughout the levels of your organization in order to fully harness the talent and energy resident there. Many companies struggle because only a few of their staff are trained and experienced in proposal management, design, and development. And often, development and knowledge management are ill-defined and poorly implemented.

Ongoing, positive communication throughout your company is a second measure of your successful business culture. That multidirectional communication extends among the acquisition/capture team and the proposal team, business development staff and technical operations staff, business development staff and your client, proposal manager and senior management, proposal manager and capture manager, proposal manager and proposal writers, and proposal manager and publications staff.

Management support must necessarily assume many different forms. But it must be manifested clearly in order to be fully effective. That support should be present in ensuring that efforts to build appropriate résumé and project summary files (part of an overall knowledge base) on your company's staff and contractual history are met with complete and timely support throughout the technical and programmatic ranks. Management support should take the shape of assisting the proposal manager in enforcing the proposal milestone schedule and of empowering the proposal manager to meet the challenges of a given proposal effort. Ensure qualified leadership with the authority to do the job. And management involvement and support also lie in committing the resources—such as human, financial, equipment, and floor space—to make every proposal your company elects to pursue a superior, client-centered sales document that tells a compelling story. Support for an apprentice system for proposal managers would be well-targeted energy.

As a leader, you are able to infuse an ethos of rational and formal planning into your company's business development and proposal development infrastructure. Expend the time and effort early in your firm's history to develop a mission statement and strategic plan. Generate and have all of your management team follow written, albeit revisable, business development and knowledge management protocol and processes. This applies particularly to bid–no bid decision making. Too many times, small and large businesses dilute their collective resources in pursuing marketing opportunities that do not support their primary lines of business. Planning also extends to developing the proposal directive prior to conducting the formal kickoff meeting. Up-front planning, analysis, and decision making yield significant dividends downstream in the proposal and oral presentation processes.

And finally, recognition and reward. People need to know in a variety of ways that they are meeting your company's business and proposal development expectations. They need to be recognized when they contribute in any one of several ways to a proposal victory. Senior management should develop and implement a definitive recognition and incentive plan that is communicated clearly to everyone in your company. Success should be noted in a big way. Those people in your firm who are thinking to win should be recognized and should receive tangible benefits. Remember that your people and their knowledge are the most important ingredients in your company's success in establishing and sustaining a recognizable brand of excellence in the marketplace.

Appendix A

Sample proposal kickoff package

A WELL-PREPARED proposal kickoff package or proposal directive—which includes important components such as a milestone schedule, annotated outline, page allocations, and key person contact list—is a valuable tool for proposal managers to maximize the value of bringing staff together for the formal kickoff meeting. The kickoff package also serves as a helpful, easy-to-follow reference tool for proposal writers, contracts staff, teaming partners, production staff, and senior management throughout the proposal response life cycle.

KICKOFF PACKAGE
GROUNDWATER TREATMENT SYSTEM
for
Any County, Virginia

Prepared by:

Acquisition Team
XYZ, Inc.

Proposal #:
Department Code:

_____

PROPRIETARY INFORMATION

Agenda

1. Proposal Summary

2. Schedule for Technical and Business Management Volumes

3. Volume Responsibilities

4. Evaluation Factors for Award

5. Technical Outline and Writing Assignments

    a. Themes and Critical Issues

    b. Page Allocations

    c. Proposal Writing Guidelines

6. Business Management Writing Assignments

7. Action Items

8. Items for Management Approval

9. Telephone List

10. Proposal Security

Proposal Summary Sheet

Title: Groundwater Treatment System

Solicitation Number:

Client: Any County, Virginia

1. Proposal Due Date and Time: Technical proposals and bids are due by 5 P.M. on Monday, in Virginia

2. Client Name (Point of Contact):   Contract Specialist
    Telephone #:

3. Period of Performance: 25 weeks + 1 month O&M

4. Copies to Client: 3 copies of the proposal and 1 copy of the bid (p. 8)[1]

---

1   Page number references are from the RFP.

Proposal Team
Procurement Name

Proposal Manager:

Proposal Team Members:

    Technical:

    Program Management:

    Cost:

    Legal:

    QA:

Evaluation Factors for Award
Procurement Name

The evaluation factors are listed below:

(1)	Project Organization	15
(2)	Project Execution	25
(3)	Schedule	15
(4)	Qualifications and Experience	30
(5)	Health and Safety	5
(6)	SB, SDB, or WOB	5
(7)	Financial Resources	5
Total		100

PROPOSAL MILESTONE SCHEDULE

Sunday	Monday	Tuesday	Wednesday	Thursday	Friday	Saturday
			December 28 Meet to discuss the contract scope and proposal team membership.	29 Develop Kickoff Package that includes a milestone schedule and outline.	30 Develop Kickoff Package that includes a milestone schedule and outline.	31
January 1	2	3	4 Develop Kickoff Package that includes a milestone schedule and outline.	5	6 9 A.M.: Meet to review Kickoff Package. 11 A.M.: Proposal Kickoff Meeting.	7 Proposal writing.
8 Proposal writing.	9 Proposal writing.	10 Proposal writing. Proposal Manager to contact all writers to assess status. Writers to provide input to him as sections are complete.	11 Proposal writing.	12 Proposal writing.	13 Proposal writing. Proposal Manager to contact all writers to assess status. Writers to provide input to him as sections are complete.	14

Sunday	Monday	Tuesday	Wednesday	Thursday	Friday	Saturday
22	23	22	25	26	27	28
Address Red Team comments.	Address Red Team comments.	Address Red Team comments. Proposal Manager to provide hardcopy MASTER markup document to Production Services by 4 P.M. Document freezes. No additional changes accepted.	Final production.	Final production.	Final production. QC, phototcopying, and assembly.	
29	30					
	Proposal due in Virginia.					

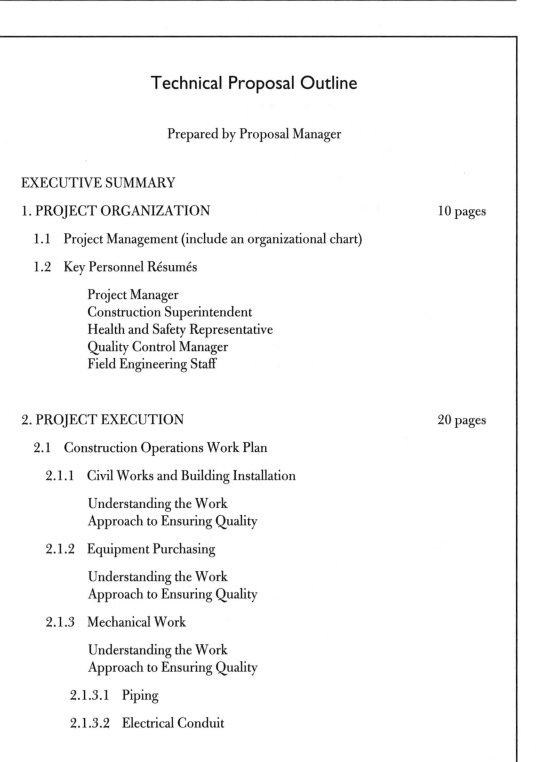

Technical Proposal Outline

Prepared by Proposal Manager

EXECUTIVE SUMMARY

1. PROJECT ORGANIZATION   10 pages

   1.1   Project Management (include an organizational chart)

   1.2   Key Personnel Résumés

         Project Manager
         Construction Superintendent
         Health and Safety Representative
         Quality Control Manager
         Field Engineering Staff

2. PROJECT EXECUTION   20 pages

   2.1   Construction Operations Work Plan

     2.1.1   Civil Works and Building Installation

         Understanding the Work
         Approach to Ensuring Quality

     2.1.2   Equipment Purchasing

         Understanding the Work
         Approach to Ensuring Quality

     2.1.3   Mechanical Work

         Understanding the Work
         Approach to Ensuring Quality

       2.1.3.1   Piping

       2.1.3.2   Electrical Conduit

4.2   Staffing Profile (by trade)

4.3   Case Histories (minimum of 3)
      References, including name, position, and phone
      Company's role and responsibility
      Contract amount
      Date completed/anticipated completion date
      Percentage of contract performed by workforce

5. HEALTH AND SAFETY   10 pages

5.1   Site Safety Plan

5.2   Site Safety Officer (name, qualifications, and résumé)

5.3   Corporate Health and Safety Program

6. BIDDER STATUS   1 page

7. FINANCIAL RESOURCES   5 pages

7.1   Overview of Financial Stability

7.2   Summary of Gross Billings (for past 10 years)

7.3   Bonding Capacity

7.4   History of Work Performance

COMPLIANCE MATRIX

TOTAL PAGE COUNT   65 pages

Action Item
Procurement Name

#	Person Responsible	Assignment	Date Assigned	Date Due	Status
1		Select appropriate résumés.	1/6/05	1/10/05	CLOSED
2		Select appropriate project summaries.	1/6/05	1/10/05	OPEN
3		Complete the bid portion of the proposal.	1/6/05	1/26/05	OPEN

Phone List
Procurement Name

Name:

    Work#:
    Home#:
    FAX#:
    Days on vacation, out of office, or otherwise unavailable:

Name:

    Work#:
    Home#:
    FAX#:
    Days on vacation, out of office, or otherwise unavailable:

Name:

    Work#:
    Home#:
    FAX#:
    Days on vacation, out of office, or otherwise unavailable:

Name:

    Work#:
    Home#:
    FAX#:
    Days on vacation, out of office, or otherwise unavailable:

Name:

    Work#:
    Home#:
    FAX#:
    Days on vacation, out of office, or otherwise unavailable:

Appendix B

Template to capture important résumé information

CRITICAL TO THE SUCCESS of your organization is the ongoing capture of information about your technical and programmatic staff's experience and knowledge base. This appendix provides a sample template for information collection for an IT firm. Having this information readily accessible in electronic form can make the proposal process easier and more cost-effective.

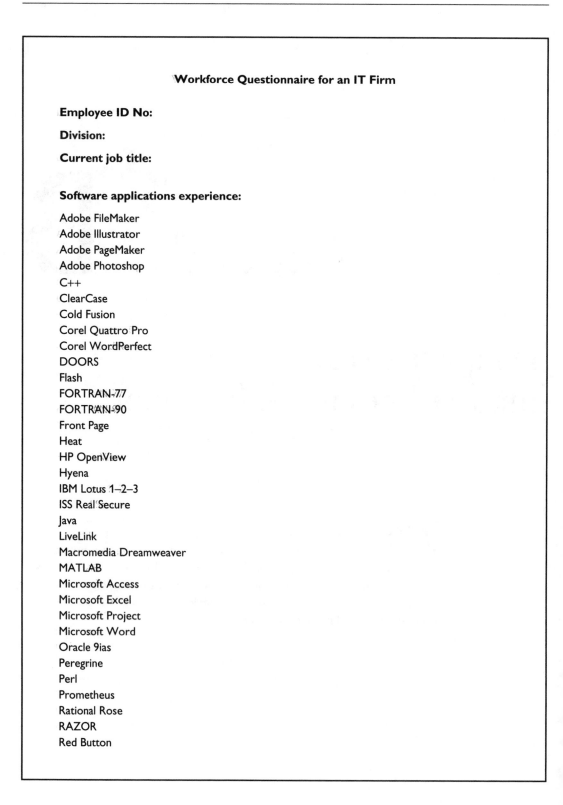

Workforce Questionnaire for an IT Firm

Employee ID No:

Division:

Current job title:

Software applications experience:

Adobe FileMaker
Adobe Illustrator
Adobe PageMaker
Adobe Photoshop
C++
ClearCase
Cold Fusion
Corel Quattro Pro
Corel WordPerfect
DOORS
Flash
FORTRAN-77
FORTRAN-90
Front Page
Heat
HP OpenView
Hyena
IBM Lotus 1–2–3
ISS Real Secure
Java
LiveLink
Macromedia Dreamweaver
MATLAB
Microsoft Access
Microsoft Excel
Microsoft Project
Microsoft Word
Oracle 9ias
Peregrine
Perl
Prometheus
Rational Rose
RAZOR
Red Button

Remedy
Samba
Statistical Analytical Systems (SAS)
Support Magic
TCP/IP
Tivoli
TK Ined
UML
UNIX
Visual Basic
VRML
Windows 2000
XML

Hardware Experience:

Supercomputers
    Cray X1
    Hitachi SR 8000
    IBM
HP workstations
IBM workstations
IBM mainframes
IBM-compatible PC microcomputers
Macintosh
SGI workstations
Sun workstations

Database experience:

Informix
Microsoft Access
Oracle
SQL server
Sybase
Visual C++

Experience supporting or working for government organizations:

Department of Agriculture (USDA)
Department of Defense (DoD)
    Army
    DISA
    DLA
    Navy
        Marine Corps
        NAVAIR

NAVSEA

Office of Naval Research (ONR)

SPAWAR

Office of the Secretary of Defense (OSD)

Department of Energy (DOE)

Department of Homeland Security (DHS)

Department of Justice (DOJ)

Department of Transportation (DOT)

General Services Administration (GSA)

NASA HQ and NASA Centers

Treasury

IRS

Technical areas of expertise:

Applications
CASE tools
Configuration management (CM)
Database Administration
Data analysis and planning
Data quality engineering
Data systems modeling
Geographic Information Systems (GIS)
Graphical user interface (GUI)
Help desk support and end-user support
Information assurance
Integration and test (I&T)
IV & V
Multimedia/graphics/documentation support
Network engineering/management
Numerical analysis
Requirements analysis
Risk assessment
Scientific computing
Software quality assurance (SQA)
Software engineering
Software development
Systems engineering and integration
Telecommunications
Total quality management (TQM)
Training
Web design and maintenance

Project or program management or technical leadership experience:

Maximum number of staff supervised:

Publications: Yes

Knowledge of and experience with industry standards (ISO 9001:2000, IEEE, SEI CMMI®, ITIL): Yes

Security clearance:

Appendix C

Marketing information and intelligence sources: federal, international, and private sector

T HERE ARE THREE MAJOR information sources for federal competitive procurements: (1) publicly available information, (2) client-derived information, and (3) community-derived information. The latter refers to friends, competitors, seminars, subcontractors, vendors, and published scientific or conference reports.

C.1 Sources of federal marketing leads and information

Acquisition forecasts

Public Law 100-656, the Business Opportunity Development Reform Act of 1988, amended the Small Business Act to place new emphasis on acquisition planning. The law requires agencies to annually compile and make available 1-year projections of contracting opportunities that small and small disadvantaged companies can perform. The NASA Acquisition Forecast, for example, includes projections of all anticipated contract actions above $25,000 that small and small disadvantaged businesses can perform under direct contracts with the government or as part of a subcontract arrangement. NASA's Acquisition Forecast is divided into sections for Headquarters in Washington, D.C., and for each Center (Goddard, Marshall, Ames, Lewis, and so forth). Retain these forecasts in your company's proposal data center or library for 3 to 5 years in order to track contracts as they become available for recompetition. Examples of online agency acquisition forecasts include the following:

U.S. Environmental Protection Agency (EPA)

http://yosemite1.epa.gov/oarm/oam/forecastdatabase.nsf

United States Military Academy (USMA)

http://www.usma.edu/doc/FY04ACQPLAN.pdf

Missile Defense Agency (MDA)

http://www.acq.osd.mil/mda/barbb/barbb.htm

NASA Dryden Flight Research Center (DFRC)

http://www.dfrc.nasa.gov/Business/Procurement/forecast.html

NASA Glenn Research Center (GRC)

http://www.lerc.nasa.gov/www/Procure/Acquisition_Forecast.xls

Department of Education

http://www.ed.gov/fund/contract/find/forecast04.pdf

Department of Energy (DOE) e-Center

http://doe-iips.pr.doe.gov

Nuclear Regulatory Commission (NRC)
http://www.nrc.gov/who-we-are/forecast.html

Department of Transportation (DOT)
http://osdbuweb.dot.gov/business/procurement/forecast.html

Department of Treasury
http://www.treas.gov/sba/getsic.html

Department of Veterans Affairs (VA)
http://www.osdbu.va.gov/cgi-bin/WebObjects/FcoPublic.woa

U.S. Department of Agriculture (USDA)
http://www.usda.gov/da/smallbus/sbonline1.htm

Department of Commerce (DOC)
http://www.osec.doc.gov/osdbu/ForecastFY04_FY05.htm

Additional Resources

Aerospace Daily & Defense Report [International Standard Serial Number (ISSN) 0193-4546/90]
1200 G Street, N.W., Suite 922
Washington, D.C. 20005
(202) 383-2350
http://www.Aviationnow.com/Avnow/news/channel_Aerospacedaily.jsp

Aviation Week & Space Technology (AW&ST)
http://www.aviationnow.com/Avnow

Baltimore Business Journal (newspaper)
American City Business Journals, Inc. (Baltimore)
111 Market Place, Suite 720
Baltimore, MD 21202
(410) 576-1161
Editor, Joanna Sullivan
Circ.: 13,500
http://baltimore.bizjournals.com/baltimore

Business Credit (journal)

National Association of Credit Management (NACM)
8840 Columbia 100 Parkway
Columbia, MD 21045-2158
(410) 740-5560
http://www.nacm.org/advertise.shtml

Competitors' newsletters and annual reports

These can be obtained by contacting public relations departments. You may need an outsider such as a college student or consultant to do this.

Contract Management (journal)

National Contract Management Association (NCMA)
8260 Greensboro Drive
Suite 200
McLean, VA 22102
(800) 344-8096
http://www.ncmahq.org/publications/jcm.asp

Contractual exposure

When your company's personnel are on the client's site to perform existing work, they can certainly ask questions.

Defense Technical Information Center (DTIC)

8725 John J. Kingman Road
Ft. Belvoir, VA 22060-6218
(800) 225-3842
http://www.dtic.mil
DTIC is the central point within the DoD for acquiring, storing, retrieving, and disseminating scientific and technical information to support the management and conduct of DoD research, development, engineering, and studies programs. DTIC services are available to DoD and its contractors, and to other U.S. government organizations and their contractors.

Defense Procurement and Acquisition Policy

http://www.acq.osd.mil/dpap/ebiz/index.htm

Department of Defense Telephone Directory

Issued three times a year, the directory contains numbers for the Army, Navy, and Air Force.

The DIALOG Corporation (a business of The Thomson Corporation)

Corporate Headquarters
11000 Regency Parkway
Suite 10
Cary, NC 27511
(800) 3 DIALOG
http://www.dialog.com
The online service called DIALOG is available for use in gathering marketing intelligence. DIALOG can be used to query any of 900 business and commercial, defense, aerospace, academic, career placement, *Federal Register*—File 180, Federal Research in Progress (FEDRIP)—File 266, Grants—File 85, and other databases. The cost for this service is determined by method of access, access time, specific database being used, and method of downloading the data. Searches can be performed, for example, on public or private companies, specific market segments, general subject areas, government policy and research forecasts, competitor acquisitions and contract award history, and general fiscal trends.

Federal Business Council, Inc. (FBC)

(A nonprofit association)
10810 Guilford Road
Suite 685
Annapolis Junction, MD 20701
(800) 878-2940; (301) 206-2940
http://www.fbcinc.com
The Federal Business Council sponsors a variety of business seminars to give your company procurement direction, contacts, and specific knowledge regarding the federal marketplace.

Federal Computer Week

3141 Fairview Park Drive
Suite 777
Falls Church, VA 22042
(703) 876-5100
Circ.: 93,000
http://www.fcw.com
Published 42 times per year, *Federal Computer Week* provides practical news, analysis, and insight on how to buy, build, and manage technology across all segments of federal, state, and local government.

The Federal Marketplace

(888) 661-4094
(208) 726-5553
Fax: (208) 726-5590
http://www.fedmarket.com

The Federal Marketplace provides assistance to companies in marketing and selling products and services to the federal government as well as state and local governments and international governments and organizations. This is a must-see Web site. Visit it regularly!

Fedmarket.com is an online community for government buyers and vendors. The site is owned and operated by Wood River Technologies, Inc., a private company located in Sun Valley, Idaho. In 1998, the company developed *Bidengine* and this led to the development of *ITBids,* a more refined government bid collection and distribution system for the IT industry. Fedmarket.com maintains and updates a database of more than 15,000 federal, state, local, and international procurement URLs and has a proprietary system in place to capture and catalog the hundreds of new bid notices posted daily at government sites. This capability puts fedmarket.com in a position to provide business opportunity content to its subscribers and vertical industry portal clients.

In 2000, fedmarket.com launched govcommerce.net, a safe, secure, and easy-to-use purchasing system for government buyers and vendors. Buyers pay nothing to use the service; winning vendors pay a 2% transaction fee. Large sales-volume vendors can have transaction fees as low as 1%. Govcommerce.net provides government sales support for your company if you do not have a government marketing staff. The organization actively promotes the Web site and your products and services to the government. Partnering with industry experts Michael Asner and Dan Safford, the company also launched Proposalworks.com. This resource center for proposal writers and evaluators includes a fully searchable, indexed library of best practices, evaluator guidelines, and actual winning proposals.

Federal Register

http://www.gpoaccess.gov/fr/index.html

Official daily newspaper that informs of proposed, interim, or final regulations issued by the federal government. It is in the *Federal Register* that the FAR receives its official publication. Changes to the FAR are published as *Federal Acquisition Circulars* (FACs). Careful attention to the *Federal Register* will help to identify program changes and new priorities.

Federal Yellow Book 2001 (ISSN 0145-6202)

Leadership Directories, Inc.
1001 G Street, NW
Suite 200 East
Washington, D.C. 20001
(202) 347-7757
http://www.leadershipdirectories.com/fyb.htm
The *Federal Yellow Book* is an organizational directory of the departments
and agencies of the executive branch of the federal government. It provides
organizational position, address, and telephone number for more than
40,000 senior people.

Freedom of Information requests

Freedom of Information Act (5 U.S.C. § 552). Consider asking for only
one item per written request to avoid delay if a particular item undergoes
further legal review or is denied.

GovCon (Web site)

http://www.govcon.com
This site offers market research reports, proposal-related books, and a
"Find a Consultant" service.

Government Information Quarterly (ISSN 0740-624X)

JAI Press
Greenwich, CT
http://www.sciencedirect.com/science

Minority business associations (for example, Women's Business Center of Northern Virginia; http://www.wbcnova.org)

The Nash & Cibinic Report

http://west.thomson.com/store/product.asp?product_id=15864305
A monthly report of analysis and advice on U.S. government contracts pre-
pared by Professors Ralph C. Nash, Jr., and John Cibinic, Jr., of George
Washington University.

National Association of Small Disadvantaged Businesses (NASDB)

Mr. Henry T. (Hank) Wilfong, Jr., President
P.O. Box 13603
Silver Spring, MD 20911
(301) 588-9312
http://www.nasdb.org/index.phtml

National Small Business United (NSBU)

1156 15th Street, NW, Suite 1100
Washington, D.C. 20005
(202) 293-8830
http://www.nsbu.org
Todd O. McCraken, President
NSBU keeps small business owners in touch with legislative and regulatory issues that affect them.

National Technical Information Service (NTIS)

Sills Building
5285 Port Royal Road
Springfield, VA 22161
(703) 605-6000
http://www.ntis.gov
NTIS is a part of the U.S. Department of Commerce Technology Administration and is an excellent source for U.S. and foreign government-sponsored research and development (R&D) results, business information, engineering solutions, and marketing reports. Complete technical reports for most government-sponsored R&D and engineering activities are provided through NTIS. The Service also makes available U.S. Army and Air Force manuals and regulations as well as a wide variety of software, data files, and databases, and information on federal laboratory inventions and technologies.

SBA's Office of Women's Business Ownership

http://www.sba.gov/financing/special/women.html

Space Business News

Phillips Business Information, Inc.
1201 Seven Locks Road, Suite 300
Potomac, MD 20854
PBI@PHILLIPS.COM
Space Business News covers global developments in civil, military, and commercial space. Provides analysis of news and trends in this competitive marketplace.

Trade association meetings

Trade association meetings are announced in trade journals and newsletters as well as on the Internet.

U.S. General Services Administration (GSA)

https://www.fpds.gov (Next Generation Federal Procurement Data System)

An *FPDS*[1] *Federal Procurement Report* is published annually and is free of charge. This type of report contains information on products and services purchased by 60 agencies within the federal government, arranged in the following manner: (I) Total Federal View, (II) Geographic View, and (III) Agency View. Section I includes federal procurement information by NAICS code as well as the top 100 federal contracting firms, ranked by contract dollars. Section II presents federal procurement spending within each state, subdivided by such elements as the type of product or service and type of business. Finally, Section III offers a detailed agency-by-agency listing of procurement activities, subdivided by such parameters as type of contract, product or service, type of business/contractor, and competitive or noncompetitive. Because the report contains statistics on the procurements of 60 federal agencies, your company can utilize this information to build information on market size. From this, you can begin to establish target values for market share. A private contactor, Global Computer Enterprises, Inc., took over the FPDS operations in 2004.

U.S. General Services Administration

Government Wide Information Systems Division
Federal Procurement Data System (MVSC)
U.S. General Services Administration
1800 F Street, NW
Room 4020
Washington, D.C. 20405
(202) 219-3416

Washington Business Journal (ISSN 0737-3147)

American City Business Publications, Inc. (Arlington)
1555 Wilson Blvd., Suite 400
Arlington, VA 22209-2405
(703) 875-2200
http://washington.bizjournals.com/washington

---

1   FPDS is an acronym that stands for Federal Procurement Data System, which was designed under the guidance of the Office of Federal Procurement Policy (OFPP). The FPDS is the central repository of detailed information on federal contracts more than $25,000.

Washington Technology (ISSN 1058-9163)
Post Newsweek Tech Media
10 G Street, NE
Washington, D.C. 20002
(202) 772-2500
http://www.washingtontechnology.com
Washington Technology is published semimonthly. A supplemental publication is *Washington Technology's 8(a) and Small Business Report*, which appears twice per year. This publication offers news and information important to minority-owned businesses, including how to establish relationships with federal agencies, 8(a) program upgrades, and financing opportunities for 8(a) companies.

Sources of international marketing leads and information

From the U.S. federal government

- Export America; http://www.export.gov;
- U.S. Agency for International Development (U.S. AID); Washington, D.C.; http://www.usaid.gov;
- U.S.-Asia Environmental Partnership (U.S.-AEP); Washington, D.C.; http://www.usaep.org;
- U.S. Environmental Protection Agency (U.S. EPA); Washington, D.C.; http://www.epa.gov
- U.S. Trade and Development Agency (USTDA); Washington, D.C.; http://www.tda.gov.

From state-level organizations (example for Maryland)

- Office of International Business (OIB) (http://www.choosemaryland. org/international/index.asp);
- World Trade Center Institute; Baltimore, Maryland (http://www. wtc.org);
- Maryland Sister States Program; Baltimore, Maryland (http://www. sos.state.md.us/International/MDSSP.htm).

From other organizations and media

- Multilateral Banks: World Bank, Asian Development Bank (ADB), and Inter-American Development Bank (IDB); Washington, D.C.;
- Border Environmental Cooperation Commission (BECC), headquartered in Chihuahua, Mexico;
- Canada-China Business Association (CCBA), headquartered in Richmond, British Columbia (http://www.ccba.bc.ca/Page_CCBA_English_Main.htm);
- Maryland/Israel Development Center (http://www.marylandisrael.org);
- Canada China Business Council (CCBC) (http://www.ccbc.com);
- U.S.-China Business Council, Washington, D.C. (http://www.uschina.org);
- *The China Business Review* (*CBR;* http://www.chinabusinessreview.com); official magazine of the U.S.-China Business Council;
- *China Daily,* published in the Peoples' Republic of China, distributed from New York (http://www.chinadaily.com.cn);
- *Diario Oficial de la Federatción*, Mexico City, Mexico (Mexico's Federal Official Journal); contains information about bidding opportunities in Mexico;
- World Bank Monthly Operational Summary (MOS), Washington, D.C.;
- U.S. Trade and Development Agency (USTDA), *Pipeline* newsletter published biweekly, Washington, D.C. (http://www.tda.gov/pipeline/index.html);
- *Bu$iness Mexico* (magazine), American Chamber Commerce/Mexico, available through Amazon.com;
- *The Economist*, London, United Kingdom;
- The Economist Intelligence Unit (EIU), London—leading provider of country-specific market trends and business strategies (http://www.eiu.com).

C.3 Sources of U.S. private-sector marketing leads and information

El Digest: Hazardous Waste & Toxic Substance Management

Environmental Information, Ltd.
P.O. Box 390266
Edina, MN 55439
(952) 831-2473
Fax: (952) 831-6550
http://www.envirobiz.com/Digest.htm
The *EI Digest* subscription-based research service is an excellent source of business research focusing on the management of hazardous waste and toxic substances.

The PEC Report

Industrial Information Resources, Inc. (IIR)
6161 Savoy Lane, Suite 150
Houston, TX 77036
(713) 783-5147
http://www.industrialinfo.com
A leading source for market intelligence on industrial projects that are being planned, engineered, and constructed in the industrial process, manufacturing, and energy markets.

Appendix D

Glossary of proposal-related terms

A-76   OMB Circular A-76, revised May 29, 2003, is a set of policies and procedures designed to help determine whether public or private sources will undertake the federal government's commercial activities and services, ranging from software consulting and research and lab work to facilities management. A-76 is, in effect, a management tool geared toward providing the best value to the American taxpayer and improving the performance of government. The Office of Federal Procurement Policy (OFPP) is the "owner" for A-76 studies, as per the May 2003 revision to the circular. The roots of Circular A-76 can be traced from the former Bureau of the Budget's Bulletin 55-4 (issued January 15, 1955), which stated that the federal government would "not start or carry on any commercial activity" that the private sector could do. Revisions have been made periodically ever since. A-76 is a federal government management tool. The A-76 competition provides an opportunity for government managers to streamline organization, implement best business practices, increase productivity, enhance quality, increase efficiency of operations, lower operational costs, and adjust IT initiatives to new regulatory drivers such as the President's Management Agenda. The A-76 study process focuses on:

- Government/customer requirements;
- Organizational structure;
- Work processes;
- Defined outcomes;
- Competition.

Historically, the government wins 50% to 84% of competitions. But regardless of who wins, 30% to more than 40% savings are achieved.

Action title   Figure caption that conveys a sales message. Captions should emphasize themes. Example: Well-Defined Interfaces at the Project Manager/COTR Level Will Ensure Mutual Understanding of Changing Requirements.

Architect-Engineer Contract Administration Support System (ACASS)   ACASS is a database of selected information on architect-engineering (A-E) firms. ACASS provides required information for DoD and federal government selection committees to aid them in their process of awarding A-E contracts. It is maintained by the Portland, Oregon, office of the U.S. Army Corps of Engineers (USACE). It contains detailed Standard Form 330 and historical SF254/255 data on contractors, DoD awards, and past performance evaluations for the last 6 years of A-E services contracts. For more information on ACASS, go to https://www.nwp.usace.army.mil/ct/I/welcome.htm.

ACASS provides an efficient and economical automated database system by which to adhere to these requirements. The use of ACASS for past performance history is mandated for all DoD agencies in the Defense Federal Acquisition Regulation Supplement (DFARS). For the Corps of Engineers, ACASS eliminated the need for 43 district offices to disseminate performance evaluations among each other and to maintain copies of SF254s. ACASS is a catalyst for overall improvement in performance of the A-E community. If the performance evaluation is to be rated "below average" or "poor," the A-E firm will receive notification of such prior to the evaluation being signed by the reviewing official. The firm then has 15 days to appeal the proposed rating.

Armed Services Procurement Act (ASPA)   Along with FPASA, this act covered most areas of government procurement (10 U.S.C. § 2301 et seq, 1948).

Base year   The first year of a given contract.

Basis of estimate (BOE)   The basic purpose of the BOE is to sell your company's technical approach, staffing, and cost to the client. The BOE is a nontechnical communication mechanism to present an argument in favor of expending a stated quantity of resources to accomplish a given task. It is documentation of how an estimate was developed and includes your company's assumptions that impact cost, a description of the methodology used in developing the estimate, and justification for using a particular methodology. BOEs must reflect an internal consistency among the Technical, Management, and Cost Volumes of your proposal. And BOEs must be traceable to task descriptions, the SOW, and the WBS. BOEs are most appropriate when the government has not specified hours or any quantitative level of effort (LOE).

Instead of merely stating that Task A will require 5 staff-months of Labor Category 1 personnel and 8 staff-months of Labor Category 4 personnel, a BOE-based approach might build on the following:

> *Task A requires 5 staff-months of Labor Category 1 personnel and 8 staff-months of Labor Category 4 personnel because, on a similar task performed in 2004 under our XYZ-11 contract with Marshall Space Flight Center, this ratio of senior to junior labor proved to be extremely effective in bringing the work to a successful, in-budget conclusion. Such a labor mix will meet all task requirements as discussed in our Technical Approach section of this proposal.*

BOEs may be substantiated by historical or current corporate contractual or commercial work, trade-off studies (particularly in the case of design-oriented procurements), vendor quotes, purchase orders, and so forth. Show your calculations, as you would on a math test, and document your references, as you would in a technical or scientific paper. Your client's source selection board needs quantifiable data to evaluate your proposal. Make sure that you would buy your products or services based on the data you have provided. (Requirements for submission of BOEs are established by FAR 15.804-6, Table 15-2.)

Best value   Best value is a process used in competitive negotiated contracting to select the most advantageous offer by evaluating and comparing factors in addition to cost or price. Best value allows offerors flexibility in selection of their best proposal strategy through trade-offs that may be made between the cost and noncost evaluation factors. It should result in an award that will give the government the greatest, or best value, for its money. Executive Order 12931 issued on October 13, 1994, directs executive agencies to "place more emphasis on past performance and promote best value rather than simply low cost in selecting sources for supplies and services" (Section 1, Item d). Best value has emerged as the centerpiece of acquisition reform.

Bidders conference   See Preproposal conference.

Bid opening date (BOD)   The final date that a bid must be received by the appropriate government office.

Blackout   Cessation of information exchange on the part of the client once the RFP is announced in FedBizOpps or is actually released.

Black Team   Low-profile, highly confidential contractor activity designed to assess the competition.

Blanket purchase agreement (BPA)   This agreement between the government and a vendor gives the government the option to purchase goods or services from the vendor when needed on an on-call basis. BPAs are a method of filling anticipated repetitive needs for supplies or services by establishing "charge accounts" with qualified sources of supply, and may include federal supply schedule contractors. Under such an agreement, the contractor and the agency agree to contract clauses applying to future orders between the parties during its term. Future orders would incorporate, by reference or attachment, clauses covering purchase limitations, authorized individuals, itemized lists of supplies or services furnished, date of delivery of shipments, billing procedures, and discounts. Under the FAR, the existence of a blanket purchase agreement does not justify purchasing from only one source or avoiding small business preferences.

Bogey   Ballpark dollar amount used for preliminary costing.

Boilerplate   Frequently used, standard materials that go into proposals, such as phase-in plans, configuration management plans, quality assurance plans, project descriptions and summaries, professional compensation plans, make-or-buy policy, and so forth. All boilerplate, however, should be tailored and customized for the specific proposal. As Tom Hewitt, founder of Federal Sources, has noted, "Boilerplate Loses!" Federal

Sources is a research firm based in McLean, Virginia, which tracks the procurement industry.

Brooks Act (ADPE Contracting) (1965) 40 U.S.C. § 759. This act (P.L. 92-582) granted exclusive purchasing authority for automatic data processing equipment (ADPE) to the Administrator of the General Services Administration (GSA), subject to the fiscal control of the Office of Management and Budget (OMB). This statute is implemented by the Federal Information Resources and Management Regulation (FIRMR), 41 CFR 201-2.001 and Part 201-23, which supplements the FAR. (Cibinic, John, and Ralph C. Nash, *Formation of Government Contracts*, 2nd ed., Washington, D.C.: George Washington University, 1986, p. 15.) The Clinger-Cohen Act rescinded the Brooks Act.

Business information center (BIC) The U.S. Small Business Administration's 80 BICs provide small businesses with a one-stop location for assistance and advice. Each BIC offers self-help software, reference materials, interactive media, and on-site counseling through SCORE. Call (800) 8-ASK-SBA for the location nearest your office or visit http://www.sba.gov/bi/bics on the Web.

Call plan Document designed to clearly indicate who within your company will be visiting the client, as well as when, why, and at what level of the client organization.

Capability Maturity Model Integration (CMMI®) CMMI® is a comprehensive process maturity framework that combines the Capability Maturity Model® (CMM®) for software with broader disciplines in systems engineering and product development. CMMI® best practices enable organizations to more explicitly link management and engineering activities to business objectives. See http://www.sei.cmu.edu/cmmi.

Capture plan (or strategy) Formalized, documented sequence of activities that includes the formation of an Acquisition/Capture Team, the generation of a call plan, and actual proposal design and development. A capture plan is directed to the acquisition of a specific business target. Prepared early in the proposal life cycle, the capture plan should include a call plan to define and orchestrate frequent contact with the client at all levels—from senior executive management to business development staff to all appropriate technical ranks. Specifically, the call plan is designed to gather intelligence about the client, the particular procurement, and the competition. Information collected must then be analyzed by members of the acquisition team, built into the proposal directive, and eventually factored into your company's proposal where appropriate.

Central Contractor Registration   The Central Contractor Registration (CCR) site (http://www.ccr.gov) is the primary vendor database for the U.S. federal government. The CCR collects, validates, stores, and disseminates data in support of agency acquisition missions. Both current and potential federal government vendors are required to register with CCR in order to be awarded contracts by the U.S. government. The CCR Web portal is also valuable for searching for small business sources.

Clinger-Cohen Act (CCA) of 1996   Public Law 104-106, signed into law on February 10, 1996, and also known as the Clinger-Cohen Act, was driven by the federal government's growing reliance on information technology (IT) and the resulting increased attention and oversight on its acquisition, management, and use. Building on FASA, the Clinger-Cohen Act provided the statutory foundation to streamline IT acquisitions and minimize layered approvals. The act emphasized accountability, outcomes-based performance, and results-based IT management. It promoted the improved performance of the civilian agency acquisition workforce. And it allowed contracting officers to select competitive contractors more efficiently. In addition, Clinger-Cohen required federal agencies to use a disciplined capital planning and investment control (CPIC) process to acquire, use, maintain, and dispose of information technology. The use of commercial-off-the-shelf (COTS) products was to be maximized. The Federal Acquisition Reform Act (FARA) and the Information Technology Management Reform Act (ITMRA) are parts of the Clinger-Cohen Act.

The Clinger-Cohen Act rescinded the Brooks Act (Public Law 92-582), also known as Qualifications Based Selection (QBS), which was enacted on October 18, 1972. This 1972 act had established the procurement process by which architects and engineers (A-Es) were selected for design contracts with federal design and construction agencies. The Brooks Act had also established a qualifications-based selection process, in which contracts for A-Es were negotiated on the basis of demonstrated competence and qualification for the type of professional services required at a fair and reasonable price.

Commercial and government entity (CAGE)   A five-character code used to support a variety of automated systems throughout the government. The Defense Logistics Information Service (DLIS) Center in Battle Creek, Michigan, is the only source for assignment and maintenance of CAGE codes. To receive a CAGE code, you must apply online at the Central Contractor Registration Web site (http://www.ccr.gov).

Competition in Contracting Act (1984)   CICA, P.L. 98-369, July 18, 1984. This act provides for increased use of competitive procedures in contracting for procurement. CICA modified both FPASA (1949) and ASPA (1947). (Cibinic, John, and Ralph C. Nash, *Formation of Government Contracts*, 2nd ed., Washington D.C.: George Washington University, 1986, pp. 13–14, 287.)

Competitive range   Proposals deemed within the competitive range are technically sufficient and exhibit a reasonable relationship between cost or price and the product or service to be procured. Proposals with a reasonable chance of being selected for contract award are deemed within the competitive range.

Consolidated contracting initiative (CCI)   A management process that emphasizes developing, using, and sharing contract resources to meet government agency objectives. CCI emphasizes development of contracts that can be used by other government procurement offices and agencies. The goals of CCI are to reduce government contracting staff's time spent on acquisition-related tasks, shorten acquisition lead times, minimize redundant contracts, and improve cooperation with other government agencies.

Contracting officer (CO)   The contracting officer is the person with authority to enter into, administer, and/or terminate contracts and make related determinations and findings. The term includes certain authorized representatives of the contracting officer operating within the limits of their authority as delegated by the contracting officer.

Contractor Performance Assessment Reporting System (CPARS) A Web-enabled application that collects and manages the library of automated CPARS. CPARS is for unclassified use only. Classified information is not to be entered into this system. A CPARS report assesses a contractor's performance and provides a record, both positive and negative, on a given contract during a specific period of time. Each assessment is based on objective facts and supported by program and contract management data, such as cost performance reports, customer comments, quality reviews, technical interchange meetings, financial solvency assessments, construction and production management reviews, contractor operations reviews, functional performance evaluations, and earned contract incentives.

Cost analysis   Cost analysis is the review and evaluation of the separate cost elements and profit/fee in an offeror's proposal (including cost or pricing data or information other than cost or pricing data). Cost analysis is also determining how well the proposed costs represent what the cost of the contract should be, assuming reasonable economy and efficiency.

Cost realism analysis   Cost realism is the process of independently reviewing and evaluating specific elements of each offeror's proposed cost estimate to determine if the estimated proposed cost elements are realistic for the work to be performed, reflect a clear understanding of the requirements, and are consistent with the unique methods of performance described in the offeror's technical proposal.

Data Universal Numbering System (DUNS)   A unique nine-digit identification sequence that is now standard for all U.S. federal government e-commerce transactions. It is the principal contractor identification code. To apply for a DUNS number, visit Dun & Bradstreet online at http://www.dnb.com.

DD Form 254   Department of Defense Contract Security Classification Specification.

Defense Contract Audit Agency (DCAA)   Performs contract audit functions required by the DoD and many civilian agencies. Determines the suitability of contractor accounting systems for government work, evaluates the validity of proposed costs, and also reviews the efficiency of contractor operations. Negotiated solicitations contain a clause that authorizes the government Contracting Officer (CO) or his representative to examine the contractor's records under circumstances prescribed in FAR 52.215-2. The DCAA auditor is generally the CO's representative. (See McVay, Barry L., *Proposals That Win Federal Contracts: How to Plan, Price, Write, and Negotiate to Get Your Fair Share of Government Business,* Woodbridge, VA: Panoptic Enterprises, 1989, p. 16.)

Desktop publishing (DTP)   Desktop publishing refers to a method by which text and graphic materials of publishable quality are produced under author control (as opposed to outside printing company control) at relatively low cost. DTP can introduce higher quality, more professional documentation in your company's publication group. Desktop publishing software applications include Corel Ventura and Adobe PageMaker.

Discriminators   Authenticated, conclusive statements that demonstrate clearly the positive differences between your company and the competition, and demonstrate the benefits of those differences to your client. Also known as competitive or unique themes.

Electronic commerce (EC)   The paperless exchange of business information, using electronic data interchange (EDI), electronic mail, electronic bulletin board systems (BBS), electronic fund transfer (EFT), and other related technologies.

Electronic Government (E-Government) Act of 2002   Public Law 107-347, 44 U.S.C. Ch. 36, the Electronic Government Act, was signed into law December 17, 2002. This act, which advocates a more citizen-focused approach to current governmentwide IT policies and programs, was designed in part to institutionalize the President's Management Agenda (PMA). This law requires federal agencies to develop performance measures for implementing e-government. This act also requires agencies to conduct governmentwide e-government initiatives and to leverage cross-agency opportunities to leverage e-government. The Federal CIO Council, originally established in 1996 by Executive Order 13011, was codified under the E-Government Act. This council is comprised of the CIOs of 29 federal agencies as well as representatives from the Office of Management and Budget. Its charter is to implement elements of GPRA, FISMA, and ITMRA. The Federal CIO Council is the principal inter-agency forum to assist CIOs in meeting the goals of the PMA.

Excluded parties list (EPLS)   List of parties excluded from participating in federal procurement programs.

Export Working Capital Program (EWCP)   An SBA program that provides short-term, transaction-specific financing for certain loans.

Federal Acquisition Reform Act (FARA) of 1996   FARA (Division D of P.L. 104-106; 40 U.S.C. 1401) was signed into law on February 10, 1996. This act enabled the federal procurement system to emulate many of the most successful buying practices used in the commercial marketplace. FARA focused on reforming the way in which the federal government makes larger dollar purchases and acquires IT. FARA increased the discretion of federal contracting officers in making competitive range determinations and lowered the approval levels for justification and approvals resulting in efficient competition. FARA also permitted the use of SAP in the acquisition of commercial items up to $5 million.

Federal Acquisition Regulations System   Established for the codification and publication of uniform policies and procedures for acquisition by all executive agencies. The Federal Acquisition Regulations System consists of the FAR, which is the primary document, and agency acquisition regulations that implement or supplement the FAR. The FAR

System is articulated in Title 48 of the CFR. The FAR is organized into Subchapters a–h, Parts (of which there are 53), Subparts, Sections, and Subsections. Example: 25.108-2, Part 25, Subpart 1, Section 08, and Subsection 2. The FAR is published by the GSA FAR Secretariat and maintained by the DoD, the General Services Administration (GSA), and NASA under several statutory authorities of those agencies. Any critical understanding of the FAR must include the protest decisions of the Comptroller General and the General Services Board of Contract Appeals (GSBCA).

The FAR is available online at http://www.arnet.gov/far/.

Federal Acquisition Streamlining Act (FASA) of 1994   FASA (Title V, FASA VP.L. 103-355) was signed into law October 13, 1994. It required federal agencies to establish cost, schedule, and measurable performance goals for all major acquisition programs and to achieve, on average, 90% of those goals. FASA focused largely on the purchase of commercial items and smaller dollar buys (those under $100,000). Importantly, the act exempted commercial items from many unique government requirements. Acquisitions of more than $2,500 but not exceeding $100,000 were reserved for small businesses. Under FASA, contracting officers were encouraged to use approaches in awarding contracts that leveraged SAP. In addition, FASA gave agencies statutory authority to access the computer records of contractors doing business with the federal government. The act also placed greater emphasis on the use of past performance when selecting a contractor. Finally, FASA encouraged the use of electronic commerce and established the statutory framework for task- and delivery-order contracting.

Federal Activities Inventory Reform Act (FAIR) of 1998   Legislation that requires federal government agencies to identify functions that could be performed by the private sector.

Federal Business Opportunities (FedBizOpps)   http://www. FedBizOpps.gov is the single point-of-entry portal for federal government procurement opportunities of more than $25,000. Government buyers are able to publicize their business opportunities by posting information directly to FedBizOpps via the Internet. Through one portal—FedBizOpps—commercial vendors seeking federal markets for their products and services can search, monitor, and retrieve opportunities solicited by the entire federal contracting community.

Federal Enterprise Architecture (FEA)   To facilitate efforts to transform the federal government into one that is citizen centered, results oriented, and market based, the Office of Management and Budget (OMB) is developing the Federal Enterprise Architecture (FEA), a business-driven and performance-based framework to support cross-agency collaboration, transformation, and governmentwide improvement. Begun on February 6, 2002, the FEA is being constructed through a collection of interrelated "reference models" designed to facilitate cross-agency analysis and the identification of duplicative investments, gaps, and opportunities for collaboration within and across federal agencies. FEA provides OMB and other federal agencies with a new way of describing, analyzing, and improving the federal government and its ability to serve the American citizen. The outcome of this effort will be a more citizen-centered, customer-focused government that maximizes technology investments to better achieve mission outcomes.

Federal Information Resources Management Regulation (FIRMR)   Governmentwide regulations that govern the purchase of computer goods and services (41 CFR Chapter 201).

Federal Property and Administrative Services Act (FPASA)   This act covered most areas of government procurement along with the Armed Services Procurement Act (41 U.S.C. § 252; 1949).

Federal Stock Number (FSN)   A code used to identify documents sold by the U.S. Government Printing Office, Superintendent of Documents.

Federal Supply Schedules   Under the Schedule program, GSA enters into indefinite-delivery, indefinite-quantity contracts with commercial firms to provide commercial goods and services governmentwide at stated prices for given periods of time. Authorized buyers at agencies place separate orders for individual requirements that specify the quantity and delivery terms associated with each order, and the contractor delivers products or services directly to the agency. The program is designed to provide federal agencies with a simplified process for obtaining millions of commonly used commercial supplies and services at prices associated with volume buying. The program consists of single award Schedules, in which GSA awards contracts to multiple companies supplying comparable services and products, often at varying prices. When agency requirements are to be satisfied through the use of multiple award schedules, the small business provisions (such as the exclusive reservation for small businesses for contracts over $2,500 up to $100,000) of the FAR do not apply.

Federal Supply Service (FSS)   An organization within the U.S. General Services Administration.

Fee determination official (FDO)   A government official who determines amount of award fee on a contract based upon contractor performance.

File Transfer Protocol (FTP)   A method of transferring files to and from remote locations.

Foldout page   An 11 × 17–inch double page, usually graphical in nature. Certain RFPs limit the number of foldout pages in a proposal. Foldouts are important space-savers for page-limited proposals.

Freedom of Information Act (FOIA)   Enacted in 1966, this act provides that select information is to be made available to the public either by publication in the *Federal Register*, providing an opportunity to read or copy records, or providing copies of records (5 U.S.C. 552, as amended by the Electronic Freedom of Information Act Amendments of 1996).

G-2   Intelligence; information on competitors, contracts, and so forth.

Gantt chart   A bar chart named after American engineer Henry L. Gantt who devised such a presentation in 1917. Sometimes referred to as project time lines, Gantt charts are commonly used scheduling charts. Gantt charts ordinarily have a list of tasks down the left side (vertical axis). A bar or line on the horizontal axis shows the date when each task begins and ends. Gantt charts are useful for envisioning an entire project over time.

Ghosting   Lines of argumentation used in proposal text to discredit or play down the competitor's strong points or emphasize weak areas. For example, suppose your competitor has demonstrated an inability to deliver technical progress reports in a timely manner. In your proposal—without naming the competitor—you would emphasize your commitment to and success with *on-time deliverables*.

Government Information Security Reform Act (GISRA) of 2000
This act brought together existing IT security requirements from the Paperwork Reduction Act of 1995 and the Clinger-Cohen Act of 1996. GISRA also codified existing OMB IT security policies found in OMB Circular A-130 and IT security budget guidance in OMB Circular A-11, "Preparation, Submission, and Execution of the Budget." Specifically, GISRA directed agency CIOs to conduct annual IT security reviews of their systems and programs. Review results are to be reported to OMB. After GISRA expired in November 2002, the Federal Information Security

Management Act (FISMA) was signed into law by President George W. Bush as Title III, part of the Electronic Government Act of 2002. FISMA permanently reauthorized the framework established by GISRA.

Government Performance Results Act (GPRA)   This act holds federal agencies accountable for achieving performance results. Enacted in 1993, it legislates that federal agencies are to be more responsive and accountable to the public/customers relative to achieving program results. GPRA requires that agencies develop comprehensive strategic, business, and performance plans documenting the organizational objectives, goals, strategies, and measures for determining results. The organizational plans and measures for achieving the goals are key elements in determining an agency's need for services (P.L. 103-62, 1993).

Government Wide Acquisition Contract (GWAC)   GWACs are critical federal contractual vehicles for small, midsized, and large businesses. They are becoming the vehicle of choice for many government agencies. Much of the procurement paperwork, price justifications, and contractor selections can be done up front in the GWAC procurement process. Individual purchases can then be made from a small cadre of qualified vendors quickly through the use of a task order competition or purchase order. Examples of important current and upcoming GWAC vehicles include GSA STARS, GSA Alliant, NIH CIO—SP2*i*, Department of Transportation ITOP II, and NASA SEWP III.

Greening   Insertion of junior-level staff in the labor mix to reduce life-cycle costs.

HUBZone   The Small Business Administration (SBA) Historically Underutilized Business Zone (HUBZone) Empowerment Contracting Program (http://www.sba.gov/hubzone) provides federal contracting assistance for HUBZone-certified small business concerns located in qualified business communities in an effort to increase employment opportunities, investment, and economic development. The program provides for set-asides, sole-source awards, and price evaluation preferences for HUBZone small business concerns. In order to be considered for a contract award, the small business concern must be certified by the SBA as a HUBZone small business concern and appear on the SBA list of HUBZone small business concerns. Nationally, HUBZones are located in more than 7,000 urban census tracts, in 900 rural counties, and on every federally recognized Native American reservation.

The HUBZone Empowerment Contracting Program refers to Historically Underutilized Business Zones and was created in 1997 as a result of legislation sponsored by Senator Christopher "Kit" Bond, the ranking member of the Senate Small Business Committee. The federal government has a 2% goal of channeling federal contracts through the HUBZone program, which equated to about $4 billion in FY2001.

For a small business to be HUBZone certified, its principal office must be located in a HUBZone and at least 35% of its employees must reside in one of these specially designated areas. In addition, the firm must be owned and controlled by U.S. citizens, a community development company, or Indian tribe. Participants in the program can receive contracts through competition limited to qualified HUBZone firms or on a sole-source basis. HUBZone firms are also given a price preference in bidding during full and open competition over non-HUBZone large firms.

Hypermedia   Richly formatted documents that contain a variety of information types, such as textual, image, movie, and audio.

Hypertext Markup Language (HTML)   Coding mechanism used to author Web pages. HTML is a subset of SGML.

Hypertext Transport Protocol (HTTP)   The protocol, or planned method, for exchanging text, graphic images, sound, video, and other multimedia files used by Web servers.

Incumbent   The contractor or contractor team that is currently performing on a contract that is approaching recompetition.

Indefinite-delivery, indefinite-quantity contract   This type of contract provides for an indefinite quantity, within stated limits, of goods and services during a fixed period of time. Agencies place separate task or delivery orders for individual requirements that specify the quantity and delivery terms associated with each order. The Federal Acquisition Regulation (FAR) expresses a preference for multiple awards of these contracts, which allows orders to be placed using a streamlined, commercial style selection process where consideration is limited to the contract awardees. The competition between the multiple awardees is designed to encourage better prices and responses than if the agency were negotiating with a single contractor. Contractors are to be afforded a fair opportunity to be considered for award of task and delivery orders but cannot generally protest the award of such orders. Indefinite-delivery, indefinite-quantity contracts include GWACs and GSA Federal Supply Schedule contracts.

Information Technology Management Reform Act (ITMRA)
IMTRA (Division E of P.L. 104-106) was signed into law on February 10, 1996. This act enables the federal procurement system to emulate many of the most successful buying practices used in the commercial marketplace. It is focused on reforming the way in which the federal government makes larger dollar purchases and acquires IT, and it establishes the role, duties, and qualifications of the chief information officer within federal agencies.

Information Technology Omnibus Procurement (ITOP) II   A multiple-award, indefinite-delivery, indefinite-quality (MA/IDIQ) contractor vehicle designed to provide federal agencies with fast and efficient total information technology solutions including technical services, hardware, and software. The ITOP II GWAC was transitioned to the GSA GWAC Center from the Department of Transportation.

IT Infrastructure Library (ITIL)   ITIL is a series of documents used to aid in the implementation of a framework for IT service management. Created by the U.K. government, ITIL has become the standard for best practice in the provision of IT service. See http://www.itil.co.uk.

Javits-Wagner-O'Day (JWOD) Act   The JWOD program creates employment and training opportunities for people who are blind or have other severe disabilities and, whenever possible, prepares them for competitive jobs. Under the JWOD Program, government employees are required to buy selected supplies and services from nonprofit agencies employing such persons. As a result, federal customers obtain quality products and services at reasonable prices, while JWOD employees are able to lead more productive, independent lives. The Committee for Purchase from People Who Are Blind or Severely Disabled is the federal agency that administers the JWOD Program and maintains a Procurement List of mandatory source items. Two national organizations, NIB (National Industries for the Blind) and NISH (serving people with a range of disabilities), have been designated to provide technical and financial support to more than 550 nonprofit agencies participating in the JWOD Program. The Federal Acquisition Streamlining Act of 1994 continues the legal requirement to buy JWOD items.

Joint Photographic Expert Group (JPEG)   A method of storing an image in digital format.

Knowledge Management   The enterprisewide application of intellectual capital to achieve an organizational mission and goals.

Labor surplus area (LSA)   A federal program to set aside certain contracts to businesses located in areas with high unemployment.

Matrix management   Management configuration in which staff are assigned to support a program or project manager outside their own line organizations or groups.

Micropurchase   A subcategory of simplified acquisitions that refers to purchases of less than $25,000.

Minority On-Line Information Service (MOLIS)   An online database that provides information on minority institutions to government and private industry to foster partnerships. This online database of more than 260 minority institutions that is used by government agencies, the private sector, and other organizations and institutions to develop partnerships with minority institutions. MOLIS is also used by government agencies to identify possible faculty at minority institutions to serve as peer reviewers.

MOLIS is a one-stop source of in-depth information about the research and educational capabilities of participating Hispanic-serving institutions (HSIs), historically black colleges and universities (HBCUs), tribal colleges and universities (TCUs), and minority postsecondary institutions (MPIs). MOLIS provides information on the minority institutions' research centers, research interests and capabilities, facilities, equipment, faculty profiles, statistics on the number of degrees awarded and enrollment figures, scholarship and fellowship information, and federal opportunity information.

MOLIS has complete searching capabilities, file transfer/document delivery, and multiple access methods, including an 800 number for minority institutions. Key federal agencies support MOLIS under a cooperative agreement from the Department of Defense, including the National Aeronautics and Space Administration, the Department of Defense (OSDBU), the U.S. Department of Agriculture, and the Agency for International Development. These agencies use MOLIS to identify potential beneficiaries of available federal resources. See http://www.molis.org.

Minority-owned business   A business that is at least 51% owned by one or more individuals classified by the U.S. government as socially and economically disadvantaged and whose management and daily business operations are controlled by one or more such individuals. Socially or economically disadvantaged individuals, by government classification, include Black Americans, Hispanic Americans, Native Americans (includes Eskimos, Aleuts, and native Hawaiians), Asian-Pacific Americans, and Asian-Indian Americans.

Modular   A proposal volume format in which each section strictly adheres to a page allocation that contains facing text and art pages, usually one page of each per section.

Modular contracting   Type of contracting that provides for the delivery, implementation, and testing of a workable IT system or solution in discrete increments, or modules. Modular contracting is one of many approaches that can be used by federal agencies to acquire major IT systems. It may be achieved by a single procurement, or multiple procurements, but is intended to ensure that the government is not obligated to purchase more than one module at a time. Modular contracting is intended to balance the government's need for fast access to rapidly changing technology and incentivized contractor performance, with stability in program management, contract performance, and risk management.

To help improve federal agencies' acquisition and management of major IT systems, Congress passed, and the president signed, the Information Technology Management Reform Act (ITMRA, P.L. 104-106, February 1996; also known as the Clinger/Cohen Act). Section 5202 of this law directs federal agencies to use modular contracting, to the maximum extent practicable, for the acquisition of major IT systems. Following the ITMRA, the president issued Executive Order No. 13011, which instructs agencies to apply modular contracting "where appropriate" and "to the maximum extent practicable."

Note that one of great enablers of modular contracting is the ready availability of task order, multiagency, and Governmentwide Acquisition Contracts (GWACs). With these tools, program managers have access to many major IT service and product providers through these contracts. In some cases, task orders can be in place within 60 days of the initial request. This allows program managers to compete modular tasks quickly and easily.

Modular contracting is one of the federal government's strategies that moves the acquisition strategy from a traditional grand design to a more manageable incremental approach. A significant benefit offered by modular contracting is its potential to refocus acquisition management attention and strategies to a more responsive and realistic systems development acquisition model.

National Stock Number (NSN)   A unique number assigned by the General Services Administration that catalogs a wide range of items by commodity, group, and class.

North American Industry Classification System   NAICS (pronouced "nakes"), which replaced the Standard Industrial Classification (SIC) code system, was developed jointly by the United States, Canada, and Mexico to provide comparable statistics about business activity across North America's three NAFTA trading partners. NAICS will make it easier to compare North American business data with the International Standard Industrial Classification System (ISIC), developed and maintained by the United Nations. The *NAICS Manual* is available through the National Technical Information Service (NTIS) at (800) 553-6847 or on the Web at http://www.ntis.gov/products/bestsellers/naics.asp?loc=4-2-0.

Office of Small and Disadvantaged Business Utilization (OSDBU)
An office within federal agencies that provides procurement assistance to small, disadvantaged, minority-8(a), HUBZone, veteran-owned, service-disabled-owned, and woman-owned businesses. The Small Business Act as amended by P.L. 95-507 established the OSDBU. The OSDBU in federal agencies acts as an advocate for small businesses.

Option years   Additional years available on a given contract if the government elects to exercise its option to renew that contract.

Oral presentation   An oral examination in which an offeror submits information bearing on its capability and proposed solution to a panel of government source selection officials. This information may include, for example, a description of, and justification for, the offeror's performance policies, processes, and plans. The oral presentation is distinct from the offer (i.e., the proposal) in that it is not itself an offer or part of an offer, and does not become a part of any resultant contract. The source selection panel uses the oral information to determine the offeror's understanding of the prospective work and, thus, its capability to perform successfully. Oral presentations may be audio or videotaped for subsequent government review.

Out years   The final years exercised on a given contract.

Past Performance Information Management System (PPIMS)   The Army's central repository for the collection and use of Army-wide contractor Past Performance Information (PPI). Available to authorized government personnel, PPIMS is used to support both the Contracting Performance Review process and future award decisions. PPIMS was developed in response to the requirements of Office of Federal Procurement Policy (OFPP) Letter 92-5 and the Federal Acquisition Streamlining Act (FASA) as implemented in Subpart 42.1500 of the Federal Acquisition Regulation (FAR). These requirements greatly expanded the number of contracts for which contractor performance evaluations must be prepared and for which PPI must be used in source selection. PPIMS is a Web-based application.

Past Performance Information Retrieval System (PPIRS)   The PPIRS database (http://www.ppirs.gov/) is available to all source selection officials across the entire federal government. Its purpose is to assist federal acquisition officials in purchasing goods and services that represent the best value for the federal government. Contractor access to PPIRS is gained through the Central Contractor Registration (CCR) (http://www.ccr.gov) process. A contractor must be registered in CCR and must have created a Marketing Partner Identification Number (MPIN) in the CCR profile to access their PPIRS information.

Formerly called the Past Performance Automated Information System (PPAIS), PPIRS is currently sponsored by the DoD E-Business Office and administered by the Naval Sea Logistics Center Detachment Portsmouth (NAVSEA). The federal PPIRS database is a Web-enabled application that allows the retrieval of contractor past performance information. It is also a central warehouse used to retrieve performance assessment reports received from the recognized federal report card collection systems.

Performance-based acquisition (PBA)   The term *performance-based acquisition* (PBA) has replaced *performance-based contracting* (PBC). Performance-based acquisition is a major trend within federal procurement. In FY2002, for example, the Department of Defense awarded more than 20% of its service requirements using performance-based specifications. According to an Office of Management and Budget (OMB) report on "Bush Administration Priorities" issued in April 2004, the federal government as a whole has a goal of achieving 40% performance-based contracts for 2005. The Bush Administration has re-energized PBA as a key initiative by emphasizing the importance of *contracting for results.* This emphasis is also an integral message in the President's Management Agenda (PMA), which calls for creating a better government that is citizen centered, results oriented, and market based. As a direct result of the Bush Administration's re-energizing of the PBA approach, the OMB in 2001 issued a mandate (OMB Memorandum M-01-11 [2/14/01 memo] and M-01-15 [3/19/01 memo]) that explicitly directs agencies to write performance-based techniques on a specific percentage of the total eligible service contracting dollars worth more than $25,000. The Services Acquisition Reform Act (SARA), which was included in the National Defense Authorization Act of 2004 (P.L. 108-136), provides incentives for performance-based services acquisitions in a manner similar to what the Federal Acquisition Streamlining Act (FASA) and the Clinger-Cohen Act did with regard to federal hardware procurement.

The principal objective of PBA is to express government needs in terms of required and measurable performance objectives, outcomes, and/or results, rather than the method of performance. This approach is to encourage industry-driven, competitive solutions and commercial best practices. The overarching focus is on the *results,* not the processes used to deliver those results. Either a performance work statement (PWS) or statement of objectives (SOO) may be used for performance-based service acquisition (PBSA). Performance-based Requests for Solutions (RFSs) describe how the contractor's performance will be evaluated in a quality assurance surveillance plan (QASP). Positive and negative incentives are identified as appropriate.

The PWS is a statement within the solicitation that identifies the technical, functional, and performance characteristics of the agency's requirements. The SOO is an alternative to the PWS. It is a summary of key agency goals and outcomes. The QASP is a plan for assessing contractor performance in order to ensure accomplishment of the government's objectives. The level of surveillance is linked to contract dollar amount, risk, and complexity of the requirement.

Performance-based service contracting (PBSC)   PBSC is the process of contracting for services by using mission-related, outcome-oriented statements of work [performance work statements (PWSs)] and quality assurance performance measures. PBSC focuses on the desired outcome and its quality measures rather than on the how of providing the required services. A PWS is the basic document used in PBSC that describes the specific requirements the contractor must meet in performance of the contract in terms of output and a measurable standard for the output.

Performace Work Statement (PWS)   A statement in a solicitation that identifies the technical, functional, and performace characteristics of the agency's requirements.

PERT chart   PERT, or Program Evaluation and Review Technique, is a "flow diagram showing the sequence of … various activities and their interdependence in terms of completion dates. In a sense it is a road map to a destination to show the intermediate points and distances between them. Activities are sequenced from left to right." (See Krathwohl, Dave R., *How to Prepare a Research Proposal: Guidelines for Funding and Dissertations in the Social and Behavioral Sciences,* 3rd ed., Syracuse, NY: Syracuse University Press, 1988, p. 75.) Developed initially by the U.S. Navy, Lockheed, and Booz Allen Hamilton in the 1950s, PERT charts are used in project management and project planning activities and are useful because task dependencies are illustrated. Tasks are scheduled, organized, and coordinated. Critical Path Method (CPM) has become synonymous with the PERT technique.

Portrait versus landscape pages   *Portrait* refers to pages of a document oriented such that the vertical dimension is 11 inches and the horizontal dimension is 8.5 inches. (This page is in portrait format.) *Landscape* refers to pages oriented such that the vertical dimension is 8.5 inches and the horizontal dimension is 11 inches. Landscape pages are often used for full-page graphics or tables.

Preinvitation notice (PIN)   A summary of a solicitation package sent to prospective bidders, who may then request the entire solicitation package.

Preproposal conference   A face-to-face meeting between the government technical and contractual contacts and the potential bidders from industry. Attendance is very important as it goes directly to your company's interest in the given procurement. Many times, the major technical requirements will be reviewed in a briefing format. A site visit may be included with the proposal conference.

President's Management Agenda (PMA)   Launched in August 2001, President George W. Bush's vision for government reform is guided by three important principles: (1) citizen centered, (2) results oriented, and (3) market based. In the PMA, Mr. Bush identified five governmentwide initiatives and nine program initiatives. Governmentwide initiatives include Strategic Management of Human Capital, Competitive Sourcing, Improved Financial Performance, Expanded Electronic Government, and Budget and Performance Integration. The PMA was launched as a strategy for improving the management and the performance of the U.S. federal government. Importantly, federal agencies have been held publicly accountable for adopting the disciplined approaches of the PMA through a government wide colorimetric scorecard system (GREEN—YELLOW—RED). GREEN indicates that a given agency has met all of the established standards of success under the PMA, or that that same agency's implementation is proceeding according to plan.

Price analysis   The process of examining and evaluating a proposed price to determine if it is fair and reasonable, without evaluating its separate cost elements and proposed profit.

Procurement Technical Assistance Centers (PTACs)   Government and privately funded organizations dedicated to helping small companies do business with the federal government.

Profit center   The smallest organizationally independent segment of a company charged by that company's management with profit and loss (P&L) responsibilities.

Proposal   An offer prepared by a contractor to perform a service or supply a product (or a combination of both) in response to an RFP or RFS issued by a procuring agency. A proposal is a sales document, knowledge product, and legal document.

Proposal Directive   The Proposal Directive is a carefully structured "living document" that serves as the "encyclopedia" of a win for a given procurement opportunity. During the course of the acquisition/capture and proposal process, the Proposal Directive may undergo 10 or more iterations as new intelligence about the competition, for example, is uncovered and other information (e.g., win themes) is further refined. The Proposal Directive is the repository that contains all of the relevant information about the customer organization, competition, specific program/project, and relevant technologies. It contains the call plan reports from customer visits; a synopsis of the hopes, fears, biases, success criteria, and critical issues of key government decision makers; the win strategy white paper for your company along with the proposal outlines for each volume; win themes and discriminators; proposal response milestone schedule; and specific proposal-related action items. Thus, the *marketing intelligence* that the Business Development staff gather is collected in the Proposal Directive. The *solution sets* that the Acquisition/Capture Team develops based on the marketing intelligence are documented in the Proposal Directive. And the *proposal development work products*—schedules, outlines, action items, and so on—are also included within the Proposal Directive. Note that the Proposal Directive is a highly competition-sensitive document and, therefore, must be safeguarded very carefully.

Quality assurance surveillance plan (QASP)   A plan for assessing contractor performance in order to ensure accomplishment of the government's performance objectives stated in the contract and compliance with the appropriate inspection clauses. The level of surveillance should be commensurate with the dollar amount, risk, and complexity of the requirement.

Risk-sharing method   Also known as share-in-savings, shared-savings, value-based, transaction-based, and shared-benefits contracting, the risk-sharing method calls for vendors and contractors to cover the up-front funding of a project in return for the promise of a share of the savings or additional benefits that result downstream. In addition to reducing costs and relieving government agencies of capital funding expenditures or major investments, share-in-savings acquisition has significant potential both to improve the performance of agency programs and to infuse new technology and new methods of doing business. Shared-savings approaches are a significant shift away from the way federal agencies traditionally have bought services: by closely specifying the how and what of each contract and paying contractors to enact their plans. Slowly, federal agencies are moving toward performance-based contracting—specifying only results, leaving companies free to devise the solutions to achieve them while closely measuring performance and basing rewards on meeting outcome goals. By adding fiscal incentives for performance exceeding what is expected and penalties for falling short, federal agencies are shifting even more risk to vendors and contractors. But shared-savings deals can also extend to reserving payment until the contractor produces tangible results.

Writing in *Government Executive* magazine, Anne Laurent notes that perhaps the best-known example of benefits-sharing contracting in government is operated by the Department of Energy to help agencies install energy-saving devices. DOE's Federal Energy Management Program (FEMP) has crafted energy-savings performance contracts under which energy service companies absorb all the up-front costs of identifying a facility's energy needs and then buying, installing, operating, and maintaining energy-efficient equipment to cut energy bills. In return, companies get a share of the energy savings generated by improvements made during the contracts, which can last as long as 25 years.

A cogent example of risk sharing was found in the major RFP released by the Department of Veterans. In Section C of the RFP, Partnering Services, it was stipulated that the prime shall construct a flexible arrangement with the VA Austin Automation Center (AAC) that focuses on a long-term partnership instead of a project-by-project relationship. The partnership goals include supporting AAC in marketing, business planning, solution development, proposal preparation, and systems implementation services. AAC desires that the contractor invest in the business relationship and assume a portion of the business risks involved with implementing new services and maintaining existing services. The RFP continued by stating that a partnership agreement shall be developed by the contractor and mutually agreed to by the government. AAC anticipated that the partnership agreement would provide for (1) agreement that each party will assume its portion of the business risk associated with the partnership, and

(2) share business risks and rewards in implementing new services and maintaining existing services.

Sealed bidding   A method of contracting that employs competitive bids, public bid openings, and awards. Award is made to the responsible bidder whose bid, conforming to the invitation for bids (IFB), will be most advantageous to the government considering only price and price-related factors included in the invitation.

Services Acquisition Reform Act (SARA) of 2003   SARA (P.L. 108-136; Title 14 of the FY2004 National Defense Authorization Act) provides incentives for use of performance-based contracting for services.

Set aside   A kind or class of procurement reserved for contenders who fit a certain category, for example, business size, region, or minority status (see FAR 19.501).

SF33   Solicitation offer and award. SF 33 is prescribed for use in soliciting bids for supplies or services and for awarding contracts that result from bids.

SF129   Standard Form 129 is the Solicitation Mailing List Application. The list is used for government planning purposes to match direct procurement opportunities with companies that may be able to provide the specific products or services required, as specified in FAR 14.205-1(d). This form is now being replaced by the EDI 838 contractor registration process.

SF330   SF330, Architect-Engineer Qualifications form, was developed to replace the long-standing SF254 and SF255. SF254, Architect-Engineer and Related Services Questionnaire, and SF255, Architect-Engineer and Related Services Questionnaire for Specific Projects, had been largely unchanged since their introduction in 1975. The A-E industry, however, has changed substantially and the requirements of federal agencies have evolved considerably. In response to these changes, SF330 was developed by an interagency committee, submitted to industry and government for comment through the *Federal Register*, and fielded.

SF330 reflects current A-E practices and provides the essential information agencies need for the selection of A-E firms in a streamlined format. The SF330 is a two-part form. Part I, Contract-Specific Qualifications, replaces SF255. Part II, General Qualifications, replaces SF254. SF254 is now essentially used as an appendix to SF255 and is rarely used alone.

Hence, SF330 was designed as a single two-part form. Similar to SF254, SF330 Part II can still be separately submitted to agencies to be maintained in their A-E qualification data files.

The objectives of the SF330 are as follows:

- Merge the SF254 and SF255 into a single streamlined form.
- Expand essential information about qualifications and experience.
- Reflect current A-E disciplines, experience types, and technology.
- Eliminate duplicate information.
- Eliminate information of marginal value.
- Facilitate electronic usage.

SF 1411   Contract Pricing Cover Sheet.

Simplified acquisition procedures (SAPs)   Streamlined techniques and guiding principles designed to reduce the administrative burden of awarding the lower dollar value procurements that account for the vast majority of DoD acquisition. They allow for informal quoting and competition procedures, encourage accepting oral quotes versus written quotations, prefer comparing quoted prices versus conducting negotiations, and provide streamlined clauses to support the award document. Saving money, improving opportunity and efficiency, and avoiding administrative burden are at the core of the SAP program. The Federal Acquisition Streamlining Act and the Federal Acquisition Reform Act were integral in the creation of simplified acquisitions and the simplified acquisition threshold of $100,000.

Six Sigma   A statistical process improvement method focusing on quality from a customer's point of view. The goal of Six Sigma is to increase profits by eliminating variability, defects, and waste that undermine customer loyalty. Six Sigma is a federally registered trademark of Motorola. See http://www.isixsigma.com.

Small business   A business that is independently owned and operated and is not dominant in its field; a business concern meeting government size standards for its own particular industry type. A small business in the manufacturing field normally employs, with its affiliates, not more than 500 persons. In the services arena, small businesses are usually defined as businesses that generate less than $5 million in gross annual receipts.

Small Business Act (SBA)   This 1953 act established the U.S. Small Business Administration. Public Law 95-507, enacted in October 1978, made major revisions to strengthen the SBA.

Small Business Administration (SBA)   The federal government agency whose function is to aid, counsel, provide financial assistance to, and protect the interests of the small business community (from McVay, *Proposals That Win Federal Contracts,* p. 324). The U.S. SBA has the responsibility of making certain that small business obtains a fair share of government contracts and subcontracts. Their mission is articulated in the Small Business Act of 1953, which established that organization.

Small Business High Technology Institute (SBHTI)   SBHTI fosters the development of R&D-based small businesses (http://www.sbhti.org).

Small Business Innovation Research (SBIR) program   SBIR is a highly competitive SBA program that encourages small businesses to explore their technological potential and provides the incentive to profit from its commercialization. Under this program, federal agencies with large R&D budgets must direct designated amounts of their R&D contracts to small businesses. Small businesses must be American-owned and independently operated, for-profit, principal researcher employed by the business, and have no more than 500 employees. Federal agencies that participate in the SBIR Program include the DoD, USDA, DOC, DOE, DHHS, DOT, EPA, NASA, NRC, NSF, and the Department of Education. These agencies designate R&D topics and accept proposals for a three-phase program. Phase I (feasibility studies) has awards up to $100,000 for 6 months, and Phase II (full-scale research) has awards up to $750,000 for up to 2 years. Phase III (commercialization) is the period during which the Phase II innovation moves from the laboratory into the marketplace. No SBIR funds support this phase. The small business must find the funding in the private sector or non-SBIR federal agency. The U.S. SBA serves as the coordinating agency and information link for SBIR. Examples of successful SBIR proposals include thin-film optical filters for intense laser light, membrane-based process for debittering citrus juice, and neonatal/infant/ fetal pump-oxygenator system. Since the inception of the SBIR Program in 1983, nearly $4 billion in competitive federal R&D awards have been made to qualified small business concerns under the program. Only Phase I winners may submit Phase II proposals.

Software Engineering Institute (SEI) The Software Engineering Institute is a federally funded research and development center sponsored by the U.S. Department of Defense and operated by Carnegie Mellon University. SEI is the current "steward" of the CMMI® model.

Sole source acquisition A contract for the purchase of supplies or services that is entered into or proposed to be entered into by the contracting officer of an agency after soliciting and negotiating with only one source (vendor, contractor). This is done on the grounds that only that one source is capable of satisfying the government's requirements.

Special Drawing Rights (SDR) An international currency unit set up by the International Monetary Fund.

Standard form (SF) SFs and optional forms may be obtained from the Superintendent of Documents, Government Printing Office (GPO), Washington, D.C. 20402, or from the prescribing agency (see FAR 53.107).

Standard International Trade Classification (SITC) A classification system devised by the United Nations for commodities used in international trade.

Statement of Objectives (SOO) An alternative to a performance work statement (PWS); it is a summary of key agency goals, outcomes, or both that is incorporated into performance-based service acquisitions in order that competitors may propose their solutions, including a technical approach, performance standards, and a quality assurance surveillance plan based on commercial business practices.

Statement of Work (SOW) A statement that defines the government's requirements in clear, concise language identifying specific work to be accomplished.

Stet Literally, "let it stand"; a proofreader's notation meaning that the indicated change should not be made.

Storyboarding Storyboarding (the associated work products are also referred to as *scenarios* and *scribble sheets*) is a critical management process that focuses on building a coherent, consistent, and compelling proposal story. Building on the marketing intelligence and solution sets for a given procurement, key stakeholders in the proposal process (your capture manager, proposal manager, business development staff, subject matter experts, and so on) meet in a series of brainstorming sessions and generate the following critical work products: (1) annotated outline for each major section of the proposal that provides an expanded architecture for each proposal section; (2) "elevator speeches" that tell the overarching story for each section; and (3) populated "pain tables" that illustrate your company's understanding of the customer's "pain points" and major issues, your approach/solution, and your company's value proposition. Your value proposition includes the tangible and intangible benefits that your solution will bring to the customer organization.

Storyboarding allows your company and your proposal and capture teams to achieve the "right dive angle off the board." Just as a championship diver cannot adjust her position relative to the water when she's a mere foot above the surface, neither can a proposal team move in a fundamentally different direction when deep into the proposal response life cycle—at least not without a lot of depleted morale, extremely long hours, wasted resources, and interpersonal challenges across the team.

Coming out of the storyboarding process, one should be able to review the elevator speeches from end to end and grasp the entire fact-based storyline of the proposal response.

Strawman RFP or proposal   Mock, simulated, or preliminary RFP or proposal.

Tag Image File Format (TIFF)   A file format used to store image files.

Technical leveling   The elimination of differences between competitors by repeated discussions and exchange of information. It is inappropriate for the government to assist offerors by suggesting modifications that improve their proposals [FAR 15.610(d)(1)]. (Cibinic, John, and Ralph C. Nash, *Formation of Government Contracts*, 2nd ed., Washington, D.C.: George Washington University, 1986, pp. 529, 625.)

Technical transfusion   The transfer of one competitor's ideas to another competitor during the course of discussions with the government. (Cibinic, John, and Ralph C. Nash, *Formation of Government Contracts*, 2nd ed., Washington, D.C.: George Washington University, 1986, pp. 529, 622.)

Themes   Authenticated, substantiated statements that clearly articulate why the client should select your company over your competition. Examples include common themes, unique themes, competitive themes. Themes are supported claims, platforms, and sound arguments.

Tiger Team   Specialized working group generally assigned to track one particular aspect of a proposal effort, such as cost strategy.

Transmission Control Protocol/Internet Protocol (TCP/IP)   A set of rules that establish the method by which data are transmitted over the Internet between computers.

Uniform resource locator (URL)   A path or pointer for locating files and sites on the Web.

Walsh-Healey Act   Enacted in 1936, "41 U.S.C. §§ 35–45, requires that, for contracts exceeding $10,000, the contractor be a 'manufacturer of' or a 'regular dealer in the materials, supplies, articles, or equipment to be manufactured or used in the performance of a contract.' " (Cibinic, John, and Ralph C. Nash, *Formation of Government Contracts*, 2nd ed., Washington, D.C.: George Washington University, 1986, p. 226.)

War room   A facility or portion of a facility dedicated to proposal operations, planning, writing, and review. The walls of a war room are often used to display proposal (and RFP) sections for review and comment. Seeing specific proposal sections in the context of other sections aids in overall continuity and flow, as well as helps eliminate unnecessary redundancy. A war room should be a highly secure area because of the open display of proposal materials.

Woman-owned business (WOB)   A business that is at least 51% owned, controlled, and operated by a woman or women. "Control" is defined as exercising the power to make policy decisions, and "operate" is defined as actively involved in day-to-day management.

World Wide Web (WWW; W3; the Web)   A distributed hypertext-based information system conceived at CERN to provide its user community with an easy way to access global information. The Web allows for the presentation and linkage of information dispersed across the Internet in an easily accessible way. The Internet 2 is projected to be 100 times faster than the Internet. More than 206 U.S. colleges and universities are participating in the upgrade process. The Internet is an international computer network of networks that connect government, academic, and business institutions.

Selected list of
acronyms and abbreviations

AAMB	American Association of Minority Businesses
AAP	acquisition analysis process
AAWBOA	African American Women Business Owners Association
ABELS	Automated Business Enterprise Locator System (online listing of minority-owned firms, maintained by the Minority Business Development Agency)
ACA	after contract award
ACASS	architect-engineer contract administration support system
ACE	American Council on Education

ACEC	American Consulting Engineers Council
ACH	automated clearing house
ACO	administrative contracting officer
ACOP	Acquisition Center Business Opportunity (U.S. Army CECOM)
ACRI	acquisition cost reduction initiatives
ACWP	actual cost of work performed
ADA	Americans with Disabilities Act (Section 508)
ADB	Asian Development Bank; African Development Bank
ADP	automated (or automatic) data processing
ADPE	automatic data processing equipment
ADR	administrative dispute resolution; alternative dispute resolution
A-E	architect-engineer
A/E/C	Architect/Engineer/Construction
AECA	U.S. Arms Export Control Act
AEP	Asian Environmental Partnership
AF	Air Force
AFAA	Air Force Audit Agency
AFARS	Army Federal Acquisition Regulation Supplement
AFB	Air Force base
AFBOP	Air Force Business Opportunities Page
AFCEE	Air Force Center for Environmental Excellence

AFFARS   Air Force Federal Acquisition Regulation Supplement

AFI   Air Force Instruction (e.g., AFI 32-4002, "Hazardous Materials Emergency Planning and Response Compliance")

AFMC   Air Force Material Command

AFMPCOE   Air Force Mentor-Protégé Center of Excellence

AFR   Air Force Regulation

AFRL   Air Force Research Laboratory

AID   Agency for International Development (http://www.usaid.gov)

AISB   American Institute of Small Business (http://www.aisb.biz)

ALC   additional labor categories

ALN   asynchronous learning network

AMC   Army Material Command

AMD   Acquisition Management Directorate (DCA)

AMS   Acquisition Management System

AMSDL   acquisition management source data list

ANSI   American National Standards Institute

AOL   America Online

APEC   Asia-Pacific Economic Cooperation

API   application programming interface

APMP   Association of Proposal Management Professionals (http://www.apmp.org)

APR   agency procurement request

AQL   acceptable quality level

AR	Army regulation; acquisition reform
ARDEC	Armament Research, Development and Engineering Center
ARNet	Acquisition Reform Network
ARO	after receipt of order
ARRT	Acquisition Reengineering and Realignment Task Force
ASBA	American Small Business Association [Rock City, IL; (800) 942-2722]
ASBCA	Armed Services Board of Contract Appeals
ASBCD	Association of Small Business Development Centers
ASC	Accredited Standards Committee (ANSI); Aeronautical Systems Center (WPAFB)
ASEAN	Association of Southeast Asian Nations
ASFI	Army Single Face to Industry (acquisition Web site)
ASPA	Armed Services Procurement Act (1948, 10 U.S.C. § 2301 et seq.)
ASPM	Armed Services Pricing Manual
ASPR	Armed Services Procurement Regulation
ASSIST	Acquisition Streamlining and Standardization Information System
ATP	Advanced Technology Program (NIST)
ATR	assistant technical representative
ATRB	Award-Term Review Board
B&P	bid and proposal
B2B	business-to-business
BA	basic agreement

BAA	broad agency announcement
BABC	British-American Business Council
BAFO	best and final offer (replaced by FPR)
BARBB	Missile Defense Agency Business Acquisition Reporting Bulletin Board
BARFO	best and revised final offer
BBS	bulletin board system
BCA	Board of Contract Appeals
BCWP	budgeted cost for work performed
BCWS	budgeted cost for work scheduled
BDG	business development group
BECC	Border Environmental Cooperation Commission
BI	SBA's Office of Business Initiatives
BIC	business information center
B/L	bill of lading
BLS	Bureau of Labor Statistics, U.S. Department of Labor
BMP	best management practices
BOA	basic ordering agreement
BOD	bid opening date
BOE	basis of estimate
BOM	bill of materials
BOS	business opportunity specialist (SBA)
BOTB	British Overseas Trade Board

BPA	blanket purchase agreement
BPCR	breakout procurement center representative
BPO	blanket purchase order
BPR	business process reengineering
BRC	business resource center
BSO	business support centers (Japan)
BT	British Telecom
BusinessLINC	Business Learning, Information, Networking, and Collaboration
BVA	best value analysis
BVS	best value selection
CAAC	Civilian Agency Acquisition Council
CAAS	contract for advisory and assistance services
CACES	computer-aided cost estimation system
CACO	corporate administrative contracting officer
CAD	computer-aided design
CAGE	commercial and government entity
CAIP	community adjustment and investment program
CAIV	cost as an independent variable
CAN	contract award notice
CANDI	commercial and nondevelopmental item
CAO	chief acquisition officer

CAR	commerce acquisition regulation
CARE	Cooperative for Assistance and Relief Everywhere, Inc.
CAS	cost accounting standard (enumerated at FAR 30); contract administration services
CASB	Cost Accounting Standards Board
CASE	computer-aided software engineering
CBD	*Commerce Business Daily* (now defunct, replaced by FedBizOpps)
CBR	*China Business Review*
CBSC	Canada Business Service Center
CCA	Clinger-Cohen Act of 1996 (P.L. 104-106; also includes FARA and ITMRA)
CCASS	construction contract appraisal support system
CCBA	Canada-China Business Association
CCBC	Canada China Business Council
CCF	Contract Cases Federal
CCH	Commerce Clearing House
CCI	consolidated contracting initiative
CCL	Commerce Control List (15 C.F.R. § 774, Suppl. 1)
CCN	Cooperating Country Nationals
CCP	current cost or pricing
CCR	Central Contractor Registry (Columbus, Ohio, and Ogden, Utah; an electronic database of government vendor firms that is maintained by the DoD; http://www.ccr.gov); Central Consultancy Register (Brussels, Belgium)

CDA	Contract Disputes Act
CDC	certified development company
CDL	contract data list
CDRD	contract data requirements document
CDRL	contract data requirements list; contractor establishment code
CEB	cost element breakdown
CEC	Center for Electronic Commerce; contract establishment code
CECOM	U.S. Army's Communications-Electronics Command
CEEBIC	Central & Eastern Europe Business Information Center
CEI	contract end item
CENDI	Commerce, Energy, NASA, and Defense Information
CEO	chief executive officer
CER	cost estimating relationship
CERN	Conseil European pour la Recherche Nucleaire (European Organization for Nuclear Research; originators of HTTP and HTML)
CEU	continuing education unit
CFE	contractor-furnished equipment; centers for excellence
CFME	Committee for External Economic Events (France)
CFP	customer-furnished property
CFR	Code of Federal Regulations (e.g., Title 41, Public Contracts)
CGLI	comprehensive general liability insurance
CI	configuration item; contractor inquiry

CICA	Competition in Contracting Act of 1984 (P.L. 98-369, 98 Stat. 1175)
CID	commercial item description
CIF	cost, insurance, freight
CIO	chief information officer
CIS	capability/interest survey; Commonwealth of Independent States (former Soviet Union)
CIT	Center for Information Technology (NIH)
CITEC	Centers for Industrial and Technological Cooperation
CITI	Center for Information Technology Innovation
CKO	chief knowledge officer
CLIN	contractor line item
CLS	contractor logistics support
CM	configuration management
CMM®	Capability Maturity Model® (Software Engineering Institute)
CMMI®	Capability Maturity Model Integration®
CO	contracting officer
COB	close of business
COC	certificate of competency (issued by SBA)
COCO	contractor owned, contractor operated; Chief of the Contracting Office (DOT)
COD	cash on delivery
CODSIA	Council of Defense and Space Industries Associates
COE	common operating environment

COG	continuity of government
COI	conflict of interest
COMSEC	communications security
CONAHEC	Consortium for North American Higher Education Collaboration
CONUS	continental United States
COOP	continuity of operations
COP	communities of practice
CORBA	Common Object Request Broken Architecture
CORDUS	Community R&D Information Service (Europe)
COTR	contracting officer's technical representative
COTS	commercial-off-the-shelf
CPAF	cost plus award fee
CPARS	contractor performance assessment reporting system
CPD	comptroller's procurement decisions
CPFF	cost plus fixed fee
CPI	consumer price index; continuous process improvement
CPIC	capital planning and investment control
CPIF	cost plus incentive fee
CPM	critical path method
CPMO	Contractor Performance Measurement Organization
CPPF	cost plus percentage fee

CPRS	contractor purchasing system review
CPU	central processing unit
CR	clarification request
CRA	continuing resolution authority
CRADA	cooperative research and development agreement
CRM	customer relationship management
CRS	Catholic Relief Services
C/SCSC	cost/schedule control system criteria
CSR	commercial market representative
CTO	chief technology officer
CTW	Contracts and Tenders Worldwide (White Plains, New York, and London)
CWBS	contract work breakdown structure
CY	calendar year
D&B	Dun & Bradstreet
D&F	determinations and findings
DABBS	Defense Communications Agency's Acquisition Bulletin Board System
DACON	Data on Consulting Firms (automated World Bank and Inter-American Development Bank systems)
DAPS	Defense Automated Printing Service (Philadelphia)
DAR	Defense Acquisition Regulation
DARO	days after receipt of order
DARPA	Defense Advanced Research Projects Agency

DBA	Davis-Bacon Act (1931); doing business as
DBE	disadvantaged business enterprise
DBMS	database management system
DCA	delegation of contracting authority
DCAA	Defense Contract Audit Agency
DCAAM	Defense Contract Audit Agency Manual
DCADS	Defense Contract Action Data System
DCAS	Defense Contract Administration Service, Defense Logistics Agency
DCI	data collection instrument
DCMA	Defense Contract Management Agency
DCMC	Defense Contract Management Command
DCSC	Defense Construction Supply Center
DDTC	U.S. Directorate of Defense Trade Controls
DEARS	Department of Energy FAR Supplement
DELTA	Defense Loan and Technical Assistance
DESC	Defense Electronics Supply Center
DFAR	Defense Federal Acquisition Regulations (also called DAR)
DFARS	Defense Federal Acquisition Regulation Supplement
DFAS	Defense Finance Accounting Service
DGR	designated government representative
DGSC	Defense General Supply Center

DHHS	Department of Health and Human Services
DHS	Department of Homeland Security
DICON	Data on Individual Consultants (automated Asian Development Bank system)
DID	data item description
DII	Defense Industry Initiative
DIRMM	DOT's IRM Manual
DISA	Defense Information Systems Agency; Data Interchange Standards Association
DISC	Defense Industrial Supply Center
DLA	Defense Logistics Agency; Direct Labor Analysis
DLIS	Defense Logistics Information Service
DLSC	Defense Logistics Services Center (now called DLIS)
DM	division manager
DMS	document management system
DN	deficiency notice
D.O.	delivery order
DOC	Department of Commerce
DoD	Department of Defense
DODD	Department of Defense Directive (e.g., DODD 5000.1)
DODI	Department of Defense Instruction (e.g., DODI 5000.2)
DODISS	Department of Defense Index of Specifications and Standards

DODSSP	Department of Defense Single Stock Point for Specifications and Standards
DOE	Department of Energy
DOI	department operating instruction
DOJ	Department of Justice
DoN	Department of the Navy
DOS	Department of State
DOSAR	Department of State Acquisition Regulation
DOSC	delivery order selection criteria
DOT	Department of Transportation
DPA	Delegation of Procurement Authority
DPAP	Defense Procurement and Acquisition Policy
DPAS	Defense Priorities and Allocations System
DPLH	direct productive labor hour
DR	deficiency report
DRD	data requirements document
DRFP	draft request for proposal
DRL	data requirements list
DROLS	Defense Research Development, Test, and Evaluation On-Line System
DSARC	Defense Systems Acquisition Review Council
DSP	Defense Standardization Program
DTC	design to cost

DT&E	development, test, and evaluation
DTI	Department of Trade and Industry (United Kingdom)
DTIC	Defense Technical Information Center
DTP	desktop publishing
DUAP	Dual-Use Applications Program (DoD)
DUNS	Data Universal Numbering System
EAC	estimate at completion (cost)
EAR	Export Administration Regulations (15 C.F.R. § 730-774)
eB	electronic business
EBRD	European Bank for Reconstruction and Development
EBS	electronic bid set
EC	electronic commerce; European Community; European Commission
ECAPMO	Electronic Commerce Acquisition Program Management Office
ECI	electronic commerce infrastructure
ECIC	Electronic Commerce Information Center
ECIP	electronic commerce interoperability process
ECM	enterprise content management
ECP	engineering change proposal
ECPN	electronic commerce processing node
ECPO	Electronic Commerce Program Office
ECRC	Electronic Commerce Resource Center
ECU	European currency unit

ED	Department of Education
EDF	European Development Fund
EDI	electronic data interchange
EDIFACT	EDI for Administration, Commerce and Transport (Europe and Asia)
EEA	European Economic Area (Norway)
EELV	evolved expendable launch vehicle
EEO	Equal Employment Opportunity
EFT	electronic funds transfer
EIB	European Investment Bank
EIC	Euro Information Centre(s)
EIN	employer identification number
EIT	electronic and information technology
EIU	Economist Intelligence Unit (London)
ELAN	Export Legal Assistance Network
EN	evaluation notice
EO	executive order
EOI	expression of interest
E&MD	engineering and manufacturing development
EPA	Environmental Protection Agency
EPAAR	Environmental Protection Agency Acquisition Regulations
EPIC	Electronic Processes Initiatives Committee

EPIN	European Procurement Information Network
EPLS	excluded parties list
EPS	equipment performance specification; electronic publishing system; electronic proposal submission; Electronic Posting System (replaced by FedBizOpps)
EPTF	Electronic Procurement Task Force (APMP)
ERM	electronic records management
ESS	electronic source selection
E-TAP	Export Trade Association Partnership
ETC	estimate to complete
ETS	Electronic Tendering System (Hong Kong)
ETTAP	Entrepreneurial Training and Technical Assistance Program
EU	European Union
EVA	earned value analysis
EVMS	earned value management system
EWCP	Export Working Capital Program
F3I	form, fit, function, and interface
FAA	Federal Aviation Administration
FAC	Federal Acquisition Circular
FACNET	Federal Acquisition Computer Network (now defunct)
FAI	Federal Acquisition Institute
FAIR	Federal Activities Inventory Reform Act of 1998
FAQ	frequently asked questions

FAR	Federal Acquisition Regulation
FARA	Federal Acquisition Reform Act (included in the 1996 Defense Authorization Act, Clinger-Cohen Act)
FAS	free along side
FASA	Federal Acquisition Streamlining Act (October 1994) (P.L. 103-355)
FAST	Federal and State Technology Partnership Program
FAT	factory acceptance test; first article testing (when DoD buys certain goods, they may perform extensive tests on the first item delivered)
FBC	faster, better, cheaper
FCCM	facilities capital cost of money
FCIA	Foreign Credit Insurance Association
FCR	federal contracts report
FDI	foreign direct investment
FDIC	Federal Deposit Insurance Corporation
FDO	fee determination official
FEA	Federal Enterprise Architecture
FedBizOpps	Federal Business Opportunities
FEDRIP	Federal Research in Progress (database)
FED-STD	federal standard
FedTeDS	Federal Technical Data Solution
FEMA	Federal Emergency Management Agency
FEMP	Federal Energy Management Program

FESMCC	Federal EDI Standards Management Coordinating Committee
FFP	firm-fixed price
FFRDC	Federally Funded Research and Development Center
FFWA	Florida Freelance Writers Association
FHWA	Federal Highway Administration
FIND	Federal Information & News Dispatch
FIP	Federal Information Processing
FIPS	Federal Information Processing Standards (Department of Commerce)
FIRMR	Federal Information Resources Management Regulation (41 CFR Chapter 201)
FISMA	Federal Information Security Management Act of 2002
FLC	Federal Laboratory Consortium
FLSA	Fair Labor Standards Act
FMS	foreign military sales
FMSS	Financial Management Systems Software
FMV	fair market value
FNS	Food and Nutrition Service (USDA)
FOB	free on board
FOCI	foreign ownership, control, or influence
FOIA	Freedom of Information Act
FOIR	Freedom of Information Request
FPASA	Federal Property and Administrative Services Act (1949, 41 U.S.C. § 251 et seq.)

FPDC	Federal Procurement Data Center
FPDS	Federal Procurement Data System
FP-EPA	fixed price with economic price adjustment
FPI	fixed-price incentive
FPIF	fixed-price incentive fee
FPLE	fixed-price level of effort
FPLH	fixed-price labor hour
FPMR	Federal Property Management Regulations
FPO	for position only (used in reference to graphics or photos)
FPR	Federal Procurement Regulations; fixed price with redetermination; final proposal revision
FPRA	forward pricing rate agreement
FR	Federal Register
FSC	Federal Supply Classification (or Class) (associated with SAACONS Vendor Information Program)
FSN	Federal Stock Number
FSS	Federal Supply Service
FTE	full-time equivalent
FTP	File Transfer Protocol
FTR	federal travel regulation
FY	fiscal year
FYDP	five-year defense plan or program; future years defense plan

G&A	general and administrative
GAAP	generally accepted accounting principles
GAAS	generally accepted auditing standards
GAO	General Accounting Office
GATT	General Agreement on Tariffs and Trade
GBL	government bill of lading
GDP	gross domestic product
GEF	Global Environmental Fund (World Bank)
GEMPC	Government Estimate of Most Probable Cost
GEMS	Australian Government Electronic Marketplace Service
GFDL	Geophysical Fluid Dynamics Laboratory (NOAA)
GFE	government-furnished equipment
GFF	government-furnished facilities
GFM	government-furnished material
GFP	government-furnished property
GFY	government fiscal year
GIF	Graphics Interchange Format
GIS	geographic information system
GISRA	Government Information Security Reform Act of 2000
GISS	Goddard Institute for Space Studies (New York)
GMSS	guaranteed maximum shared saving
GOCO	government owned, contractor operated

GPA	grade point average
GPE	government point of entry
GPEA	Government Paperwork Elimination Act
GPN	General Procurement Notice (*UN Development Business*)
GPO	Government Printing Office
GPRA	Government Performance and Results Act (P.L. 103-62, 1993)
GRA&I	Government Reports Announcements and Index
GRC	NASA Glenn Research Center
GSA	General Services Administration
GSBCA	General Services Board of Contract Appeals
GSD	Government Supplies Department (Hong Kong, China)
GSFC	Goddard Space Flight Center
GST	goods and services tax (Canada)
GTN	Global Technology Network
GWAC	governmentwide acquisition contract
HBCU	historically black colleges and universities
HCA	head of contracting activity
HHS	Health and Human Services
HIA	high-impact agencies
HMSO	Her Majesty's Supply Office (United Kingdom)
HOA	head of the operating administration

HRIS	Human Resource Information System
HS	harmonized system (international trade)
H&S	health and safety
HSI	Hispanic-serving institution
HTML	Hypertext Markup Language
HTRW	hazardous, toxic, and radioactive waste
HTTP	Hypertext Transport Protocol
HUBZone	historically underutilized business zone
HUD	Department of Housing and Urban Development
IA	information assurance
IACWBE	Interagency Committee on Women's Business Enterprises
IAE	Integrated Acquisition Environment
IAPSO	Interagency Procurement Services Office (United Nations)
IAW	in accordance with
IBRD	International Bank for Reconstruction and Development (part of the World Bank Group)
ICAR	Individual Contract Action Report (Standard Form 279)
ICB	international competitive bidding
ICE	independent cost estimate
ICR	intelligent character recognition (scanning technology)
ICSB	International Council for Small Business (http://www.icsb.org)
ICSID	International Center for Settlement of Investment Disputes (part of the World Bank Group)

IDA	International Development Agency (part of the World Bank Group)
IDB	Inter-American Development Bank; Industrial Development Board
IDC	indefinite delivery contract
IDE	Institute of Developing Economies (Japan)
ID-IQ	indefinite delivery/indefinite quantity (contract type, examples of which are the Air Force's Desktop V and the Army's Small Multiuser Computer Contracts)
IE	individual experience (SF254, Block 11)
IEEE	Institute of Electrical and Electronics Engineers
IEI	invitation for expression of interest
IESC	International Executive Service Corps
IEW	intelligence and electronic warfare
IFB	invitation for bid
IFC	International Finance Corporation (part of the World Bank Group)
IG	inspector general
IGCE	independent government cost estimate
IK	Indigenous Knowledge (World Bank initiative)
ILA	integrated logistics assessment
ILS	integrated logistics support
IMAS	International Marketing Assistance Service (SUNY-Albany program)
IMC	International Medical Corps
IMD	International Institute for Management Development (in Switzerland)

IMF	International Monetary Fund (World Bank)
I.N.F.O.	Information Network Fully Online
INGO	international nongovernmental organization
IOC	initial operating capability
IOT&E	initial operational test and evaluation
IPD	integrated logistics assessment
IPPA	Integrated Public Procurement Association (Europe)
IPPD	integrated product and process development
IPT	integrated product team
IQC	indefinite quantity contract
IRAD	independent research and development
IR&D	internal (independent) research and development
IRM	information resources management
IRS	Internal Revenue Service
ISBC	International Small Business Consortium (Norman, Oklahoma; http://www.isbc.com/isbc); International Small Business Congress (http://www.isbc.or.jp)
ISDN	Integrated services digital network (128 Kbps)
ISIC	International Standard Industrial Classification system
ISO	International Standards Organization
ISP	Internet service provider (current top commercial ISPs are AOL, the Microsoft Network, and Prodigy)
ISSA	Interservice Support Agreement

ISSAA	Information Systems Selection and Acquisition Agency (U.S. Army)
ISSN	International Standard Serial Number
I&T	integration and test
IT	information technology
ITA	International Trade Administration
ITAR	International Traffic in Arms Regulations (22 C.F.R. Parts 120–130)
ITIL	Information Technology Infrastructure Library (British standards for managing information technology services)
ITL	International Trade Law
ITMRA	Information Technology Management Reform Act (1996)
ITOP	Information Technology Omnibus Procurement
ITT	invitation(s) to tender
IV&V	independent verification and validation
J&A	justification and approval
JCALS	Joint Computer-Aided Acquisition and Logistics Support
JDAM	Joint Direct Attack Munition
JECPO	Joint Electronic Commerce Program Office
JETRO	Japanese Government Procurement Database (Japan External Trade Organization)
JN	job number
JOFOC	justification for other than full-and-open competition
JPEG	Joint Photographic Expert Group

JSC	Johnson Space Center
JTR	joint travel regulations
JV	joint venture
JWOD Act	Javits-Wagner-O'Day Act
Kbps	kilobits per second
KISS	keep it short and simple
KM	knowledge management
KME	knowledge management environment
KSC	Kennedy Space Center
LAN	local-area network
LCC	life-cycle costs
LCOTR	lead COTR
LEDU	Local Enterprise Development Unit (Northern Ireland)
L-H	labor-hour
LIB	limited international bidding
LLNL	Lawrence Livermore National Laboratory
LNGO	local nongovernmental organization
LOB	line of business
LOE	level of effort
LOI	letter of interest (Canadian equivalent of RFI); letter of intent
LOO	letter of obligation
LOSP	Liaison Outreach and Services Program (DOT)

LPO	local project overhead
LPTA	lowest price, technically acceptable
LRAE	long-range acquisition estimate (Air Force)
LS	lump sum
LSA	labor surplus area
LTOP	lease-to-ownership program
LWOP	lease with option to purchase
MAC	multiple award contracts; multiagency contract
MACOM	Major Army Command
MA/IDIQ	multiple award/indefinite delivery, indefinite quantity
MAN	metropolitan area network
MBDA	Minority Business Development Agency (U.S. Department of Commerce)
MBDC	Minority Business Development Center (DOT)
MBE	minority business enterprise
MBIRD	Minority Business Information Resources Directory
MBOC	Minority Business Opportunity Committee
Mbps	megabits per second
MBRC	Minority Business Resource Center
MCTL	Military Critical Technology List
MDA	Missile Defense Agency; Multilateral Development Agencies
MED	Minority-/female-/disabled-owned business; Minority Enterprise Development

MEGA	Minority Enterprise Growth Assistance
MEO	most efficient organization
MERX	Internet-based national electronic tendering service (Canada)
MESBIC	Minority Enterprise Small Business Investment Company
MI	minority institutions
M&IE	meals and incidental expenses
MIGA	Multilateral Investment Guarantee Agency (part of the World Bank Group; established in 1988)
MIL-HBK	Military Handbook
MILSPEC	Military Specification
MIL-STD	Military Standard
MIS	management information system
MMO	Materials Management Office
MNC	multinational corporation
M&O	maintenance and operations
MOA	memorandum of agreement
MOLIS	Minority On-Line Information Service
MOS	monthly operational summary (World Bank)
MOU	memorandum of understanding
MP	Mentor-Protégé (DoD)
MPC	most probable cost
MPCG	most probable cost to the government

MPI	minority postsecondary institution
MPIN	marketing password identification numbers
MPT	modular proposal technique
MR	modification request
MRO	maintenance, repair, and operating supplies
MSE	major source of employment
MSFC	Marshall Space Flight Center
MYP	multiyear procurement
NABDC	Native American Business Development Center
NAF	nonappropriated fund
NAFTA	North American Free Trade Agreement
NAICS	North American Industry Classification System (NAFTA, 1997) (pronounced "nakes")
NAIS	NASA Acquisition Internet Service
NAPM	National Association of Purchasing Management
NAPS	Navy Acquisition Procedures Supplement
NARA	National Archives and Records Administration
NASA	National Aeronautics and Space Administration
NASAPR	NASA Procurement Regulation
NAVSEA	Naval Sea Systems Command
NAVSUP	Naval Supply Systems Command
NAWBO	National Association of Women Business Owners

NBA	National Business Association
NCB	national competitive bidding
NCMA	National Contract Management Association
NCMB	National Coalition of Minority Businesses
NCSA	National Center for Supercomputing Applications (at the University of Illinois, Urbana-Champaign)
NDI	nondevelopmental items
N.E.C.	not elsewhere classified (related to SIC codes)
NECO	Navy's Electronic Commerce Online
NEH	National Endowment for the Humanities
NEP	network entry point (EC/EDI; two NEPs in Ohio and Utah)
NFAR	NASA Federal Acquisition Regulations
NFS	NASA FAR Supplement
NFWBO	Center for Women's Business Research (formerly the National Foundation for Women Business Owners; http://www.womensbusinessresearch.org) (Silver Spring, Maryland; http://www.womensbusinessresearch.org/about.html)
NGO	nongovernmental organization
NHB	NASA Handbook
NIB	National Industries for the Blind
NIC	newly industrializing countries
NICRA	Negotiated Indirect Cost Rate Agreement
NIGP	National Institute of Governmental Purchasing (Canada)
NIH	National Institutes of Health

NIS	New Independent States (of the former Soviet Union)
NIST	National Institute of Standards and Technology
NITAAC	National Information Technology Acquisitions and Assessment Center
NLT	no later than
NMCARS	Navy Marine Corps Acquisition Regulation Supplement
NMBC	National Minority Business Council (http://www.nmbc.org)
NMI	NASA Management Instruction
NMSO	National Master Standing Offers (Canada)
NMVC	New Market Venture Capital program (SBA)
NNSA	National Nuclear Security Administration (DOE)
NOAA	National Oceanic and Atmospheric Administration
NPP	notice of proposed procurement
NPPPU	National Public Procurement Policy Unit (Ireland)
NPR	National Partnership for Reinventing Government (formerly National Performance Review)
NRA	NASA Research Announcement
NRC	Nuclear Regulatory Commission
NSA	National Security Agency
NSBU	National Small Business United
NSF	National Science Foundation
NSN	National Stock Number
NSNA	no stock number assigned

NSP	not separately priced
NSS	National Security Systems
NTDB	National Trade Data Bank
NTE	not to exceed
NTIS	National Technical Information Service
NTP	notice to proceed
NWBOC	National Women Business Owners Corporation
o/a	on or about
O&M	operations and maintenance
O&S	operations and support
OA	office automation
OBS	organizational breakdown structure; Open Bidding Service (Canada)
OCI	organizational conflict of interest
OCR	optical character recognition (scanning technology)
ODC	other direct cost
ODMA	Open Document Management API
OEM	original equipment manufacturer
OFCC	Office of Federal Contract Compliance
OFPP	Office of Federal Procurement Policy
OGAS	other government agency system
O/H	overhead
OIB	Office of International Business (Maryland)

OICC	officer in charge of construction
OIRM	Office of Information Resources Management
OJ or OJEC	*Official Journal of the European Communities*
OLE	object linking and embedding
OMB	Office of Management and Budget
OMG	object management group
OPBA	Ontario Public Buyers Association (Canada)
OPIC	overseas private investment corporation
OPM	Office of Personnel Management
OPW	Office of Public Works (Ireland)
OSCS	One Stop Capital Shop
OSD	Office of the Secretary of Defense
OSDBU	Office of Small and Disadvantaged Business Utilization
OSHA	Occupational Safety and Health Administration (established in 1970)
OSTA	Optical Storage Technology Association (Santa Barbara, California)
OTA	Office of Technology Assessment
OT&E	operational test and evaluation
OTS	off-the-shelf; Overseas Trade Services
OUPMA	Ontario University Purchasing Management Association
P&L	profit and loss
P&S	product and service code

P³I	preplanned product improvement
PAD	Project Approval Document (part of NASA acquisition process)
PALT	procurement administrative lead time
PAR	proposal analysis report
PARC	principal assistant responsible for contracting
PASS	procurement automated source system (now defunct; replaced by PRO-*Net*)
PAT	Process Action Team (DoD)
PBA	performance-based acquisition
PBBE	performance-based business environment
PBC	performance-based contracting (now called PBA)
PBO	performance-based organization
PBSA	performance-based service acquisition
PBSC	performance-based service contracting
PC	personal computer
PCO	procuring (or procurement or principal) contracting officer
PCP	Potential Contractor Program (DTIC)
PCR	procurement center representative
PDF	personnel data form; Portable Document Format (.pdf)
PEA	Procurement Executives Association
PEAG	proposal evaluation analysis (or advisory) group
PEB	performance evaluation board
PEDS	program element descriptive summary

PEN   program element number

PEP   project execution plan; performance evaluation plan

PER   proposal evaluation report

PERT   program evaluation and review techniques (diagrams)

PI   principal investigator

PID   procurement initiation document (German procurement law)

PIID   procurement instrument identifier (required of all U.S. federal agencies by FAC 2001–16; each procurement must have a unique identifier)

PIIN   procurement instrument identification number

PIN   preinvitation notice; prior indicative notice

PIO   procurement information online

PIP   procurement improvement plan

PIXS   Preaward Information Exchange System (Wright-Patterson AFB)

PL   public law

PLI   professional liability insurance

PM   project manager, program manager

PMA   President's Management Agenda (2001)

PMC   Procurement Management Council (within DOT)

PMO   Program Management Office

P/N   part number

PO   purchase order

POA   plan of action

POC	point of contact
POG	paperless order generator
PON	program opportunity notice
PPAIS	DoD Past Performance Automated Information System (now PPIRS)
PPBS	planning, programming, and budgeting system
PPI	proposal preparation instruction; pixels per inch; past performance information
PPIMS	Past Performance Information Management System (U.S. Army)
PPIRS	Past Performance Information Retrieval System
PPM	principal period of maintenance
PPT	performance-price trade-off (U.S. Air Force evaluation strategy); past performance tool (U.S. Army)
PQ^{2}D	price quantity quality delivery
PQS	personnel qualification sheet
PR	purchase request or public relations
PRA	Paperwork Reduction Act of 1995
PRAG	Performance Risk Assessment Group (government evaluators)
PRDA	program R&D announcements
PRO-*Net*	Procurement Marketing and Access Network
PROCNET	Procurement Network (ARDEC)
PSA	Presolicitation Announcement (for SBIR and STTR opportunities, published quarterly by SBA); professional services agreement
PSC	product service code

PSD	private sector development
PSPQ	potential supplier profile questionnaire
PTAC	Procurement Technical Assistance Center
PWBS	program (or project) work breakdown structure
PWS	performance work statement
QA	quality assurance
QASP	quality assurance surveillance plan
QBL	qualified bidders list
QBS	qualifications-based selection
QC	quality control
QCBS	quality- and cost-based selection
QML	qualified manufacturers list
QPL	qualified products list
QSL	qualified suppliers list
QVL	qualified vendors list
RAFV	risk adjusted functional value
RAM	random access memory; reliability and maintainability
RAM-D	reliability, availability, maintainability, or durability
R&D	research and development
RCAS	Reserve Component Automation System
RDT&E	research, development, test, and evaluation

RFA	request for application (grants)
RFC	request for comment
RFI	request for information
RFO	request for offer
RFP	request for proposal
RFQ	request for quotation [normally prepared on Standard Form 18, FAR 53.301-18. Unlike bids or proposals following an IFB or RFP, quotations submitted in response to an RFQ are not considered to be offers. Standing alone, a quotation cannot be accepted by the government to form a contract for goods or services. (Cibinic, John, and Ralph C. Nash, *Formation of Government Contracts*, 2nd ed., Washington D.C.: George Washington University, 1986, p. 158.)]; request for qualifications
RFS	request for solution
RFTP	request for technical proposals
RIP	raster image processing
RM	risk management
RM&S	reliability, maintainability, and supportability
ROI	return on investment
ROICC	resident officer in charge of construction
ROM	rough order of magnitude (cost)
ROS	return on sales
RTF	Rich Text Format
RTN	routing and transfer number (electronic payments)
SA	supplemental agreement
SAACONS	Standard Army Automated Contracting System

SADBUS	small and disadvantaged business utilization specialist
SAME	Society of American Military Engineers
SAP	Simplified Acquisition Procedures (FAR Part 13)
SARA	Service Acquisition Reform Act of 2003
SAT	Simplified Acquisition Threshold
SBA	Small Business Administration
SBANC	Small Business Advancement National Center (University of Central Arkansas; http://www.sbaer.uca.edu)
SBD	Standard Bidding Documents (e.g., the World Bank)
SBDC	Small Business Development Center (there are 58 SBDCs)
SBE	small business enterprise
SBHTI	Small Business High Technology Institute
SBIC	Small Business Investment Company
SBIR	Small Business Innovation Research program
SBLO	small business liaison officer
SBS	Small Business Service (United Kingdom)
SBSA	small business set aside
SBTC	Small Business Technology Coalition
SCORE	Service Corps of Retired Executives (SBA program with more than 400 offices)
SCI	sensitive compartmentalized information
SDB	small and disadvantaged business

SDR	Special Drawing Rights
SDRL	subcontract data requirements list
SEB	Source Evaluation Board
SEC	Securities and Exchange Commission
SEDB	Socially and Economically Disadvantaged Business
SEED	Support for East European Democracy
SEI	Software Engineering Institute (Carnegie Mellon University, Pittsburgh, Pennsylvania)
SEMP	system engineering management plan
SF	standard form
SFAS	Statement of Financial Accounting Standard
SGML	Standard Generalized Markup Language
SIC	Standard Industrial Classification (replaced by NAICS codes)
SIG	special interest group
SIMAP	Système d'Information pour les Marchés Public (European Commission)
SIR	screening information request
SITC	Standard International Trade Classification
SLA	service-level agreement
SLIN	subcontractor line item
SMC	Space and Missile Systems Center (Air Force)
SMDC	U.S. Army's Space and Missile Defense Command
SME	small and medium-sized enterprises (European Commission); subject matter expert

SML	solicitation mailing list
SOC	statement of capability
SOHO	small office, home office
Sol.	solicitation
SON	statement of need (equivalent of SOW)
SOO	statement of objectives
SOQ	statement of qualifications
SOW	statement of work
SP&BDG	Strategic Planning and Business Development Group
SPE	senior procurement executive
SPI	single process initiative
SPN	Specific Procurement Notice (*UN Development Business*)
SQA	software quality assurance
SRD	system requirements document
SSA	source selection authority; Social Security Administration
SSAC	Source Selection Advisory Council
SSBIC	Specialized Small Business Investment Company
SSCASS	Service and Supply Contract Appraisal Support System
SSDD	source selection decision document
SSEB	Source Selection Evaluation Board
SSET	source selection evaluation team

SSF	source selection facility
SSO	source selection official
SSP	source selection plan
SSQAAP	Small Supplier Quality Assurance Assistance Program
STARS	Streamlined Technology Acquisition Resources for Services (GSA GWAC)
STOP	sequential thematic organization of proposals (technique developed at Hughes Ground Systems)
STR	senior technical review; senior technical representative
STRICOM	U.S. Army Simulation Training and Instrumentation Command
STTR	Small Business Technology Transfer Pilot Program
SUNY	State University of New York
SWOT	Strengths, Weaknesses, Opportunities, and Threats
Ts & Cs	terms and conditions
TABD	Transatlantic Business Dialogue
TAC	Transportation Acquisition Circular; Technology Assistance Centers
TADSBAT	Training and Development of Small Businesses in Advanced Technologies (OSDBU program under NASA)
TAM	Transportation Acquisition Manual
TAR	Transportation Acquisition Regulation
TASBI	Transatlantic Small Business Initiative (United States and European Union)
TBD	to be determined
TBIC	Tribal Business Information Center

TBN	to be negotiated
TBP	to be proposed
TCN	Third County Nationals
TCO	termination contracting officer; total cost of ownership
TCP/IP	Transmission Control Protocol/Internet Protocol
TCS	Technical Consulting Services (part of USAID)
TCU	tribal colleges and universities
TD	task description
TDA	U.S. Trade and Development Agency
TDI	technical data interchange
TDO	term-determining official
TDP	technical data packages (U.S. Army)
TDY	temporary duty
T&E	test and evaluation
TED	*Tenders Electronic Daily*
TEP	total evaluated price
TEV	total evaluated value
TIFF	Tag Image File Format
TILO	Technical Industrial Liaison Office (Army)
TIN	Taxpayer Identification Number; contact the IRS to obtain a TIN
TINA	Truth in Negotiations Act [P.L. 87-653 (1962); extended in 1984]

T&M	time and materials
TM	technical monitor
TM Online	Trade Mission Online
TMIS	technical management information system
T.O.	task order
TOA	total obligation authority
TOPS	Total Operating Paperless System (CECOM)
TORFP	task order RFP
TPCC	Trade Promotion Coordinating Committee
TPCR	traditional procurement center representatives
TPIN	trading partner identification number
TQM	total quality management
TRCO	technical representative of the contracting officer
TRD	technical requirement document
TTPP	Trade Tie-up Promotion Program (Japan)
U2	unclassified/unlimited
UCA	undefinitized contract action
UCC	uniform commercial code
UCF	uniform contract format
UCLINS	uniform contract line item numbering system
UN	United Nations
UNCITRL	United Nations Commission on International Trade Law

UNCSD	United Nations Common Supply Database
UNDB	*UN Development Business*
UNDP	United Nations Development Program
URL	uniform resource locator
URN	unique reference number
USACE	United States Army Corps of Engineers
USAFE	United States Air Force–Europe
USAREU	United States Army–Europe
USASBE	U.S. Association for Small Business and Entrepreneurship
USB	universal serial bus
U.S.C.	United States Code
USCG	U.S. Coast Guard
USDA	U.S. Department of Agriculture
USEAC	U.S. Export Assistance Centers (SBA is a key partner)
USEPA	U.S. Environmental Protection Agency
USGS	U.S. Geological Survey
USITC	U.S. International Trade Commission
USML	United States Munitions List
USPS	U.S. Postal Service
USTDA	U.S. Trade and Development Agency
VA	Department of Veterans Affairs

VAAR	Veterans Administration Acquisition Regulation
VAN	value-added network
VAS	value-added service
VE	value engineering
VGA	video graphics adapter
VECP	value engineering change proposal
VEP	value engineering proposal
VIBES	virtual international business and economic sources
VP	vice president
VPC	virtual proposal center
VPN	virtual private network
VSB	very small business
VUSME	Virtual University for Small and Medium-Sized Enterprises
W2W	Work to Welfare Program
WASME	World Association of Small and Medium-Sized Enterprises
WAIS	Wide Area Information Server (a database)
WBE	woman-owned business enterprise
WBENC	Women's Business Enterprise National Council
WBS	work breakdown structure
WEIC	Wales Euro Info Centre
WinPET	Windows proposal evaluation tool
WISBIS	Women in Small Business Information Site

WISE	Web-integrated solicitation elements
WMDVBE	Women, Minority, and Disabled-Veteran Business Enterprise
WNET	Women's Network for Entrepreneurial Training
WOB	woman-owned business
WOBREP	women-owned business representative (federal agencies such as the Department of Agriculture, Commerce, Defense, Education, and Veterans' Affairs have WOBREPs)
WTO	World Trade Organization
WTO GPA	World Trade Organization Agreement on Government Procurement
WWW	World Wide Web

Selected bibliography

Abramson, Neil R., and Janet X. Ai, "You Get What You Expect in China," *Business Quarterly*, Winter 1996, pp. 37–44.

Anderson, Tania, "8(a) Companies Learn the Secret of Partnering," *Washington Technology*, Vol. 10, October 1996, p. 22.

Angus, Jeff, Jeetu Patel, and Jennifer Harty, "Knowledge Management: Great Concept . . . But What Is It?" *Information Week*, Vol. 673, March 16, 1998, pp. 58–70.

Annual Report on the State of Small Business, The White House to the Congress of the United States, June 6, 1996.

Anthes, Gary H., "Procurement Horror Stories Draw Feds' Scrutiny," *Computerworld*, February 21, 1994.

Barquin, Ramon C., Alex Bennett, and Shereen G. Remez, (eds.), *Knowledge Management: The Catalyst for Electronic Government,* Vienna, VA: Management Concepts, Inc., 2001.

Beazley, Hamilton, Jeremiah Boenisch, and David Harden, *Continuity Management: Preserving Corporate Knowledge and Productivity When Employees Leave*, New York: John Wiley, 2002.

Beveridge, James M., and Edward J. Velton, *Positioning to Win: Planning and Executing the Superior Proposal*, Radnor, PA: Chilton Book Company, 1982.

Blankenhorn, Dana, "11 Lessons from Intranet Webmasters," *Netguide*, October 1996, pp. 82–90.

BMRA Procurement Alert Notice, Fairfax, VA: Business Management Research Associates, Inc., Vol. 8, No. 3, Fall 2003.

Bowman, Joel P., and Bernadine P. Branchaw, *How to Write Proposals That Produce*, Phoenix, AZ: Oryx Press, 1992.

"A Brief History of the National Performance Review," September 1996, http://www.defenselink.mil/nii/bpr/bprcd/4128.htm.

Brockmeier, Dave, "Help Shape Federal Acquisition Regulations," *Business Credit*, Vol. 97, March 1995, pp. 40–42.

Brusaw, C. T., G. J. Alred, and W. E. Oliu, *Handbook of Technical Writing*, New York: St. Martin's Press, 1976.

Budget of the United States Government Fiscal Year 2004, Washington, D.C.: U.S. Government Printing Office, 2003.

Caudron, Shari, "Spreading Out the Carrots," *Industry Week*, May 19, 1997, pp. 20, 22, 24.

The Chicago Manual of Style, 15th ed., Chicago, IL: The University of Chicago Press, 2003.

Cibinic, John, and Ralph C. Nash, *Formation of Government Contracts*, 2nd ed., Washington, D.C.: George Washington University, 1986.

Clemons, Erid K., and Bruce W. Weber, "Using Information Technology to Manage Customer Relationships: Lessons for Marketing in Diverse Industries," *Proceedings of 26th Hawaii International Conference on System Sciences*, January 1993.

Clipsham, Neil B., "Appraising Performance Appraisal Systems," *The Military Engineer*, No. 591, April–May 1998, pp. 31–32.

Cohen, Sacha, "Knowledge Management's Killer App," *The Knowledge Management Yearbook 1999–2000,* James W. Cortada and John A. Woods, (eds.), Boston, MA: Butterworth-Heinemann, 1999, pp. 394–403.

"Community KM Glossary," http://knowledgecreators.com/km/kes/glossary.htm.

"Complying with the Anti-Kickback Act," *Developments in Government Contract Law*, No. 10, September 1990.

Cranston, H. S., and Eric G. Flamholtz, "The Problems of Success," *Management Decision*, Vol. 26, September 1988, p. 17.

Cremmins, Edward T., *The Art of Abstracting*, Philadelphia, PA: ISI Press, 1982.

Daum, Juergen H., *Intangible Assets and Value Creation,* New York: John Wiley, 2002.

Davenport, Thomas H., and Laurence Prusak, *Working Knowledge: How Organizations Manage What They Know,* Boston, MA: Harvard Business School Press, 1997.

Davenport, Tom, and Larry Prusak, "Know What You Know," http://www.brint.com/km/davenport/cio/know.htm, 2000.

Davidson, Paul, "U.S. Air Force SMC Innovations in Electronic Procurement," *APMP Perspective*, Vol. I, July/August 1996, pp. 7, 12, 13, 15.

Day, Robert A., *How to Write and Publish a Scientific Paper*, Philadelphia, PA: ISI Press, 1979.

December, John, and Neil Randall, *The World Wide Web Unleashed,* 2nd ed., Indianapolis, IN: Samsnet, 1995.

Delphos, William A., *Inside the World Bank Group: The Practical Guide for International Business Executives,* Washington, D.C.: Delphos International, 1997.

DerGurahian, Jean, "Albany MBA Candidates Earn Real-World Success," *The Business Review*, March 31, 2000.

Desktop Guide to Basic Contracting Terms, 2nd ed., Vienna, VA: National Contract Management Association, 1990, 1989.

DiGiacomo, John, and James Klekner, *Win Government Contracts for Your Small Business,* Riverwoods, IL: CCH, 2001.

"Disadvantaged Firms to Get Edge in Bids for Federal Contracts," *Baltimore Sun,* June 25, 1998, p. 2A.

"DOD Acquisition Chief Pushes COTS/Dual-Use, Wants More Suppliers," *Technology Transfer Week*, Vol. 30, April 1996.

"DOD's Airborne Recon Plan Pushes Sensors, Image Recognition," *Sensor Business News*, May 8, 1996.

Doebler, Paul D., "Who's the Boss? Sorting Out Staffing and Workflow Issues," *EP&P*, April 1988, pp. 21–26.

Do-It-Yourself Proposal Plan, TRW Space & Defense Proposal Operations, Rev. 3, February 1989.

Douglas, Susan P., C. Samuel Craig, and Warren J. Keegan, "Approaches to Assessing International Marketing Opportunities for Small- and Medium-Sized Companies," *Columbia Journal of World Business*, Fall 1982, pp. 26–32.

Edvinsson, Leif, and Michael S. Malone, *Intellectual Capital: Realizing Your Company's True Value by Finding Its Hidden Brainpower,* New York: HarperBusiness, 1997.

Edwards, Vernon J., *Competitive Proposals Contracting*, Falls Church, VA: Educational Services Institute, 1989 (in association with The George Washington University School of Government and Business Management).

Edwards, Vernon J., *Federal Contracting Basics*, Falls Church, VA: Educational Services Institute, 1990 (in association with The George Washington University School of Government and Business Management).

Elbert, Bruce, and Bobby Martyna, *Client/Server Computing: Architecture, Applications, and Distribution Systems Management*, Norwood, MA: Artech House, 1994.

Erramilli, M. Krishna, and C. P. Rao, "Choice of Foreign Market Entry Modes by Service Firms: Role of Market Knowledge," *Management International Review*, Vol. 30, 1990, pp. 135–150.

Farrell, Michael, "Many Firms Want to Do Business Internationally But Lack the Wherewithal," *Capital District Business Review*, Vol. 22, September 1995, p. 15.

Federal Acquisition Regulation (as of May 1, 1990), Chicago, IL: Commerce Clearing House, 1990.

"Federal Acquisition Streamlining Act Enacted," *Business Credit*, Vol. 97, March 1995, p. 6.

Flagler, Carolynn, "The ANY Aspects of International Proposals," *Contract Management*, March 1995, pp. 11–19.

Fox, Harold W., "Strategic Superiorities of Small Size," *Advanced Management Journal*, Vol. 51, 1986, p. 14.

Frank, C., "Uncle Sam Helps Business Get High Tech to Market," *Missouri Tech Net News,* November 4, 1997; http://www.umr.edu/~tscsbdc/sbirsumm.html.

Freed, Richard C., Shervin Freed, and Joe Romano, *Writing Winning Business Proposals: Your Guide to Landing the Client, Making the Sale, and Persuading the Boss*, New York: McGraw-Hill, 1995.

Frey, Robert S., "Beyond Compliance: Toward Solution and Storyline Development as Valuable Proposal Management Core Competencies," *Journal of the Association of Proposal Management Professionals,* Spring/Summer 2003, pp. 8–11.

Frey, Robert S., "Build Tangible and Intangible Benefits into Your Proposals," *Kansas City Small Business Monthly*, February 2001, pp. 34, 39.

Frey, Robert S., "Contracting with the Federal Government," *St. Louis Small Business Monthly*, Volume 15, Issue VI, July 2002, pp. 1, 3.

Frey, Robert S., "Effective Small Business Response Strategies to Federal Government Competitive Procurements," *The Journal of Business and Management*, Vol. 4, No. 1, Summer 1997, pp. 40–74.

Frey, Robert S., "Envisioning Your Future Contracting with the Federal Government," *Kansas City Small Business Monthly*, March 2002, pp. 19, 47.

Frey, Robert S., "Federal Contracting and Small Business," *Kansas City Small Business Monthly*, Vol. 13, Issue 2, February 2004, p. 21.

Frey, Robert S., "Federal Performance-Based Contracting: What Is It, and What Does It Mean for Your Small Business," *Corporate Corridors*, Vol. 2, No. 3, July/August 2003, pp. 60–61.

Frey, Robert S., "ISO-Driven Proposal and Business Development Excellence," *APMP Perspective*, Vol. 13, Fall 2003, No. 4, pp. 3–6.

Frey, Robert S., "Knowledge Management Benefits Federal Contracting Initiatives," *Kansas City Small Business Monthly*, April 2002, pp. 21, 47.

Frey, Robert S., "Knowledge Management, Proposal Development, and Small Business," *Journal of Management Development,* Vol. 20, No. 1, 2001, pp. 38–54.

Frey, Robert S., "Leveraging Business Complexity in a Knowledge Economy," *Journal of Business,* Vol. XI, Spring 1999, pp. 4–21.

Frey, Robert S., "Making the Leap: Graduating Successfully from the U.S. SBA 8(a) Program," *Corporate Corridors*, Vol. 1, No. 8, November 2002, pp. 54–55.

Frey, Robert S., "Once-Upon-a-Time Proposals: Winning New Government Contracts Through the Art of Customer-Focused, Fact-Based Storytelling," *Kansas City Small Business Monthly*, Vol. 13, Issue 7, July 2004, p. 21.

Frey, Robert S., "Past Performance: It's More Important Than You Think!" *Corporate Corridors*, Vol. 3, No. 1, February 2004, pp. 60–61.

Frey, Robert S., "Performance-Based Contracting: What Does It Mean for Small Business?" *Kansas City Small Business Monthly*, Vol. 12, Issue 2, February 2003, pp. 21, 43.

Frey, Robert S., "Proactive Strategic Partnering and Subcontracting Initiatives," *Corporate Corridors*, Vol. 2, No. 1, January/February 2003, pp. 54–55.

Frey, Robert S., "Resources for International Success: Information and Partnerships Can Jumpstart Global Business Opportunities," *St. Louis Small Business Monthly*, Vol. 16, Issue X, November 2003, p. 10.

Frey, Robert S., "Small Business Knowledge Management Success Story—*This Stuff Really Works!*" *Knowledge and Process Management: The Journal of Corporate Transformation,* Vol. 9, No. 3, July–September 2002, pp. 172–177.

Frey, Robert S., "Stop Selling Your Product! Build Tangible and Intangible Benefits Into Your Proposals," *Texas Small Business News*, Vol. 4, Issues 6/7, June/July 2002, pp. 7–8.

Frey, Robert S., "Strategic Partnering and Subcontracting," *Kansas City Small Business Monthly*, June 2001, pp. 34, 39.

Frey, Robert S., "Submitting Successful Private-Sector Proposals," *Kansas City Small Business Monthly Entrepreneur's Guide*, 2003–2004, p. 72.

Frey, Robert S., "Understanding and Responding to a Request for Proposal (RFP)," *Kansas City Small Business Monthly*, Vol. 11, Issue 8, August 2002.

Frey, Robert S., "Your Company's Federal Report Card," *Corporate Corridors*, Vol. 2, No. 2, April/May 2003, pp. 60–61.

Gibson, Rowan, (ed.), *Rethinking the Future: Rethinking Business, Principles, Competition, Control, Leadership, Markets and the World*, London, England: Nicholas Brealey Publishing, 1997.

Goldberg, Mim, "Listen to Me," *Selling Power*, July/August 1998, pp. 58–59.

Goretsky, M. Edward, "When to Bid for Government Contracts," *Industrial Marketing Management*, Vol. 16, Feburary 1987, p. 25.

Gouillart, Francis J., and Frederick D. Sturdivant, "Spend a Day in the Life of Your Customers," *Harvard Business Review*, January/February 1994.

Graham, John R., "Getting to the Top, and Staying There," *The New Daily Record*, Baltimore, MD, July 26, 1997, pp. 7A–7B.

Grundstein, Michel, "Companies & Executives in Knowledge Management," http://www.brint.com/km/cko.htm.

Guide to Doing Business in All 50 States for A/E/P and Environmental Consulting Firms; Natick, MA: Zweig White and Associates, 1998.

A Guide for Preparing BOEs (Basis of Estimates), TRW Space & Defense Sector, 1989.

Gurak, Laura J., *Oral Presentations for Technical Communications*, Boston, MA: Allyn and Bacon, 2000.

Hackeman, Calvin L., "Best Value Procurements: Hitting the Moving Target," McLean, VA: Grant Thornton, 1993 (a white paper prepared in connection with Grant Thornton's Government Contractor Industry Roundtables).

Hager, Peter J., and Howard J. Scheiber, *Designing & Delivering Scientific, Technical, and Managerial Presentations,* New York: John Wiley & Sons, 1997.

Hall, Dane, "Electronic Proposal Evaluation: How One U.S. Government Agency Does It," *The Executive Summary: The Journal of the APMP's National Capital Area Chapter,* June 1996, pp. 6, 8–9.

Hamilton, Martha A., "Managing the Company Mind: Firms Try New Ways to Tap Intangible Assets Such as Creativity, Knowledge," *Washington Post,* August 18, 1996, pp. H1, H5.

Hand, John R. M., and Baruch Lev, *Intangible Assets: Values, Measures, and Risks,* New York: Oxford University Press, 2003.

Hartman, Curtis, and Steven Pearlstein, "The Joy of Working," *INC.,* November 1987, pp. 61–71.

Heiman, Stephen E., and Diane Sanchez, *The New Strategic Selling, Revised and Updated for the 21st Century,* New York: Warner Books, 1998.

Heisse, John R., " 'Best Value' Procurement: How Federal and State Governments Are Changing the Bidding Process," April 29, 2002, http://www.constructionweblinks.com/Resources/Industry_Reports_Newsletters/April_29_2002/best_value_procurement.htm.

Helgeson, Donald V., *Engineer's and Manager's Guide to Winning Proposals,* Norwood, MA: Artech House, 1994.

Henry, Shannon, "The 8(a) Dating Game: Primes and Small Business Find Each Other Through Hard Work and Happenstance," *Washington Technology,* Vol. 12, September 1996, p. 24.

Hesselbein, Frances, Marshall Goldsmith, and Richard Beckhard, (eds.), *The Organization of the Future,* San Francisco, CA: Jossey-Bass Publishers, 1997.

Hevenor, Keith, "Storage Product Buyers Guide," *Electronic Publishing,* August 1, 1996, p. 10.

Hewitt, Tom, "Preparing Winning Proposals and Bids," Virginia: Federal Sources.

Hill, James W., and Timothy Whalen, (eds.), *How to Create and Present Successful Government Proposals: Techniques for Today's Tough Economy,* New York: IEEE Press, 1993.

Hinton, Henry L., *Federal Acquisition: Trends, Reforms, and Challenges,* Washington, D.C.: United States General Accounting Office, March 16, 2000. Testimony before the Subcommittee on Government Management, Information, and Technology, House Committee on Government Reform, GAO/T-OGC-007.

History at NASA [NASA HHR-50], Washington, D.C.: NASA Headquarters, 1986.

Hoft, Nancy L., *International Technical Communication: How to Export Information About High Technology*, New York: John Wiley & Sons, 1995.

Holtz, Herman, *Proven Proposal Strategies to Win More Business,* Chicago, IL: Upstart Publishing Co., 1997.

Holtz, Herman, and Terry Schmidt, *The Winning Proposal: How to Write It*, New York: McGraw-Hill, 1981.

"How the Pros See the Global Business Scene," *Rochester Business Journal*, Vol. 12, June 21, 1996, p. 10.

Howard, James S., and John Emery, "Strategic Planning Keeps You Ahead of the Pack," *D & B Reports*, Vol. 33, March/April 1985, p. 18.

Hoyt, Brad, "What Is KM?" *Knowledge Management News,* 1998; http://www.kmnews.com/Editorial/whatkm.htm.

Huseman, Richard C., and Jon P. Goodman, "Realm of the Red Queen: The Impact of Change on Corporate Education, and the Emergence of Knowledge Organizations," The Corporate Knowledge Center @ EC^2, Annenberg Center for Communications, University of Southern California, 1998.

"IBM, U.S. Chamber of Commerce Announce Results of New Study on U.S. Small Business and Technology," http://216.239.39.104/search?q= cache:v86sqMvVWmQJ:americanbusinessplans.com/1.pdf+%22IBM,+ U.S.+Chamber+of+Commerce+Announce+Results%22hl=en.

Ingram, Thomas N., Thomas R. Day, and George H. Lucas, "Dealing with Global Intermediaries: Guidelines for Sales Managers," *Journal of Global Marketing*, Vol. 5, 1992.

Ink, Dwight, "Does Reinventing Government Have an Achilles Heel?" *The Public Manager: The New Bureaucrat*, Vol. 24, Winter 1995, p. 27.

Introduction to the Federal Acquisition Regulation Training Course, Vienna, VA: Management Concepts, Inc.

Jain, Subash C., *International Marketing Management*, Boston, MA: Kent Publishing Company, 1993.

James, Geoffrey, "It's Time to Free Dilbert," *New York Times*, September 1, 1996, p. F-11.

Jennings, Robert W., *Make It Big in the $100 Billion Outsource Contracting Industry*, Westminster, CO: Westfield Press, 1997.

Jones, Stephen, "Navigating the Federal Bureaucracy Labyrinth," *Computerworld*, November 14, 1988, p. 145.

Joss, Molly W., "Authoring Alchemy: Ingredients for Brewing Up a Multimedia Masterpiece," *Desktop Publishers*, January 1996, pp. 56–65.

Kasser, Joe, *Applying Total Quality Management to Systems Engineering*, Norwood, MA: Artech House, 1995.

Kelleher, Kevin, "Feds, State Go On Line with Contracts," *San Francisco Business Times*, Vol. 9, April 1995, p. 3.

Kelman, Steven, "Remaking Federal Procurement," Working Paper No. 3, The John F. Kennedy School of Government Visions of Governance in the 21st Century, January 3, 2002, http://www.ksg.harvard.edu/visions/publication/kelman.pdf.

Keninitz, Donald, "The Government Contractor's Dictionary: A Guide to Terms You Should Know," http://www.kcilink.com/govcon/contractor/gcterms.html.

"Knowledge Management: Collaborating on the Competitive Edge," *CIO White Paper Library*, 2000; http://www.cio.com/sponsors/0600_km.

Knudsen, Dag, "The Proposal Game: Two Strikes and You're Out," *American Consulting Engineer*, May/June 1998, p.13.

Krathwohl, David R., *How to Prepare a Research Proposal: Guidelines for Funding and Dissertations in the Social and Behavioral Sciences*, 3rd ed., Syracuse, NY: Syracuse University Press, 1988.

LaFlash, Judson, "The Target Proposal Seminar," Woodland Hills, CA: Government Marketing Consultants, June 1981.

Laurent, Anne, "Shifting the Risk," *Government Executive Magazine,* August 30, 1999; http://www.govexec.com/features/99top/08a99s1.htm.

League, V. C., and Odessa Bertha, (eds.), *The Proposal Writer's Workshop: A Guide to Help You Write Winning Proposals,* Sacramento, CA: Curry-Co Publishing, 1998.

Leibfried, Kate H. J., and Joe Oebbecke, "Benchmarking: Gaining a New Perspective on How You Are Doing," *Enterprise Integration Services Supplement,* October 1994, pp. 8–9.

Lev, Baruch, "Accounting Needs New Standards for Capitalizing Intangibles," *ASAP: Forbes Supplement on the Information Age,* April 7, 1997; http://www.forbes.com/asap/97/0407/034.htm.

Lev, Baruch, *Intangibles: Management, Measurement, and Reporting,* Washington, D.C.: Brookings Institution, 2001.

Lissack, Michael R., "Complexity Metaphors and the Management of a Knowledge Based Enterprise: An Exploration of Discovery," http://www.lissack.com/writings/proposal.htm.

Loring, Roy, and Harold Kerzner, *Proposal Preparation and Management Handbook,* New York: Van Nostrand Reinhold, 1982.

Maher, Michael C., "The DOD COTS Directive—What About Radiation Hardness?" *Defense Electronics,* Vol. 26, October 1994, pp. 29–33.

Malone, Thomas W., and John K. Rockart, "Computers, Networks, and the Corporation," *Scientific American,* September 1991, pp. 128–136.

Mariotti, John, "Nursery-Rhyme Management," *Industry Week,* May 5, 1997, p. 19.

Markie, Tracy, and Mark Tapscott, "COTS: The Key to a Tougher Military Marketplace," *Defense Electronics,* Vol. 25, November 1993, pp. 29–33.

Marshall, Colin, "Competing on Customer Service: An Interview with British Airways' Sir Colin Marshall," *Harvard Business Review,* November/December 1995.

Martin, James A., "Team Work: 10 Steps for Managing the Changing Roles in Your Desktop Publishing Work Group," *Publish!*, December 1988, pp. 38–43.

McCartney, Laton, "Getting Smart About Knowledge Management," *Industry Week*, May 4, 1998, pp. 30, 32, 36–37.

McConnaghy, Kevin V., and Robert J. Whitehead, "RFPs: Winning Proposals," *Environmental Lab*, May/June 1995, pp. 14–19.

McCubbins, Tipton F., "Three Legal Traps for Small Businesses Engaged in International Commerce," *Journal of Small Business Management*, Vol. 32, July 1994, p. 95–103.

McVay, Barry L., *Proposals That Win Federal Contracts: How to Plan, Price, Write, and Negotiate to Get Your Fair Share of Government Business*, Woodbridge, VA: Panoptic Enterprises, 1989 (Panoptic Federal Contracting Series).

McVey, Thomas W., "The Proposal Specialist as Change Agent," *APMP Perspective*, May/June 1997, pp. 1, 3, 13.

Meador, R., *Guidelines for Preparing Proposals: A Manual on How to Organize Winning Proposals for Grants, Venture Capital, R&D Projects, and Other Proposals*, Chelsea, MI: Lewis Publishers, 1985.

Michaelson, Herbert B., *How to Write and Publish Engineering Papers and Reports*, Philadelphia, PA: ISI Press, 1982.

Minds at Work: How Much Brainpower Are We Really Using? A Research Report, Princeton, NJ: Kepner-Tregoe, 1997.

Miner, Lynn E., and Jeremy T. Miner, *Proposal Planning and Writing*, 3rd ed., Westport, CT: Greenwood Press, 2003.

Miner, Lynn E., and Jeremy T. Miner, *Proposal Planning and Writing*, 2nd ed., Phoenix, AZ: Oryx Press, 1998.

"The Modular Proposal Technique" (Volume I: A Review of Its Need, Origin, & Benefits; Volume II: The Step-By-Step Guide to Implementation; Volume III: Proposal Preparation Checklist, Categorical/Topical Outline Examples, Storyboards & Finished Modules), Van Nuys, CA: Litton Systems.

Mulhern, Charlotte, "Round 'Em Up," *Entrepreneur*, August 1998, Vol. 28, No. 8, pp. 117–122.

Munro, Neil, "8(a) Program Survives Republican Attacks," *Washington Technology's 8(a) and Small Business Report*, September 12, 1996, pp. S-6, S-8.

Munro, Neil, "Clinton Set-Aside Plan Becomes Election-Year Pawn," *Washington Technology*, April 11, 1996, pp. 5, 93.

Murray, Bill, "Looking Beyond Graduation," *Washington Technology's 8(a) and Small Business Report*, March 7, 1996, pp. S-4, S-6, and S-8.

Muzio, David, "Bush Administration Priorities," Office of Management and Budget, April 8, 2004; http://www.amci.web.com/con301_oe/ppt_pres/HHS/2004%20con%20301%20_%20April%208.ppt.

NASA Contractor Financial Management Reporting System: A Family of Reports, Washington, D.C.: National Aeronautics and Space Administration (circa 1986).

Nash, Ralph C., and John Cibinic, *Administration of Government Contracts*, Washington, D.C.: George Washington University Government Contracts, 1985.

Nash, Ralph C., and John Cibinic, *Federal Procurement Law*, 2 vols., Washington, D.C.: George Washington University Government Contracts, 1977, 1980 (Contract Formation Series).

Nasseri, Touraj, "Knowledge Leverage: The Ultimate Advantage," *Kognos: The E-Journal of Knowledge Issues*, Summer 1996.

Newton, Fred J., "Restoring Public Confidence in Government Contractors," *Management Accounting*, Vol. 67, June 1986, p. 51.

O'Guin, Michael, "Competitive Intelligence and Superior Business Performance: A Strategic Benchmarking Study," *Competitive Intelligence Review*, Vol. 5, 1994, pp. 4–12.

Osborne, David, and Ted Gaebler, *Reinventing Government: How the Entrepreneurial Spirit Is Transforming the Public Sector*, Reading, MA: Addison-Wesley, 1992.

"Outstanding Information Technology," General Services Administration, February 1998: http://www.itmweb.com/essay528.htm.

Paine, Lynn Sharp, "Managing for Organizational Integrity," *Harvard Business Review*, April/May 1994.

Peterson, Ralph R., "Plotting a Safe Passage to the Millennium," *Civil Engineering News*, September 1997.

Phelan, Steven E., "From Chaos to Complexity in Strategic Planning," Presented at the 55th Annual Meeting of the Academy of Management, Vancouver, British Columbia, August 6–9, 1995; http://www.aom.pace.edu.bps/Papers/chaos.html.

Pine, B. Joseph, *Mass Customization: The New Frontier in Business Competition News*, Boston, MA: Harvard Business School Press, 1993.

Piper, Thomas S., "A Corporate Strategic Plan for General Sciences Corporation," Spring 1989 (unpublished).

Porter, Kent, "Usage of the Passive Voice," *Technical Communication*, First Quarter 1991, pp. 87–88.

Porter-Roth, Bud, *Proposal Development: How to Respond and Win the Bid*, Central Point, OR: Oasis Press/PSI Research, 1998.

Preparing Winning Proposals and Bids, U.S. Professional Development Institute (USPDI), Washington, D.C., September 16–17, 1987.

"Presenting Your Proposal Orally and Loving It!" TRW, 1979.

Proposal Preparation Manual, 2 vols., Covina, CA: Procurement Associates, 1989 (revised).

Prusak, Laurence, (ed.), *Knowledge in Organizations*, Boston, MA: Butterworth-Heinemann, 1997.

Quinn, James Brian, Philip Anderson, and Sydney Finkelstein, "Managing Professional Intellect: Making the Most of the Best," *Harvard Business Review*, March/April 1996.

Quinn, Judy, "The Welch Way: General Electric CEO Jack Welch Brings Employee Empowerment to Light," *Incentive*, September 1994, pp. 50–52, 54, 56.

Ramsey, L. A., and P. D. Hale, *Winning Federal Grants: A Guide to the Government's Grantmaking Process*, Alexandria, VA: Capital Publications, 1996.

Ray, Dana, "Filling Your Funnel: Six Steps to Effective Prospecting and Customer Retention," *Selling Power*, July/August 1998, pp. 44–45.

Reis, Al, and Jack Trout, *The 22 Immutable Laws of Marketing: Violate Them at Your Own Risk,* New York: HarperBusiness, 1994.

Ricks, David A., *Blunders in International Business*, Cambridge, MA: Blackwell Publishers, 1993.

Rifkin, Jeremy, *The End of Work: The Decline of the Global Labor Force and the Dawn of the Post-Market Era,* New York: G.P. Putnam and Sons, 1995.

Riordan, John, and Lillian Hoddeson, *Crystal Fire: The Birth of the Information Age,* New York: W. W. Norton & Co., 1997.

Roberts, J. B., *The Art of Winning Contracts: Proposal Development for Government Contractors,* Ridgecrest, CA: J. Melvin Storm Co., 1995.

Roos, Johan, "Intellectual Performance: Exploring an Intellectual Capital System in Small Companies," October 30, 1996, pp. 4, 9.

Roos, Johan, and Georg von Krogh, "What You See Depends on Who You Are," *Perspectives for Managers,* No. 7, September 1995.

Root, Franklin R., *Entry Strategies for International Markets*, revised and expanded, New York: Lexington Books, 1994.

Root, Franklin R., *Foreign Market Entry Strategies*, New York: AMACOM, 1987.

Ruggles, Rudy L., *Knowledge Management Tools,* Boston, MA: Butterworth-Heinemann, 1997.

Sabo, Bill, "Orals: Your Worst Nightmare or Your Greatest Competitive Advantage?" *APMP Perspective*, September/October 1996, pp. 1, 11.

Safford, Dan, *Proposals: On Target, On Time,* Washington, D.C.: American Consulting Engineers Council, 1997.

SBA Budget Request & Performance Plan, FY2004 Congressional Submission, Fourth Printing, February 10, 2003.

Scarborough, Norman M., and Thomas W. Zimmerer, "Strategic Planning for Small Business," *Business*, Vol. 37, April/June 1987, p. 11.

Schillaci, William C., "A Management Approach to Placing Articles in Engineering Trade Journals," *Journal of Management in Engineering*, September/October 1995, pp. 17–20.

"Selected Viewgraphs from Judson LaFlash Seminars on Government Marketing and Proposals," *Government Marketing Consultants*, July 1980.

Selling to the Military, DOD 4205.1-M, Washington, D.C.: Department of Defense.

Shuman, Jeffrey C., and John A. Seeger, "The Theory and Practice of Strategic Planning in Smaller Rapid Growth Firms," *American Journal of Small Business*, Vol. 11, 1986, p. 7.

Slaughter, Jeff, "New Way of Doing Business with Uncle Sam: Electronic Commerce/Electronic Data Interchange," *Mississippi Business Journal*, Vol. 18, February 1996, p. 10.

Smith, P. R., and Dave Chaffey, *eMarketing eXcellence: The Heart of eBusiness*, Oxford, England: Butterworth-Heinemann, 2003.

Soat, Douglas M., *Managing Engineers and Technical Employees: How to Attract, Motivate, and Retain Excellent People*, Norwood, MA: Artech House, 1996.

Sole, D., and D. G. Wilson, "Storytelling in Organizations: The Power and Traps of Using Stories to Share Knowledge in Organizations;" http://lila.pz.harvard.edu/_upload/lib/ACF14F3.pdf.

Solomon, Robert, "International Effects of the Euro," Brookings Institution Policy Brief #42, Washington, D.C.: The Brookings Institution, 1999.

Source Evaluation and Selection, Vienna, VA: Management Concepts, Inc.

Source Selection: Greatest Value Approach, Document #KMP-92-5-P, Washington, D.C.: U.S. General Services Administration, May 1995.

Stanberry, Scott A., *Federal Contracting Made Easy*, Vienna, VA: Management Concepts, Inc., 2001.

Standard Industrial Classification Manual, Executive Office of the President, Office of Management and Budget, 1987.

Statements of Work Handbook [NHB 5600.2], Washington, D.C.: Government Printing Office, February 1975.

Steiner, Richard, *Total Proposal Building: An Expert System Dedicated to One Result: Winning Grants & Contracts from Government, Corporations & Foundations*, 2nd ed., Albany, NY: Trestleetree Publications, 1988.

Stewart, Rodney D., and Ann L. Stewart, *Proposal Preparation*, New York: John Wiley & Sons, 1984.

Stewart, Thomas A., *Intellectual Capital: The New Wealth of Organizations,* New York: Doubleday, 1997.

Stewart, Thomas A., "Mapping Corporate Brainpower," *Fortune,* Vol. 132, No. 9, 1995, p. 209.

Sullivan, Robert, *United States Government New Customers Step by Step Guide,* Information International, 1997.

Sutton, Michael H., "Dangerous Ideas: A Sequel," Remarks delivered during the American Accounting Association 1997 Annual Meeting in Dallas, Texas, August 18, 1997, p. 3; http://www.corpforms99.com/83.html.

Sveiby, Karl E., "The Intangible Assets Monitor," http://www.sveiby.com/articles/IntangAss/CompanyMonitor.html, December 20, 1997.

Sveiby, Karl E., "What Is Knowledge Management?" http:www.sveiby.com/articles/KnowledgeManagement.html, April 2001.

Tapscott, Mark, "COTS Regulation Reforms Sought by Industry Coalition," *Defense Electronics*, Vol. 25, October 1993, pp. 8–11.

Ten Steps to a Successful BOE: Basis of Estimate Estimator's Checklist, TRW Space & Defense Sector, 1987.

Tepper, Ron, *How to Write Winning Proposals for Your Company or Client*, New York: John Wiley & Sons, 1990, 1989.

Terpstra, Vern, *International Marketing*, 4th ed., New York: The Dryden Press, 1987.

"Test Your Global Mindset," *Industry Week,* November 2, 1998, p. 12.

"Thriving on Order" (an interview of Steve Bostic), *INC.*, December 1989, pp. 47–62.

Thompson, Arthur A., and A. J. Strickland, *Strategy Formulation and Implementation: Tasks of the General Manager*, 4th ed., Homewood, IL and Boston, MA: BPI/IRWIN, 1989.

Training and Validation in Basis of Estimates, TRW Space & Defense Sector, 1987 [vugraphs].

Twitchell, James B., *AdcultUSA: The Triumph of Advertising in American Culture*, New York: Columbia University Press, 1996.

Unisys Defense Systems Proposal Development Digest, McLean, VA: Unisys, 1989.

Ursey, Nancy J., *Insider's Guide to SF254/255 Preparation*, 2nd ed., Natick, MA: Zweig White & Associates, 1996.

U.S. Small Business Administration, "BSA Streamlines 8(a) Contracting," Press Release Number 98-24, June 4, 1998; http://www.sba.gov/news/current98-24.html.

U.S. Small Business Administration, "Lockheed Martin Division, Nova Group Earn SBA's Top Awards for Subcontracting," Press Release Number 98-43, June 4, 1998; http://www.sba.gov/news/current98-43.html.

U.S. Small Business Administration, "Vice President and SBA Administrator Announce Pact with Big Three Automakers," Press Release Number 98-09, February 19, 1998; http://www.sba.gov/news/current98-09.html.

U.S. Small Business Administration, "Vice President Gore Praises SBA for Expanding Economic Opportunity: Congratulates Agency on 45th Birthday," Press Release Number 98-68, July 30, 1998; http://www.sba.gov/news.

Verespej, Michael A., "Drucker Sours on Terms," *Industry Week*, April 6, 1998, p. 16.

Verespej, Michael A., "The Old Workforce Won't Work," *Industry Week*, September 21, 1998, pp. 53, 54, 58, 60, 62.

Verespej, Michael A., "Only the CEO Can Make Employees King," *Industry Week*, November 16, 1998, p. 22.

Vivian, Kaye, *Winning Proposals: A Step-by-Step Guide to the Proposal Process*, Jersey City, NJ: American Institute of Certified Public Accountants, 1993.

Wall, Richard J., and Carolyn M. Jones, "Navigating the Rugged Terrain of Government Contracts," *Internal Auditor*, April 1995, pp. 32–36.

Weil, B. H., I. Zarember, and H. Owen, "Technical-Abstracting Fundamentals," *Journal of Chemical Documentation*, Vol. 3, 1963, pp. 132–136.

Whitely, Richard, and Diane Hessan, *Customer-Centered Growth: Five Proven Strategies for Building Competitive Advantage,* Reading, MA: Addison-Wesley, 1996.

Whitman, Marina, and Rosabeth Moss Kanter, "A Third Way? Globalization and the Social Role of the American Firm," *Washington Quarterly,* Spring 1999.

Wodaski, Ron, "Planning and Buying for Multimedia: Effective Multimedia for the Small Office," *Technique: How-To Guide to Business Communication*, October 1995, pp. 16–25.

Words into Type, 3rd ed., Englewood Cliffs, NJ: Prentice-Hall, 1974.

The World Bank Annual Report, Washington, D.C.: The World Bank, 1994.

Worthington, Margaret, "The Ever Changing Definition of Allowable Costs," *Government Accountants Journal*, Vol. 35, Spring 1986, p. 52.

Worthington, Margaret M., Louis P. Goldsman, and Frank M. contra Alston, *Contracting with the Federal Government,* 4th ed., New York: John Wiley & Sons, 1988.

Writing Winning Proposals, Farmington, UT: Shipley Associates, 1988.

Zimmerman, Jan, *Doing Business with the Government Using EDI: A Guide for Small Businesses,* New York: Van Nostrand Reinhold, 1996.

Zuckman, Saul, "Is There Life After 8(a)?" *The Columbia (Md.) Business Monthly*, October 1994, p. 37.

About the author

ROBERT S. FREY has successfully performed proposal management; end-to-end proposal development, design, and production; capture management; and business planning and development for 17 years. That successful experience is coupled with more than 25 years of writing and publication-related activities. He has conceptualized and established effective proposal development infrastructures for five federal contracting firms in Virginia and Maryland. Four of those firms were small businesses operating under the Small Business Administration's 8(a) program. In addition, Mr. Frey has provided end-to-end proposal development, design, and production support for more than 1,500 government, private-sector, and international proposals. He has coordinated all phases of up to 12 scientific and engineering proposals concurrently.

Of particular note is Mr. Frey's hands-on experience with most facets of proposal development. In the process of establishing and implementing proposal infrastructures for various companies, he has actually performed the myriad tasks associated with responding to government and private-sector requests for proposals, including graphics conceptualization, electronic-text editing, computer troubleshooting, and photocopying, along with solution development, original writing, and review. During various periods, he was the only full-time staff person in a business-development capacity. He has developed a very thorough understanding of both the big picture and the day-to-day, hour-to-hour logistical, organizational, and interpersonal activities and processes associated with successful proposalmanship.

Mr. Frey also develops and presents innovative proposal training curricula and has delivered numerous interactive seminars nationwide to businesses and academe. These seminars have included a 2-day short course at the University of California at Los Angeles (UCLA), a daylong seminar for Microsoft Corporation in Redmond, Washington, and a work-shop at Gallaudet University in Washington, D.C., sponsored by the Deaf and Hard of Hearing Entrepreneurship Council. He possesses a detailed understanding of marketing-intelligence resources, as well as collection and analysis techniques. Mr. Frey participates in business strategizing and planning for numerous federal, civilian, defense, law enforcement, and intelligence agencies under a broad umbrella of acquisition vehicles, such as governmentwide agency contracts (GWACs), ID-IQs, and cost-reimbursable, firm-fixed-price, and time and materials contracts.

He has prepared proposals for defense, civilian, law enforcement, and intelligence agencies, including the U.S. Army Corps of Engineers; Air Force Center for Environmental Excellence; NASA; National Oceanic and Atmospheric Administration; Air Force Space Command; Environmental Protection Agency; Department of Transportation; Department of Energy; National Institutes of Health; Department of Veterans Affairs; Department of the Treasury; Bureau of Alcohol, Tobacco, and Firearms; Defense Logistics Agency; and the Defense Information Systems Agency.

Mr. Frey is an active member in the nationwide Association of Proposal Management Professionals (APMP, http://www.apmp.org), serving as a member of the editorial review board for the *APMP Professional Journal: Journal of the Association of Proposal Management Professionals*. He was selected to deliver a presentation titled "Managing Critical Proposal Information Effectively," at the APMP annual conference held in Colorado Springs, Colorado, in 1998. In addition, Mr. Frey published an article entitled "Effective Small Business Response Strategies to Federal Government Competitive Procurements" in the Summer 1997 issue of *The Journal of Business and Management* (Western Decision Sciences Institute/California State University, Dominguez Hills).

Successful Proposal Strategies for Small Businesses: Using Knowledge Management to Win Government, Private-Sector, and International

Contracts, Fourth Edtion, is the eighth book that Mr. Frey has written since 1985. He holds a B.S. in biology (cum laude) and an M.A. in history (with highest honors). Currently, Mr. Frey serves as vice president of knowledge management and proposal development for RS Information Systems (RSIS; http://www.rsis.com), a highly successful midtier federal systems integrator with core competencies in information technology, engineering, telecommunications, and scientific consulting, headquartered in McLean, Virginia. RSIS graduated from the SBA's 8(a) program and has grown to stand among the 50 largest federal IT contractors in the United States. The company has sustained an incredible 6-year proposal win rate of 66%. RSIS has catapulted from 140 staff professionals in 1998 to more than 1,800 currently. Revenues expanded from $15.5 million in 1998 to $325 million in 2004.

You can contact Mr. Frey at BRIDGES23@AOL.com or by calling (703) 734-7800 x 213. He can also be reached by U.S. mail at P.O. Box 3075, Oakton, VA 22124.

Index

Preparing and Delivering Effective Technical Presentations, Second Edition,
  David Adamy

Reengineering Yourself and Your Company: From Engineer to Manager to Leader,
  Howard Eisner

The Requirements Engineering Handbook, Ralph R. Young

Running the Successful Hi-Tech Project Office, Eduardo Miranda

Successful Marketing Strategy for High-Tech Firms, Second Edition, Eric Viardot

*Successful Proposal Strategies for Small Businesses: Using Knowledge Management
  to Win Government, Private-Sector, and International Contracts, Fourth Edition,*
  Robert S. Frey

Systems Approach to Engineering Design, Peter H. Sydenham

Systems Engineering Principles and Practice, H. Robert Westerman

Systems Reliability and Failure Prevention, Herbert Hecht

Team Development for High-Tech Project Managers, James Williams

For further information on these and other Artech House titles,
including previously considered out-of-print books now available through our
In-Print-Forever® (IPF®) program, contact:

Artech House	Artech House
685 Canton Street	46 Gillingham Street
Norwood, MA 02062	London SW1V 1AH UK
Phone: 781-769-9750	Phone: +44 (0)20 7596-8750
Fax: 781-769-6334	Fax: +44 (0)20 7630-0166
e-mail: artech@artechhouse.com	e-mail: artech-uk@artechhouse.com

Find us on the World Wide Web at: www.artechhouse.com